EUKARYOTIC TRANSCRIPTION FACTORS

Third Edition

Cover illustration: Binding of the a1 (yellow)/α2 (blue) homeodomain heterodimer to DNA. α-helices are shown as cylinders. Figure kindly provided by Professor Cynthia Wolberger and Dr Thomas Li.

Eukaryotic Transcription Factors

Third Edition

David S. Latchman

*Professor of Molecular Pathology, and Director,
Windeyer Institute of Medical Sciences,
University College London Medical School, London*

ACADEMIC PRESS
San Diego London Boston New York
Sydney Tokyo Toronto

ACADEMIC PRESS
525 B Street, Suite 1900, San Diego,
California 92101–4495, USA
http://www.apnet.com

ACADEMIC PRESS
24–28 Oval Road
LONDON NW1 7DX
http://www.hbuk.co.uk/ap/

A catalogue record for this book is available from the British Library

ISBN 0–12–437176–0 (Hbk)
ISBN 0–12–437177–9 (Pbk)

Library of Congress Catalog Card Number: 98–87659

Typeset by LaserScript, Mitcham, Surrey
Printed in Great Britain by the University Printing House, Cambridge
98 99 00 01 02 03 CU 9 8 7 6 5 4 3 2 1

Contents

COLOUR PLATES ARE LOCATED BETWEEN PP. 110–11

To Maurice and Vivienne
In admiration

Preface to the third edition

As in previous years, the period between the publication of the second and third editions of this book has been marked by a considerable further accumulation of information about individual transcription factors and the manner in which they act. This new edition has, therefore, been extensively updated to reflect this and several sections have been completely rewritten.

As well as such increased general understanding of transcription factors, a major new theme unifying much of this information has emerged. This involves the role of co-activator molecules such as CBP in the action of a number of different activating transcription factors as well as the finding that such co-activators frequently possess histone acetyltransferase activity, indicating that they may act by modulating chromatin structure. In addition to discussion of co-activators in the appropriate sections on individual transcription factors, the new edition of this work now includes specific new sections dealing with this important topic. Thus the role of chromatin structure and histone acetylation in the regulation of gene expression is now introduced in Chapter 1 (Section 1.4), the role of CBP in cyclic AMP-mediated gene activation, where it was originally discovered, is discussed in Chapter 4 together with other aspects of this signalling pathway (Section 4.3) and the interaction of transcriptional activators with co-activators is discussed in a separate section of Chapter 9 (Section 9.2.4).

In addition to these new sections on this aspect, other new sections have been added describing topics which are now of sufficient importance to merit a separate section. These are the methods used to determine the DNA binding specificity of an uncharacterized transcription factor (Chapter 2, Section 2.3.4), the Pax family transcription factors (Chapter

6, Section 6.3.2), anti-oncogenic transcription factors other than p53 or Rb (Chapter 7, section 7.3.4) and the regulation of transcription factor activity by protein degradation and processing (Chapter 10, Section 10.3.5). Similarly, Chapter 7 now includes an extensive discussion of the role of transcription factors in diseases other than cancer and its title has therefore been changed to 'Transcription factors and human disease' (from 'Transcription factors and cancer').

As well as these changes in the text, we have been able to include, for the first time, a special section of colour illustrations illustrating various aspects of transcription factor structure which are being progessively elucidated. It is hoped that all these changes will allow this new edition, like its predecessors, to provide an up-to-date overview of the important area of transcription factors and their vital role in regulating transcription in different cell types, during development and in disease.

Finally, I would like to thank Mrs Sarah Franklin for her efficiency in producing the text and dealing with the need to make numerous changes from the previous edition as well as Mrs Jane Templeman for continuing to use her outstanding skills in the preparation of the numerous new illustrations in this edition. Thanks are also due to Tessa Picknett and the staff at Academic Press for producing this new edition with their customary efficiency.

David S. Latchman

Preface to the second edition

In the four years since the first edition of this work was published, the explosion of information about transcription factors has continued. The genes encoding many more transcription factors have been cloned and this information used to analyse their structure and function culminating in many cases with the use of inactivating mutations to prepare so-called 'knock-out' mice, thereby testing directly the role of these factors in development. None the less, the examples used in the first part of this book to illustrate the role of transcription factors in processes as diverse as inducible gene expression and development still remain amongst the best understood. The discussion of these factors has therefore been considerably updated to reflect the progress made in the last few years. In addition new sections have been added on topics such as TBP; the *myc* oncogene and anti-oncogenes where the degree of additional information now warrants a separate section.

Even greater changes have been necessary in the second part of the book, which deals with the mechanisms by which transcription factors act. Thus, for example, the sections in Chapter 9 on the mechanisms of transcriptional activation and on transcriptional repression have been completely rewritten. In addition the increasing emphasis on transcriptional repression discussed in Chapter 9 has led to a change in the title of Chapter 10 to 'What regulates the regulators?' (from 'What activates the activators?'). Moreover, this chapter now includes a much more extensive section on the interaction between different factors, which is another major theme to have emerged in the last few years. It is hoped that these changes will allow the new edition to build on the success of the first edition in providing an overview of these vital factors and the role they play in gene regulation.

Finally I would like to thank Jane Templeman who has prepared a large number of new illustrations to complement the excellent ones she provided for the first edition and Sarah Chinn for coping with the necessity of adding, deleting or amending large sections of the first edition. I am also grateful to Tessa Picknett and the staff at Academic Press for commissioning this new edition and their efficiency in producing it.

David S. Latchman

Preface to the first edition

In my previous book, *Gene Regulation: A Eukaryotic Perspective* (Unwin-Hyman Ltd, 1990), I described the mechanisms by which the expression of eukaryotic genes is regulated during processes as diverse as steroid treatment and embryonic development. Although some of this regulation occurs at the post-transcriptional level, it is clear that the process of gene transcription itself is the major point at which gene expression is regulated. In turn this has focused attention on the protein factors, known as transcription factors, which control both the basal processes of transcription and its regulation in response to specific stimuli or developmental processes. The characterization of many of these factors and in particular the cloning of the genes encoding them has resulted in the availability of a bewildering array of information on these factors, their mechanism of action and their relationship to each other. Despite its evident interest and importance, however, this information could be discussed only relatively briefly in *Gene Regulation*, whose primary purpose was to provide an overview of the process of gene regulation and the various mechanisms by which this is achieved.

It is the purpose of this book, therefore, to discuss in detail the available information on transcription factors, emphasizing common themes and mechanisms to which new information can be related as it becomes available. As such it is hoped the work will appeal to final-year undergraduates and postgraduate students entering the field as well as to those moving into the area from other scientific or clinical fields who wish to know how transcription factors may regulate the gene in which they are interested.

In order to provide a basis for the discussion of transcription factors, the first two chapters focus respectively on the DNA sequences with which the factors interact and on the experimental methods which are used to study these factors and obtain the information about them provided in subsequent chapters. The remainder of the work is divided into two distinct portions. Thus Chapters 3 to 7 focus on the role of transcription factors in particular processes. These include constitutive and inducible gene expression, cell type-specific and developmentally regulated gene expression and the role of transcription factors in cancer. Chapters 8 to 10 adopt a more mechanistic approach and consider the features of transcription factors which allow them to fulfil their function. These include the ability to bind to DNA and modulate transcription either positively or negatively as well as the ability to respond to specific stimuli and thereby activate gene expression in a regulated manner.

Although this dual approach to transcription factors from both a process-oriented and mechanistic point of view may lead to some duplication, it is the most efficient means of providing the necessary overview both of the nature of transcription factors and the manner in which they achieve their role of modulating gene expression in many diverse situations.

Finally I would like to thank Mrs Rose Lang for typing the text and coping with the continual additions necessary in this fast-moving field and Mrs Jane Templeman for her outstanding skill in preparing the illustrations.

David S. Latchman

Acknowledgements

I would like to thank all the colleagues, listed below, who have given permission for material from their papers to be reproduced in this book and have provided prints suitable for reproduction.

Figure 4.2, photograph kindly provided by Dr C. Wu from Zimarino and Wu, *Nature* **327**, 727 (1987), by permission of Macmillan Magazines Ltd. Figure 4.17, photograph kindly provided by Professor M. Beato from Willmann and Beato, *Nature* **324**, 688 (1986) by permission of Macmillan Magazines Ltd. Figures 5.10 and 5.14, photographs kindly provided by Dr R.L. Davis from Davis *et al.*, *Cell* **51**, 987 (1987) by permission of Cell Press. Figures 6.1 and 10.2, photographs kindly provided by Professor W.J. Gehring from Gehring, *Science* **236**, 1245 (1987) by permission of the American Association for the Advancement of Science. Figure 6.15, photograph kindly provided by Dr P. Holland from Holland and Hogan, *Nature* **321**, 251 (1986) by permission of Macmillan Magazines Ltd. Figures 6.17 and 6.18, photographs kindly provided by Dr R. Krumlauf from Graham *et al.*, *Cell* **57**, 367 (1989) by permission of Cell Press. Figure 8.8, redrawn from Redemann *et al.*, *Nature* **332**, 90 (1988) by kind permission of Dr H. Jackle and Macmillan Magazines Ltd. Figures 8.12 and 8.16, redrawn from Schwabe *et al.*, *Nature* **348**, 458. (1990) by kind permission of Dr D. Rhodes and Macmillan Magazines Ltd. Figure 8.21, redrawn from Abel and Maniatis, *Nature* **341**, 24 (1989) by kind permission of Professor T. Maniatis and Macmillan Magazines Ltd.

I am also especially grateful to the colleagues who have provided colour prints of transcription factor structures, allowing us to include this new feature in the third edition. Colour plate 1, kindly provided by Dr J.H. Geiger. Colour plate 2 kindly provided by Dr R.J. Fletterick. Colour plate 3 kindly provided by Professor D. Moras. Colour plate 4 kindly provided by Dr T. Li and Professor C. Wolberger. Colour plate 5 kindly provided by Professor P.E. Wright from Lee *et al.*, *Science* **245**, 635 (1989) by permission of the American Association for the Advancement of Science. Colour plate 6 kindly provided by Professor R. Kaptein from Hard *et al.*, *Science* **249**, 157 (1990) by permission of the American Association for the Advancement of Science. Colour plate 7 kindly provided by Dr D. Rhodes from Schwabe *et al.*, *Cell* **75**, 567 (1993) by kind permission of Cell Press.

List of tables

CHAPTER ONE

DNA sequences, transcription factors and chromatin structure

1.1 THE IMPORTANCE OF TRANSCRIPTION

The fundamental dogma of molecular biology is that DNA produces RNA which in turn produces protein. Hence if the genetic information which each individual inherits as DNA (the genotype) is to be converted into the proteins which produce the corresponding characteristics of the individual (the phenotype), it must first be converted into an RNA product. The process of transcription, whereby an RNA product is produced from the DNA, is therefore an essential element in gene expression. The failure of this process to occur will obviously render redundant all the other steps which follow the production of the initial RNA transcript in eukaryotes, such as RNA splicing, transport to the cytoplasm or translation into protein (for reviews of these stages, see Nevins, 1983; Latchman, 1998).

The central role of transcription in the process of gene expression also renders it an attractive control point for regulating the expression of genes in particular cell types or in response to a particular signal. Indeed, it is now clear that, in the vast majority of cases, where a particular protein is produced only in a particular tissue or in response to a particular signal, this is achieved by control processes which ensure that its corresponding gene is transcribed only in that tissue or in response to such a signal (for reviews, see Darnell, 1982; Latchman, 1998). For example, the genes encoding the immunoglobulin heavy and light chains of the antibody molecule are transcribed at high level only in the antibody-producing B cells, whilst the increase in somatostatin production in response to treatment of cells with cyclic AMP is mediated by increased transcription of the corresponding gene. Therefore, while post-transcriptional regulation affecting, for example, RNA splicing or stability, plays some role in the regulation of gene expression (for reviews, see Ross, 1996; Wang and Manley, 1997), the major control point lies at the level of transcription.

1.2 DNA SEQUENCE ELEMENTS

1.2.1 The gene promoter

The central role of transcription both in the basic process of gene expression and its regulation in particular tissues has led to considerable study of this process. Initially such studies focused on the nature of the DNA sequences within individual genes which were essential for either basal or regulated gene expression. In prokaryotes, such sequences are found immediately upstream of the start site of transcription and form part of the promoter directing expression of the genes. Sequences found at this position include both elements found in all genes which are involved in the basic process of transcription itself and those found in a more limited number of genes which mediate their response to a particular signal (for reviews, see Travers, 1993; Muller-Hill, 1996).

Early studies of cloned eukaryotic genes, therefore, concentrated on the region immediately upstream of the transcribed region where, by analogy, sequences involved in transcription and its regulation should be located. Putative regulatory sequences were identified by comparison between different genes and the conclusions reached in this way confirmed either by destroying these sequences by deletion or mutation, or by transferring them to another gene in an attempt to alter its pattern of regulation.

This work, carried out on a number of different genes encoding specific proteins, identified many short sequence elements involved in transcriptional control (for reviews, see Davidson *et al.*, 1983; Jones *et al.*, 1988). The elements of this type present in two typical examples, the human gene encoding the 70-kDa heat-inducible (heat-shock) protein (Williams *et al.*, 1989) and the human metallothionein IIA gene (Lee *et al.*, 1987) are illustrated in Figure 1.1.

Comparisons of these and many other genes revealed that, as in bacteria, their upstream regions contain two types of elements: firstly, sequences found in very many genes exhibiting distinct patterns of regulation which are likely to be involved in the basic process of transcription itself; and, secondly, those found only in genes transcribed

a)

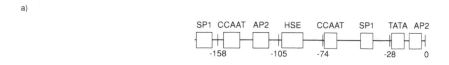

b)

Figure 1.1 Transcriptional control elements upstream of the transcriptional start site in the human genes encoding *hsp70* (a) and methallothionein IIA (b). The TATA, Sp1 and CCAAT boxes bind factors which are involved in constitutive transcription (see also Chapter 3), whilst the glucocorticoid response element (GRE), metal response element (MRE), heat-shock element (HSE), and the AP1 and AP2 sites bind factors involved in the induction of gene expression in response to specific stimuli (see also Chapter 4 and Section 7.2.1).

in a particular tissue or in response to a specific signal which are likely to produce this specific pattern of expression. These will be discussed in turn.

1.2.2 Sequences involved in the basic process of transcription

Although they are regulated very differently, the *hsp70* and metallothionein genes both contain a TATA box. This is an AT-rich sequence (consensus TATAA/TAA/T) which is found about 30 base pairs upstream of the transcriptional start site in very many but not all genes. Mutagenesis or relocation of this sequence has shown that it plays an essential role in accurately positioning the start site of transcription (Breathnach and Chambon, 1981). The region of the gene bracketed by the TATA box and the site of transcriptional initiation (the Cap site) has been operationally defined as the gene promoter or core promoter (Goodwin *et al.*, 1990). It is likely that this region binds several proteins essential for transcription, as well as RNA polymerase II itself, which is the enzyme responsible for transcribing protein coding genes (Sentenac, 1985).

Although the TATA box is found in most eukaryotic genes, it is absent in some genes, notably housekeeping genes expressed in all tissues and in some tissue specific genes (for a review, see Weis and Reinberg, 1992). In these promoters, a sequence known as the initiator element, which is located over the start site of transcription, itself appears to play a critical role in determining the initiation point and acts as a minimal promoter capable of producing basal levels of transcription.

In promoters which contain a TATA box and in those which lack it, the very low activity of the promoter itself is dramatically increased by other elements located upstream of the promoter. These elements are found in a very wide variety of genes with different patterns of expression indicating that they play a role in stimulating the constitutive activity of promoters. Thus inspection of the *hsp70* and metallothionein IIA genes reveals that both contain one or more copies of a GC-rich sequence known as the Sp1 box which is found upstream of the promoter in many genes both with and without TATA boxes (for a review, see Dynan and Tjian, 1985).

In addition, the *hsp70* promoter but not the metallothionein promoter contains another sequence, the CCAAT box, which is also found in very many genes with disparate patterns of regulation. Both the CCAAT box and the Sp1 box are typically found upstream of the TATA box as in the metallothionein and *hsp70* genes. Some genes, as in the case of *hsp70*, may have both of these elements, whereas others such as the metallothionein gene have single or multiple copies of one or the other (for a review, see McKnight and Tjian, 1986). In every case, however, these elements are essential for transcription of the genes, and their elimination by deletion or mutation abolishes transcription. Hence these sequences play an essential role in efficient transcription of the gene and have been termed upstream promoter elements (UPE) (Goodwin *et al.*, 1990). The role of the promoter and UPE sequences and the protein factors which bind to them are discussed further in Chapter 3.

1.2.3 Sequences involved in regulated transcription

Inspection of the *hsp70* promoter (Figure 1.1) reveals several other sequence elements which are only shared with a much more limited number of other genes and which are interdigitated with the upstream promoter elements discussed above. Indeed, one of these, which is located approximately 90 bases upstream of the transcriptional start site, is shared only with other heat-shock genes whose transcription is increased in response to elevated temperature. This suggests that this heat-shock element may be essential for the regulated transcription of the *hsp70* gene in response to heat.

To prove this directly, however, it is necessary to transfer this sequence to a non-heat-inducible gene and show that this transfer renders the recipient gene heat inducible. Pelham (1982) successfully achieved this by linking the heat-shock element to the non-heat-inducible thymidine kinase gene of the eukaryotic virus herpes simplex. This hybrid gene could be activated following its introduction into mammalian cells by raising the temperature (Figure 1.2). Hence the

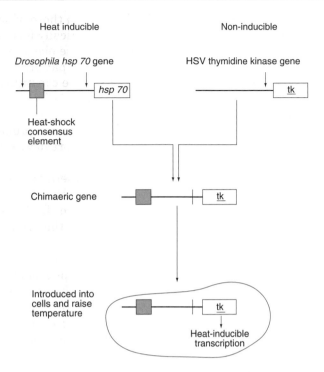

Figure 1.2 Demonstration that the heat-shock element mediates heat inducibility. Transfer of this sequence to a gene (thymidine kinase) which is not normally inducible renders this gene heat inducible.

heat-shock element can confer heat inducibility on another gene, directly proving that its presence in the *hsp* gene promoters is responsible for their heat inducibility.

Moreover, although these experiments used a heat-shock element taken from the *hsp70* gene of the fruit fly *Drosphila melanogaster*, the hybrid gene was introduced into mammalian cells. Not only does the successful functioning of the fly element in mammalian cells indicate that this process is evolutionarily conserved, it also permits a further conclusion about the way in which the effect operates. Thus, in the cold-blooded *Drosophila*, 37°C represents a thermally stressful temperature and the heat-shock response would normally be active at this temperature. The hybrid gene was inactive at 37°C in the mammalian cells, however, and was only induced at 42°C, the heat-shock temperature characteristic of the cell into which it was introduced. Hence this sequence does not act as a thermostat set to go off at a particular temperature, since this would occur at the *Drosophila* heat-shock temperature (Figure 1.3a). Rather, this sequence must act by being recognized by a cellular protein which is activated only at an elevated temperature characteristic of the mammalian cell heat-shock response (Figure 1.3b).

This experiment, therefore, not only directly proves the importance of the heat-shock element in producing the heat inducibility of the *hsp70* gene but also shows that this sequence acts by binding a cellular protein which is activated in response to elevated temperature. The binding of this transcription factor then activates transcription of the *hsp70* gene. The manner in which this factor activates transcription of the *hsp70* gene and the other heat-shock genes is discussed further in Section 4.2.

The presence of specific DNA sequences, which can bind particular proteins, will therefore confer on a specific gene the ability to respond to particular stimuli. Thus the lack of a heat-shock element in the metallothionein IIA gene (Figure 1.1) means that this gene is not heat inducible. In contrast, however, this gene, unlike the *hsp70* gene, contains a glucocorticoid response element (GRE). Hence it can bind the complex of the glucocorticoid receptor and the hormone itself which forms following treatment of cells with glucocorticoid. Its transcription is therefore activated in response to glucocorticoid, whereas that of the *hsp70* gene is not (see Chapter four, Section 4.4). Similarly, only the

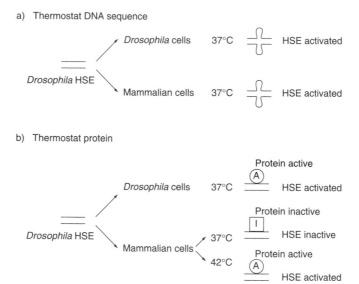

Figure 1.3 Predicted effects of placing the *Drosophila* heat-shock element (HSE) in a mammalian cell if the element acts as a thermostat detecting elevated temperature directly (a) or if it acts by binding a protein which is activated by elevated temperature (b). Note that only possibility (b) can account for the observation that the *Drosophila* HSE activates transcription in mammalian cells only at the mammalian heat-shock temperature of 42°C and not at the *Drosophila* heat-shock temperature of 37°C.

metallothionein gene contains metal response elements (MRE), allowing it to be activated in response to treatment with heavy metals such as zinc and cadmium (Thiele, 1992). In contrast, both genes contain binding sites for the transcription factor AP2 which mediates gene activation in response to cyclic AMP and phorbol esters. The manner in which the binding of specific transcription factors to different regulatory sequences modulates gene expression in response to specific inducing factors is discussed further in Chapter 4.

Similar DNA sequence elements in the promoters of tissue-specific genes play a critical role in producing their tissue specific pattern of expression by binding transcription factors which are present in an active form only in a particular tissue where the gene will be activated. For example the promoters of the immunoglobulin heavy- and light-chain genes contain a sequence known as the octamer motif (ATG-CAAAT), which can confer B cell specific expression on an unrelated promoter (Wirth *et al.*, 1987) (see Section 5.2.2 for further details). Similarly, the related sequence ATGAATAA/T is found in genes expressed specifically in the anterior pituitary gland such as the prolactin gene and the growth hormone gene, and binds a transcription factor known as Pit-1 which is expressed only in the anterior pituitary (for a review, see Andersen and Rosenfeld, 1994). If this short sequence is inserted upstream of a promoter, the gene is expressed only in pituitary cells. In contrast the octamer motif which differs by only two bases will direct expression only in B cells when inserted upstream of the same promoter. (Elsholtz *et al.*, 1990) (Figure 1.4). Hence small differences in control element sequences can produce radically different patterns of gene expression.

The role of transcription factors in producing tissue specific gene expression is discussed further in Chapter 5, whilst their role in development is discussed in Chapter 6.

1.2.4 Enhancers

One of the characteristic features of eukaryotic gene expression is the existence of sequence elements located at great distances from the start site of transcription which can influence the level of gene expression. These elements can be located upstream, downstream or within a transcription unit, and function in either orientation relative to the start site of transcription (Figure 1.5). They act by increasing the activity of a promoter, although they lack promoter activity themselves and are hence referred to as enhancers (for reviews, see Hatzopoulos *et al.*, 1988; Muller *et al.*, 1988). Some enhancers are active in all tissues and increase the activity of a promoter in all cell types whilst others function as tissue-

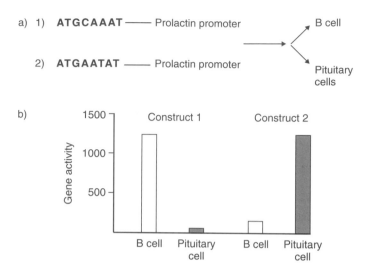

Figure 1.4 Linkage of the octamer binding motif ATGCAAAT (1) and the related Pit-1-binding motif ATGAATAT (2) to the prolactin promoter and introduction into B cells and pituitary cells (a). Only the octamer containing construct 1 directs a high level of activity in B cells, whereas only construct 2, containing the Pit-1-binding site, directs a high level of gene activity in pituitary cells (b). Data from Elsholtz *et al.* (1990).

specific enhancers which activate a particular promoter only in a specific cell type. Thus the enhancer located in the intervening region of the immunoglobulin genes is active only in B cells and the B-cell-specific expression of the immunoglobulin genes is produced by the interaction of this enhancer and the immunoglobulin promoter, which, as we have previously seen, is also B-cell specific (Garcia *et al.*, 1986) (Section 5.2, for further discussion).

As with promoter elements, enhancers contain multiple binding sites for transcription factors which interact together (for a review, see Carey, 1998). In many cases these elements are identical to those contained immediately upstream of gene promoters. Thus the immunoglobulin heavy-chain enhancer contains a copy of the octamer sequence (Sen and Baltimore, 1986) which is also found in the immunoglobulin promoters (Section 1.2.3). Similarly, multiple copies of the heat-shock consensus element are located far upstream of the start site in the *Xenopus hsp70* gene and function as a heat-inducible enhancer when transferred to another gene (Bienz and Pelham, 1986).

Enhancers, therefore, consist of sequence elements which are also present in similarly regulated promoters and may be found within the enhancer associated with other control elements or in multiple copies.

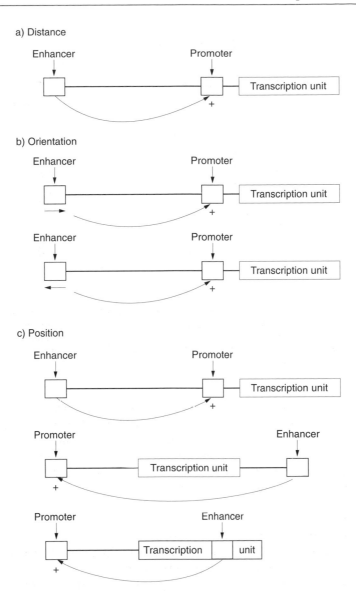

Figure 1.5 Characteristics of an enhancer element which can activate a promoter at a distance (a); in either orientation relative to the promoter (b), and when positioned upstream, downstream, or within a transcription unit (c).

1.3 INTERACTION BETWEEN FACTORS BOUND AT VARIOUS SITES

The typical eukaryotic gene will therefore consist of up to four distinct transcriptional control elements (Figure 1.6). These are: firstly, the promoter itself; secondly, upstream promoter elements located close to it which are required for efficient transcription in any cell type; thirdly, other elements adjacent to the promoter which are interdigitated with the UPEs and which activate the gene in particular tissues or in response to particular stimuli, and, lastly, more distant enhancer elements which increase gene activity either in all tissues or in a regulated manner.

Such sequences often act by binding positively acting transcription factors which then stimulate transcription (Figure 1.7a). Interestingly however, although most sequences act in such a positive way, some sequences do appear to act in a negative manner to inhibit transcription. Such silencer elements have been defined in a number of genes including the cellular oncogene c-*myc* (Section 7.2.3) and those encoding proteins such as growth hormone or collagen type II. As with activating sequences, some silencer elements are constitutively active whilst others display cell-type specific activity. Thus, for example, the silencer in the gene encoding the T-lymphocyte marker CD4 represses its expression in most T cells where CD4 is not expressed but is inactive in a subset of T cells allowing these cells to express the CD4 protein actively (Sawada *et al.*, 1994). In many cases silencer elements have been shown to act by binding transcription factors which then act to reduce the rate of transcription (Figure 1.7b). The mechanisms by which this is achieved are discussed in Section 9.3.

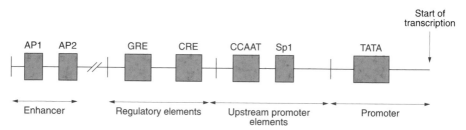

Figure 1.6 Structure of a typical gene with a TATA-box-containing promoter, upstream promoter elements such as the CCAAT and Sp1 boxes, regulatory elements inducing expression in response to treatment with substances such as glucocorticoid (GRE) and cyclic AMP (CRE) and other elements within more distant enhancers. Note that, as discussed in the text and illustrated in Figure 1.1, the upstream promoter elements are often interdigitated with the regulatory elements, whilst the same regulatory elements can be found upstream of the promoter and in enhancers.

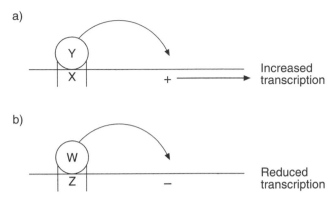

Figure 1.7 (a) A specific DNA sequence (X) can act to stimulate transcription by binding a positively acting factor (Y). (b) In contrast, binding of the negatively acting factor (W) to the DNA sequence (Z) inhibits transcription.

Obviously the balance between positively and negatively acting transcription factors which bind to the regulatory regions of a particular gene will determine the rate of gene transciption in any particular situation. In some cases, binding of the RNA polymerase and associated factors to the promoter and of other positive factors to the UPEs will be sufficient for transcription to occur, and the gene will be expressed constitutively. In other cases, however, such interactions will be insufficient and transcription of the gene will occur only in response to the binding, to another DNA sequence, of a factor which is activated in response to a particular stimulus or is present only in a particular tissue. These regulatory factors will then interact with the constitutive factors allowing transcription to occur. Hence their binding will result in the observed tissue-specific or inducible pattern of gene expression.

Such interaction is well illustrated by the metallothionein IIA gene. As illustrated in Figure 1.1, this gene contains a binding site for the transcription factor AP1 which produces induction of gene expression in response to phorbol ester treatment. The action of AP1 on the expression of the metallothionein gene is abolished, however, both by mutations in its binding site and by mutations in the adjacent Sp1 motif which prevent this motif binding its corresponding transcription factor Sp1 (Lee *et al.*, 1987). Although these mutations in the Sp1 motif do not abolish AP1 binding they do prevent its action indicating that the inducible AP1 factor interacts with the constitutive Sp1 factor to activate transcription.

Clearly such interactions between bound transcription factors need not be confined to factors bound to regions adjacent to the promoter but can also involve the similar factors bound to more distant enhancers. It is likely that this is achieved by a looping out of the intervening DNA

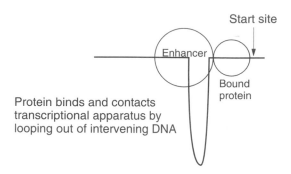

Figure 1.8 Contact between proteins bound at the promoter and those bound at a distant enhancer can be achieved by looping out of the intervening DNA.

allowing contact between factors bound at the promoter and those bound at the enhancer (Figure 1.8) (for further discussion, see Latchman, 1998).

This need for transcription factors to interact with one another to stimulate transcription means that transcription can also be stimulated by a class of factors which act indirectly by binding to the DNA and bending it so that other DNA-bound factors can interact with one another (Figure 1.9), Thus, the LEF-1 factor, which is specifically expressed in T lymphocytes, binds to the enhancer of the T-cell receptor α gene and bends the DNA so that other constitutively expressed transcription factors can interact with one another thereby allowing them to activate transcription. This results in the T-cell-specific expression of the gene even though the directly activating factors are not expressed in a T-cell-specific manner (for a review, see Werner and Burley, 1997).

1.4 CHROMATIN STRUCTURE AND ITS REMODELLING

1.4.1 Chromatin structure and gene regulation

The DNA in eukaryotic cells is packaged by association with specific proteins such as the histones into a structure known as chromatin (for reviews, see Wolffe, 1995; Latchman, 1998). The fundamental unit of this structure is the nucleosome in which the DNA is wrapped twice around a unit of eight histone molecules (two each of histones H2A, H2B, H3 and H4) (for a review see Rhodes, 1997). This structure is compacted further in genes which are not transcriptionally active or about to become active, whereas active or potentially active genes exist in the simple nucleoso-

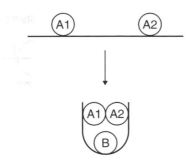

Figure 1.9 A factor which bends the DNA (B) can indirectly activate transcription by facilitating the interaction of two activating transcription factors (A1 and A2).

mal structure. Moreover, in the promoter or enhancer regions of these genes, nucleosomes are either removed altogether or undergo a structural alteration which facilitates the binding of specific transcription factors to their binding sites in these regions (Figure 1.10).

Hence the access of a transcription factor to its appropriate binding site will be affected by the manner in which that site is packaged within the chromatin. Evidently, therefore, genes which are about to be transcribed must undergo changes in chromatin structure which facilitate such transcription by allowing the access of activating transcription factors to their binding sites. Although a detailed discussion of these changes is beyond the scope of this book (for reviews see Wolffe, 1995; Felsenfeld, 1996; Felsenfeld *et al.*, 1996; Latchman, 1998), at least two mechanisms which can alter chromatin structure are of particular importance in terms of transcription factor regulation. These will be discussed in turn.

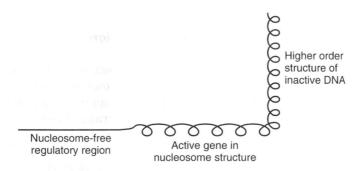

Figure 1.10 Levels of chromatin structure in active or inactive DNA.

1.4.2 Chromatin remodelling factors

A number of studies have identified protein complexes which are capable of binding to DNA, hydrolysing ATP and using the energy generated to disrupt the nucleosomal structure. The best characterized of these is the SWI/SNF complex which contains a number of different polypeptides. It was originally defined in yeast but has now been identified in a range of organisms including humans (for a review, see Pazin and Kadonaga, 1997a; Tsukiyama and Wu, 1997). The critical role of this complex in regulating gene expression is indicated by the phenotype of the *brahma* mutation in *Drosophila* which inactivates the SWI2 component of the complex. Thus, in this mutant, the genes encoding several homeobox-containing genes which control the correct patterning of the body (see Section 6.2) remain in an inactive chromatin structure and are hence not transcribed. This results in a mutant fly with a grossly abnormal body structure (for a review, see Simon, 1995).

Interestingly, the change in chromatin structure brought about by the SWI/SNF complex is maintained even after the complex has dissociated from the chromatin. Hence SWI/SNF can alter nucleosomal structure, exposing the binding site for an activating transcription factor, and then dissociate, with the activator then binding and activator transcription (Figure 1.11).

Evidently, this mechanism begs the question of how the SWI/SNF complex is itself recruited to the genes which need to be activated. This can occur via its association with the RNA polymerase complex or by its association with other transcription factors which can bind to their specific DNA binding sites even in tightly packed, non-remodelled chromatin. These processes are discussed in subsequent chapters.

1.4.3 Histone acetylation

The histone molecules which play a key role in chromatin structure are subject to a number of post-translational modifications such as phosphorylation, ubiquitination or acetylation (for a review, see Turner, 1993). In particular the addition of an acetyl group to a free amino group in lysine residues in the histone molecule reduces its net positive charge. Such acetylated forms of the histones have been found preferentially in active or potentially active genes where the chromatin is less tightly packed. Moreover, treatments which enhance histone acetylation, such as the addition of sodium butyrate to cultured cells, result in a less tightly packed chromatin structure and the activation of previously silent cellular genes. This suggests that hyperacetylation of histones could play a causal role in producing the more open chromatin structure

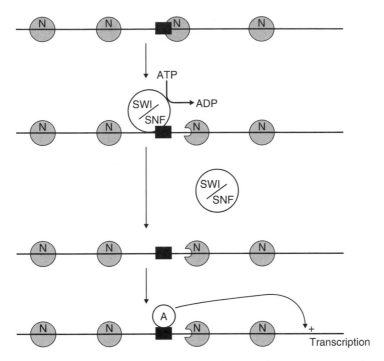

Figure 1.11 Activation of a gene in which the binding site for a transcriptional activator (solid box) is masked by the nucleosome (N) structure of chromatin. The SWI/SNF complex binds to the chromatin and uses the energy of ATP hydrolysis to displace a nucleosome or alter its structure. This change in chromatin structure is maintained after SWI/SNF dissociates, allowing the activator (A) to bind to its binding site and activate transcription.

characteristic of active or potentially active genes (for a review, see Grunston, 1997).

Hence, activation of gene expression could be achieved by factors with histone acetyltransferase activity which were able to acetylate histones and hence open up the chromatin structure, whereas inhibition of gene expression would be achieved by histone deacetylases which would have the opposite effect (Figure 1.12). Most interestingly, recent studies have identified both components of the basal transcriptional complex and specific activating transcription factors with histone acetyltransferase activity as well as specific inhibitory transcription factors with histone deacetylase activity (for reviews, see Tsukiyama and Wu, 1997; Pazin and Kadonaga, 1997b; Wu, 1997). These findings which link studies on modulation of chromatin structure with those on activating and inhibitory transcription factors are discussed further in later chapters.

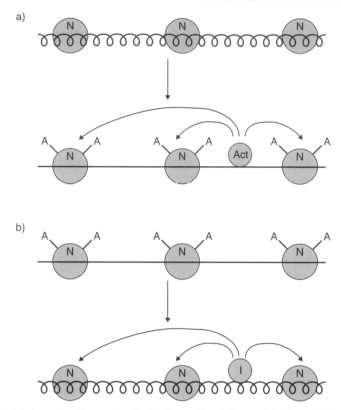

Figure 1.12 (a) An activating molecule (Act) can direct the acetylation of histones in the nucleosome (N), thereby resulting in a change in chromatin structure from a tightly packed (wavy line) to a more open (solid line) configuration. (b) An inhibitory molecule can direct the deacetylation of histones thereby having the opposite effect on chromatin structure.

1.5 CONCLUSIONS

It is clear that both the process of transcription itself and its regulation in particular tissues or in response to particular signals are controlled by short DNA sequence elements located adjacent to the promoter or in enhancers. In turn, such sequences act by binding proteins which are either active constitutively or are present in an active form only in a specific tissue or following a specific inducing signal. Such DNA-bound transcription factors then interact with each other and the RNA polymerase itself in order to produce constitutive or regulated transcription. The nature of these factors, the manner in which they function and their role in different biological processes forms the subject of this book.

REFERENCES

Andersen, B. and Rosenfeld, M.G. (1994). Pit-1 determines cell types during the development of the anterior pituitary gland. *Journal of Biological Chemistry* **269**, 29335–29338.

Bienz, M. and Pelham, H.R.B. (1986). Heat-shock regulatory elements function as an inducible enhancer when linked to a heterologous promoter. *Cell* **45**, 753–760.

Breathnach, R. and Chambon, P. (1981). Organization and expression of eukaryotic split genes coding for proteins. *Annual Review of Biochemistry* **50**, 349–383.

Carey, M. (1998). Enhanceosome and transcriptional synergy. *Cell* **92**, 5–8.

Darnell, J.E. (1982). Variety in the level of gene control in eukaryotic cells. *Nature* **297**, 365–371.

Davidson, E.H., Jacobs, H.T. and Britten, R.J. (1983). Very short repeats and coordinate induction of genes. *Nature* **301**, 468–470.

Dynan, W.S. and Tjian, R. (1985). Control of eukaryotic messenger RNA synthesis by sequence-specific DNA-binding proteins. *Nature* **316**, 774–778.

Elshotz, H.P., Albert, V.R., Treacy, M.N. and Rosenfeld, M.G. (1990). A two-base change in a POU factor binding site switches pituitary-specific to lymphoid-specific gene expression. *Genes and Development* **4**, 43–51.

Felsenfeld, G. (1996). Chromatin unfolds. *Cell* **86**, 13–19.

Felsenfeld, G., Boyes, J., Chung, J., Clark, D. and Studitsky, V. (1996). Chromatin structure and gene expression. Proceedings of the National Academy of Sciences USA **93**, 9384–9388.

Garcia, J.V., Bich-Thuy, L., Stafford, J. and Queen, C. (1986). Synergism between immunoglobulin enhancers and promoters. *Nature* **322**, 383–385.

Goodwin, G.H., Partington, G.A. and Perkins, N.D. (1990). Sequence specific DNA binding proteins involved in gene transcription. In: *Chromosomes: Eukaryotic, Prokaryotic and Viral* (Adolph, K.W., ed.), Vol. 1, pp. 31–85. Boca Raton, Florida: CRC Press.

Grunstein, M. (1997). Histone acetylation in chromatin structure and transcription. *Nature* **389**, 349–352.

Hatzopoulous, A.K., Schlokat, U. and Gruss, P. (1988). Enhancers and other cis-acting sequences. In: *Transcription and Splicing* (Hames, B.D. and Glover, D.M., eds), pp. 43–96. Oxford: IRL Press.

Jones, N.C., Rigby, P.W.J. and Ziff, E.B. (1988). Trans-acting protein factors and the regulation of eukaryotic transcription. *Genes and Development* **2**, 267–281.

Latchman, D.S. (1998). *Gene Regulation: A Eukaryotic Perspective*, 3rd edn. London: Chapman and Hall.

Lee, W., Haslinger, A., Karin, M. and Tjian, R. (1987). Activation of transcription by two factors that bind promoter and enhancer sequences of the human metallothionein gene and SV40. *Nature* **325**, 369–372.

McKnight, S. and Tjian, R. (1986). Transcriptional selectivity of viral genes in mammalian cells. *Cell* **46**, 795–805.

Muller, M.M., Gerster, T. and Schaffner, W. (1988) Enhancer sequences and the regulation of gene transcription. *European Journal of Biochemistry* **176**, 485–495.

Muller-Hill, B.W. (ed.) (1996). The *lac* operon: a short history of a genetic paradigm. Berlin: de Grayler Co.

Nevins, J.R. (1983). The pathway of eukaryotic mRNA transcription. *Annual Review of Biochemistry* **52**, 441–446.

Pazin, M.J. and Kadonaga, M. (1997a) SWI/SNF2 and related proteins: ATP-driven motors that disrupt protein-DNA interations? *Cell* **88**, 737–740.

Pazin, M.J. and Kadonaga, M. (1997b). What's up and down with histone acetylation and transcription. *Cell* **89**, 325–328.

Pelham, H.R.B. (1982). A regulatory upstream promoter element in the Drosophila *hsp70* heat-shock gene. *Cell* **30**, 517–528.

Rhodes, D. (1997). The nucleosome core all wrapped up. *Nature* **389**, 231–233.

Ross, J. (1996). Control of messenger RNA stability in high eukaryotes. *Cell* **74**, 413–421.

Sawada, S., Scarborough, J.D., Kileen, N. and Littman, D.R. (1994) A lineage-specific transcriptional silencer regulates CD4 gene expression during T lymphocyte development. *Cell* **77**, 917–929.

Sen, R. and Baltimore, D. (1986). Multiple nuclear factors interact with the immunoglobulin enhancer sequences. *Cell* **46**, 705–716.

Sentenac, A. (1985) Eukaryotic RNA polymerases. *CRC Critical Reviews in Biochemistry* **1**, 31–90.

Simon, J. (1995). Locking in stable states of gene expression: transcriptional control during *Drosophila* development. *Current Opinion in Cell Biology* **7**, 376–385.

Thiele, D.J. (1992). Metal regulated transcription in eukaryotes. *Nucleic Acids Research* **20**, 1183–1191.

Travers, A. (1993) *DNA–Protein Interactions*. London: Chapman and Hall.

Tsukiyama, T. and Wu, C. (1997). Chromatin remodelling and transcription. *Current Opinion in Genetics and Development* **7**, 182–191.

Turner, B.M. (1993). Decoding the nucleosome. *Cell* **75**, 5–8.

Wang, J. and Manley, J.L. (1997). Regulation of pre-mRNA splicing in metazoa. *Current Opinion in Genetics and Development* **7**, 205–211.

Weis, L. and Reinberg, D. (1992) Transcription by RNA polymerase II initiator directed formation of transcription competent complexes. *FASEB Journal* **6**, 3300–3309.

Werner, M.H. and Burley, S.K. (1997). Architectural transcription factors: proteins that remodel DNA. *Cell* **88**, 733–736.

Williams, G.T., McClanahan, T.K. and Morimoto, R.I. (1989). E1a transactivation of the human *hsp70* promoter is mediated through the basal transcriptional complex. *Molecular and Cellular Biology* **9**, 2574–2587.

Wirth, T., Staudt, L. and Baltimore, D. (1987). An octamer oligonucleotide upstream of a TATA motif is sufficient for lymphoid specific promoter activity. *Nature* **329**, 174–178.

Wolffe, A. (1995). *Chromatin: Structure and Function*, 2nd edn. London, San Diego. Academic Press.

Wu, C. (1997). Chromatin remodelling and the control of gene expression. *Journal of Biological Chemistry* **272**, 28171–28174.

CHAPTER TWO

Methods for studying transcription factors

2.1 INTRODUCTION

The explosion in the available information on transcription factors that has occurred in recent years has arisen primarily because of the availability of new or improved methods for studying these factors. Initially such studies frequently focus on identifying a factor that interacts with a particular DNA sequence and characterizing this interaction. Subsequently the protein identified in this way is further characterized and purified, and its corresponding gene isolated for further study. The methods involved in these two types of study will be considered in turn in Sections 2.2 and 2.3, respectively, together with the methods for determining the DNA-binding site of a transcription factor that is intially identified by other means (Section 2.3.4) For details of the methodologies involved, see Latchman (1998.)

2.2 METHODS FOR STUDYING DNA–PROTEIN INTERACTIONS

2.2.1 DNA mobility shift assay

As discussed in Section 1.2, the initial stimulus to identify a transcription factor frequently comes from the identification of a particular DNA sequence that confers a specific pattern of expression on a gene which carries it. The next step, therefore, following the identification of such a sequence will be to define the protein factors that bind to it. This can be readily achieved by the DNA mobility shift or gel retardation assay (Fried and Crothers, 1981; Garner and Revzin, 1981).

This method relies on the obvious principle that a fragment of DNA to which a protein has bound will move more slowly in gel electrophoresis than the same DNA fragment without bound protein. The DNA mobility shift assay is carried out, therefore, by first radioactively labelling the specific DNA sequence whose protein-binding properties are being investigated. The labelled DNA is then incubated with a nuclear (Dignam *et al.*, 1983) or whole cell (Manley *et al.*, 1980) extract of cells prepared in such a way as to contain the DNA-binding proteins. In this way DNA–protein complexes are allowed to form. The complexes are then electrophoresed on a non-denaturing polyacrylamide gel and the position of the radioactive DNA visualized by autoradiography. If no protein has bound to the DNA, all the radioactive label will be at the

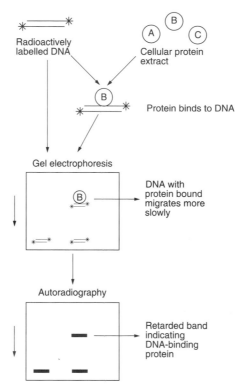

Figure 2.1 DNA mobility shift or gel retardation assay. Binding of a cellular protein (B) to the radioactively labelled DNA causes it to move more slowly upon gel electrophoresis and hence results in the appearance of a retarded band upon autoradiography to detect the radioactive label.

bottom of the gel, whereas if a protein–DNA complex has formed, radioactive DNA to which the protein has bound will migrate more slowly and hence will be visualized near the top of the gel (Figure 2.1). (For methodological details, see Smith *et al.* (1998).)

This technique can be used, therefore, to identify proteins which can bind to a particular DNA sequence in extracts prepared from specific cell types. Thus, for example, in the case of the octamer sequence discussed in Section 1.2.3, a single retarded band is detected when this sequence is mixed, for example, with a fibroblast extract. In contrast, when an extract from immunoglobulin-producing B cells is used, two distinct retarded bands are seen (Figure 2.2). Since each band is produced by a

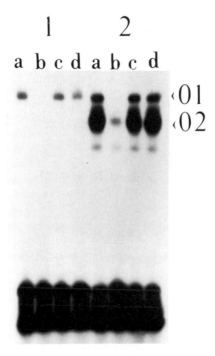

Figure 2.2 DNA mobility shift assay using a radioactively labelled probe containing the binding site for octamer binding proteins (ATGCAAAT) and extracts prepared from fibroblast cells (1) or B cells (2). Note that fibroblast cells contain only one protein Oct-1 (O1) capable of producing a retarded band, whereas B cells contain both Oct-1 and an additional tissue-specific protein Oct-2 (O2). The complexes formed by Oct-1 and Oct-2 on the labelled oligonucleotide in the absence of unlabelled oligonucleotide (track a) are readily removed by a 100-fold excess of unlabelled octamer oligonucleotide (track b). They are not removed, however, by a similar excess of a mutant octamer oligonucleotide (ATAATAAT), which is known not to bind octamer binding proteins (track c) (Lenardo *et al.*, 1987) or of the binding site for the unrelated transcription factor Sp1 (track d) (Dynan and Tjian, 1983). This indicates that the retarded bands are produced by sequence-specific DNA-binding proteins which bind specifically to the octamer motif and not to mutant or unrelated motifs.

distinct protein binding to the DNA, this indicates that, in addition to the ubiquitous octamer binding protein Oct-1 which is present in most cell types, B cells also contain an additional octamer binding protein, Oct-2, which is absent in many other cells. The role of the Oct-2 protein in immunoglobulin gene expression is discussed further in Section 5.2.2.

As well as defining the proteins binding to a particular sequence, the DNA mobility shift assay can also be used to investigate the precise sequence specificity of this binding. This can be done by including in the binding reaction a large excess of a second DNA sequence which has not been labelled. If this DNA sequence can also bind the protein bound by the labelled DNA, it will do so. Moreover, binding to the unlabelled DNA will predominate since it is present in large excess. Hence the retarded band will not appear in the presence of the unlabelled competitor since only protein–DNA complexes containing labelled DNA are visualized on autoradiography (Figure 2.3b). In contrast, if the competitor cannot bind the same sequence as the labelled DNA, the complex with the labelled DNA will form and the labelled band will be visualized as before (Figure 2.3c).

Thus, by using competitor DNAs which contain the binding sites for previously described transcription factors, it can be established whether the protein detected in a particular mobility shift experiment is identical or related to any of these factors. Similarly, if competitor DNAs are used which differ in only one or a few bases from the original binding site, the effect of such base changes on the efficiency of the competitor DNA and hence on binding of the transcription factor can be assessed. Figure 2.2 illustrates an example of this type of competition approach, showing that the octamer binding proteins Oct-1 and Oct-2 are efficiently competed away from the labelled octamer probe by an excess of identical unlabelled competitor but not by a competitor containing three base changes in this sequence which prevent binding (ATGCAAAT to ATAATAAT) (Lenardo *et al.*, 1987). Similarly no competition is observed, as expected, when the binding site of an unrelated transcription factor Sp1 (see Section 3.3.2) is used as the competitor DNA.

The DNA mobility shift assay therefore provides an excellent means of initially identifying a particular factor binding to a specific sequence and characterizing both its tissue distribution and its sequence specificity.

2.2.2 DNAseI footprinting assay

Although the mobility shift assay provides a means of obtaining information on DNA–protein interaction, it cannot be used directly to localize the area of the contact between protein and DNA. For this purpose, the DNAseI footprint assay is used (Galas and Schmitz, 1978; Dynan and Tjian, 1983).

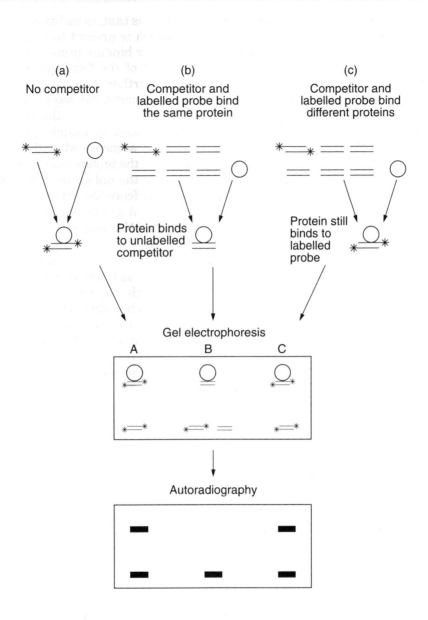

Figure 2.3 Use of unlabelled competitor DNAs in the DNA mobility shift assay. If an unlabelled DNA sequence is capable of binding the same protein as is bound by the labelled probe, it will do so (b) and the radioactive retarded band will not be observed, whereas if it cannot bind the same protein (c), the radioactive retarded band will form exactly as in the absence of competitor (a).

In this assay, DNA and protein are mixed as before, the DNA being labelled, however, only at the end of one strand of the double-stranded molecule. Following binding, the DNA is treated with a small amount of the enzyme deoxyribonuclease I (DNAseI), which will digest DNA. The digestion conditions are chosen, however, so that each molecule of DNA will be cut once or a very few times by the enzyme. Following digestion the bound protein is removed and the DNA fragments separated by electrophoresis on a polyacrylamide gel capable of resolving DNA fragments differing in size by only one base. This produces a ladder of bands representing the products of DNAseI cutting either one or two or three or four etc., bases from the labelled end. Where a particular piece of the DNA has bound a protein, however, it will be protected from digestion

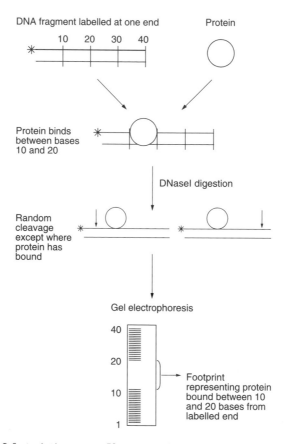

Figure 2.4 DNaseI footprinting assay. If a protein binds at a specific site within a DNA fragment labelled at one end, the region of DNA at which the protein binds will be protected from digestion with DNaseI. Hence this region will appear as a footprint in the ladder of bands produced by the DNA being cut at all other points by DNaseI.

and hence the bands corresponding to cleavage at these points will be absent. This will be visualized on electrophoresis as a blank area on the gel lacking labelled fragments and is referred to as the footprint of the protein (Figure 2.4). Similar labelling of the other strand of the DNA molecule will allow the interaction of the protein with the other strand of the DNA to be assessed.

The footprinting technique, therefore, allows a visualization of the interaction of a particular factor with a specific piece of DNA. By using a sufficiently large piece of DNA, the binding of different proteins to different DNA sequences within the same fragment can be assessed. An analysis of this type is shown in Figure 2.5. This shows the footprints (A and B) produced by two cellular proteins binding to two distinct sequences within a region of the human immunodeficiency virus (HIV) control element which has an inhibiting effect on promoter activity (Orchard *et al.*, 1990). Interestingly some insights into the topology of the DNA–protein interaction are also obtained in this experiment since bands adjacent to the protected region appear more intense in the presence of the protein. These regions of hypersensitivity to cutting are likely to represent a change in the structure of the DNA in this region when the protein has bound, rendering the DNA more susceptible to enzyme cleavage.

As with the mobility shift assay, unlabelled competitor sequences can be used to remove a particular footprint and determine its sequence specificity. In the HIV case illustrated in Figure 2.5, short DNA competitors containing the sequence of one or other of the footprinted areas were used to specifically remove each footprint without affecting the other, indicating that two distinct proteins produce the two footprints.

As well as footprinting using DNAseI, other footprinting techniques have been developed which rely on the protection of DNA which has bound protein from cleavage by other reagents that normally cleave the DNA. These include hydroxyl radical footprinting and phenanthroline–copper footprinting which, like DNAseI footprinting, rely on the ability of the reagents to cleave the DNA in a non-sequence specific manner (for further details see Kreale, 1994; Papavassilou, 1995).

Of greater interest, however, is the technique of dimethyl sulphate (DMS) protection footprinting since it can provide information on the exact bases within the binding site that are contacted by the protein. Thus this method relies on the ability of DMS specifically to methylate guanine residues in the DNA. These methylated G residues can then be cleaved by exposure to piperidene, whereas no cleavage occurs at unmethylated G residues (Maxam and Gilbert, 1980). A protein bound to the DNA will protect the guanine residues which it contacts from methylation and hence they will not be cleaved upon subsequent

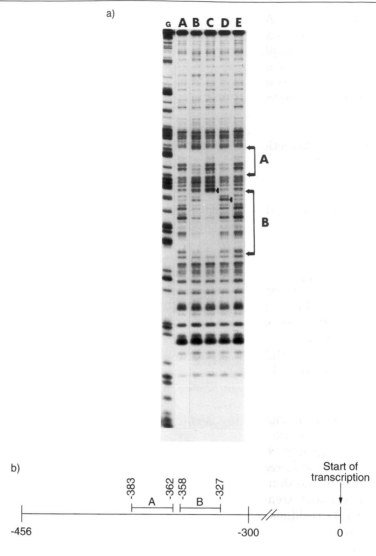

Figure 2.5 (a) DNaseI footprinting assay carried out on a region of the human immunodeficiency virus (HIV) control element. The two footprints (A and B) are not observed when no cell extract is added to the reaction (track A) but are observed when cellular extract is added in the absence of competitor (track B). Addition of unlabelled oligonucleotide competitor containing the DNA sequence of site A removes the site A footprint without affecting site B (track C), whilst an unlabelled oligonucleotide containing the site B DNA sequence has the opposite effect (track D). Both footprints are removed by a mixture of unlabelled site A and B oligonucleotides (track E). Arrows indicate the position of sites at which cleavage with DNaseI is enhanced in the presence of protein bound to an adjacent site indicating the existence of conformational changes induced by protein binding. The track labelled G represents a marker track consisting of the same DNA fragment chemically cleaved at every guanine residue. (b) Position of sites A and B within the HIV control element. The arrow indicates the start site of transcription.

piperidene treatment. As in the other footprinting techniques, therefore, specific bands produced by such treatment of naked DNA are absent in the protein–DNA sample (see Lassar *et al.* (1989) for an example of this approach). Unlike the other methods, however, because cleavage occurs at specific guanine residues, this method identifies specific bases within the DNA that are contacted by the transcription factor protein.

These footprinting techniques, therefore, offer an advance on the mobility shift assay, allowing a more precise visualization of the DNA protein interaction. (For methodological details see Spiro and McMurrary (1998).)

2.2.3 Methylation interference assay

The pattern of DNA–protein interaction can also be studied in more detail using the methylation interference assay (Siebenlist and Gilbert, 1980). Like methylation protection, this method relies on the ability of DMS to methylate G residues which can then be cleaved with piperidene. However, methylation interference is based on assessing whether the prior methylation of specific G residues in the target DNA affects subsequent protein binding.

Thus, the target DNA is first partially methylated using DMS so that, on average, only one G residue per DNA molecule is methylated (Maxam and Gilbert, 1980). Each individual DNA molecule will, therefore, contain some methylated G residues with the particular residues which are methylated being different in each molecule. These partially methylated DNAs are then used in a DNA mobility shift experiment with an appropriate cell extract containing the DNA-binding protein. Following electrophoresis the band produced by the DNA, which has bound protein, and that produced by the DNA, which has not, are excised from the gel and treated with piperidine to cleave the DNA at the methylated G residues and not at unmethylated G residues. Clearly, if methylation of a particular G prevents protein binding, then cleavage at this particular methylated G will be observed only in the DNA which failed to bind the protein. Conversely, if a particular G residue plays no role in binding, then cleavage at this G residue will be observed equally in both the DNA which bound the protein and that which failed to do so (Figure 2.6).

Figure 2.7 shows this type of analysis applied to the protein binding to site B within the negatively acting element in the HIV promoter (for the footprint produced by the binding of this protein, see Figure 2.5). In this case the footprinted sequence was palindromic (Figure 2.7), suggesting that the DNA–protein interaction may involve similar binding to the two halves of the palindrome. The methylation interference analysis of site B

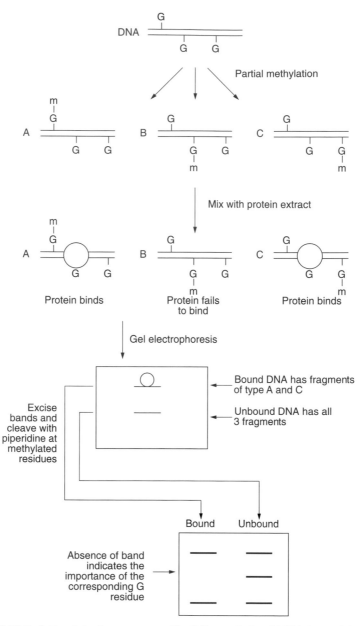

Figure 2.6 Methylation interference assay. Partially methylated DNA is used in a DNA mobility shift assay and both the DNA which has failed to bind protein and that which has bound protein and formed a retarded band are subsequently cleaved at methylated G residues with piperidine. If methylation at a specific G residue has no effect on protein binding (types A and C), the bound and unbound DNA will contain equal amounts of methylated G at this position. In contrast, if methylation at a particular G prevents binding of the protein (type B), only the unbound DNA will contain methylated G at this position.

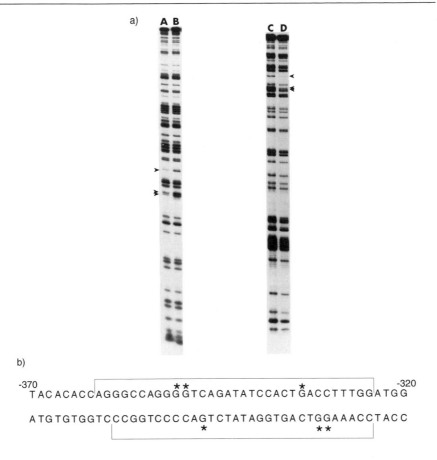

a)

b)

-370   * *   *   -320
　T AC ACAC C AGGGC C AGGG G G TC AGATA TC CACT G ACC TTTGG A TG G

　A TG TGTGG TC C CGG TC C C CAGT C TAT AGG TGA C TGG A A AC C TAC C
　　　　　　　　　　　　*   * *

Figure 2.7 (a). Methylation interference assay applied to the DNA of site B in the HIV control element, as defined in the footprinting experiment shown in Figure 2.5. Both the upper (tracks A and B) and lower (tracks C and D) strands of the double-stranded DNA sequence were analysed. Tracks B and C show the methylation pattern of the unbound DNA, which failed to bind protein, whereas tracks A and B show the methylation pattern of DNA, which has bound protein. The arrows show G residues whose methylation is considerably lower in the bound compared to the unbound DNA and which are therefore critical for binding the specific cellular protein, which interacts with this DNA sequence. (b) DNA sequence of site B. The extent of the footprint region is indicated by the square brackets and the critical G residues defined by the methylation interference assay in (a) are marked with an asterisk. Note the symmetrical pattern of critical G residues within the palindromic DNA sequence.

confirms this by showing that methylation of equivalent G residues in each half of the palindrome interferes with binding of the protein, indicating that these residues are critical for binding.

Although the DMS method only studies contacts of the protein with G residues, interference analysis can also be used to study the interaction of DNA-binding proteins with A residues in the binding site. This can be done either by methylating all purines to allow study of interference at A and G residues simultaneously (see, for example, Ares *et al.*, 1987) or by using diethylpyrocarbonate to specifically modify A residues (probably by carboxyethylation), rendering them susceptible to piperidine cleavage (see, for example, Sturm *et al.*, 1988). These techniques are of particular value when studying sequences such as the octamer motif in which there are relatively few G residues, hence limiting the information which can be obtained by studying interference at G residues alone (Sturm *et al.*, 1987; Baumruker *et al.*, 1988).

Chemical interference techniques can, therefore, be used to supplement footprinting methodologies and identify the precise DNA–protein interactions within the footprinted region. (For methodological details, see Spiro and McMurray (1998).)

2.2.4 *In vivo* footprinting assay

Although the methods described so far can provide considerable information about DNA–protein contacts they all suffer from the deficiency that the DNA – protein interaction occurs *in vitro* when cell extract and the DNA are mixed. Hence they indicate what factors can bind to the DNA rather then whether such factors actually do bind to the DNA in the intact cell where a particular factor may be sequestered in the cytoplasm or where its binding may be impeded by the association of DNA with other proteins such as histones.

These problems are overcome by the technique of *in vivo* footprinting which is an extension of the *in vitro* DMS protection footprinting technique described in Section 2.2.2. Thus intact cells are freely permeable to DMS, which can therefore be used to methylate the DNA within its native chromatin structure in such cells. Exactly as in the *in vitro* technique, G residues to which a protein has bound will be protected from such methylation and will therefore not be cleaved when the DNA is subsequently isolated and treated with piperidene. Hence the bands produced by cleavage at these residues will be absent when the pattern produced by intact chromatin is compared to that produced by naked DNA (Figure 2.8).

Obviously the amounts of any specific DNA sequence obtained from total chromatin in this procedure are vanishingly small compared to when a cloned DNA fragment is used in the *in vitro* procedure. It is hence necessary to amplify the DNA of interest from within total chromatin by the polymerase chain reaction in order to obtain sufficient material for analysis by this method. When this is done, however, *in vivo* footprinting provides an excellent means for analysing DNA–protein contacts within intact cells *in vivo* as well as determining the changes in such contacts which occur in response to specific treatments (see Herrera *et al.* (1989) and Mueller *et al.* (1990) for examples of this approach, and Spiro and McMurray (1998) for a full description of the methodologies involved).

Taken together, therefore, the three methods of DNA mobility shift, footprinting and methylation interference can provide considerable information on the nature of the interaction between a particular DNA sequence and a transcription factor. They serve as an essential prelude to a detailed study of the transcription factor itself.

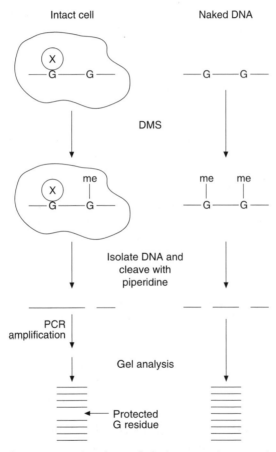

Figure 2.8. In vivo footprinting using the methylation protection assay in which specific G residues are protected by bound protein (X) from methylation by dimethyl sulphate (DMS) treatment of intact cells. Hence, following DNA isolation, cleavage of methylated G residues with piperidine and subsequent amplification by the polymerase chain reaction (PCR), the band corresponding to cleavage at this protected residue will be absent. In contrast, cleavage at this position will be observed in naked DNA where no protein protects this residue from methylation.

2.3 METHODS FOR STUDYING THE TRANSCRIPTION FACTOR ITSELF

2.3.1 Protein purification

As discussed above, once a particular DNA sequence has been shown to be involved in transcriptional regulation, a number of techniques are available for characterizing the binding of transcription factors to this sequence. Although such studies can be carried out on crude cellular extracts containing the protein, ultimately they need to be supplemented by studies on the protein itself. This can be achieved by purifying the transcription factor from extracts of cells containing it. Unfortunately, however, conventional protein purification techniques such as conventional chromatography and high-pressure liquid chromatography (HPLC) result in the isolation of transcription factors at only 1–2% purity (Kadonaga and Tjian, 1986).

To overcome this problem and purify the transcription factor Sp1, Kadonaga and Tjian (1986) devised a method involving DNA affinity chromatography. In this method (Figure 2.9), a DNA sequence containing a high affinity binding site for the transcription factor is synthesized and the individual molecules joined to form a multimeric molecule. This very high-affinity binding site is then coupled to an activated sepharose support on a column and total cellular protein passed down the column. The Sp1 protein binds specifically to its corresponding DNA sequence whilst all other cellular proteins do not bind. The bound Sp1 can be eluted simply by raising the salt concentration. Two successive affinity chromatography steps of this type successfully resulted in the isolation of Sp1 at 90% purity, 30% of the Sp1 in the original extract being recovered, representing a 500–1000-fold purification (Kadonaga and Tjian, 1986).

Although this simple one-step method was successful in this case, it relies critically on the addition of exactly the right amount of non-specific DNA carrier to the cell extract. Thus, this added carrier acts to remove proteins which bind to DNA in a non-sequence-specific manner and which would hence bind non-specifically to the Sp1 affinity column and contaminate the resulting Sp1 preparation. This contamination will occur if too little carrier is added. If too much carrier is added, however, it will bind out the Sp1, since, like all sequence-specific proteins, Sp1 can bind with low affinity to any DNA sequence. Hence in this case no Sp1 will bind to the column itself (Figure 2.10).

To overcome this problem, Rosenfeld and Kelley (1986) devised a method in which proteins capable of binding to DNA with high affinity in a non-sequence-specific manner are removed prior to the affinity column. To do this, the bulk of cellular protein was removed on a Biorex 70 high-

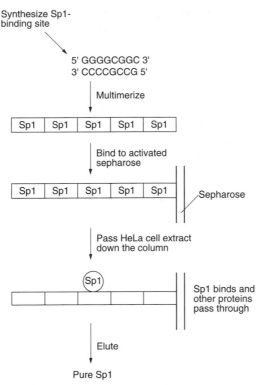

Figure 2.9 Purification of transcription factor Sp1 on an affinity column in which multiple copies of the DNA sequence binding Sp1 have been coupled to a sepharose support (Kadonaga and Tjian, 1986).

capacity ion-exchange column and proteins which can bind to any DNA with high affinity were then removed on a cellulose column to which total bacterial DNA had been bound. Subsequently the remaining proteins, which had bound to non-sequence-specific DNA only with low affinity, were applied to a column containing a high-affinity binding site for transcription factor NF1 (Figure 2.11). NF1 bound to this site with high affinity and could be eluted in essentially pure form by raising the salt concentration (Table 2.1). It should be noted that in this and other purification procedures the fractions containing the transcription factor can readily be identified by carrying out a DNA mobility shift or footprinting assay with each fraction using the specific DNA-binding site of the transcription factor.

The purified protein obtained in this way can obviously be used to characterize the protein, for example, by determining its molecular weight or by raising an antibody to it to characterize its expression pattern in different cell types. Similarly the activity of the protein can be

a) Correct amount of carrier

b) Too little carrier

c) Too much carrier

Figure 2.10 Consequences of adding different amounts of non-specific carrier DNA to the protein passing through the Sp1 affinity column. If the correct amount of non-specific carrier is added, it will bind proteins which interact with DNA in a non-sequence-specific manner, allowing Sp1 to bind to the column (a). However, addition of too little carrier will result in non-sequence-specific proteins binding to the column, thereby preventing the binding of Sp1 (b), whereas in the presence of too much carrier both the non-specific proteins and Sp1 will bind to the carrier (c).

assessed by adding it to cellular extracts and assessing its effect on their ability to transcribe an exogenously added DNA in an *in vitro* transcription assay.

Unfortunately, however, because of the very low abundance of transcription factors in the cell, these purification procedures yield very small amounts of protein. For example Treisman (1987) succeeded in purifying only 1.6 μ-grams of the serum response factor starting with 2×10^{10} cells or 40 g of cells. Such difficulties clearly limit the experiments which can be done with purified material. Indeed, the primary use of purified factor in most cases has simply been to provide material to isolate the gene encoding the protein. This gene can then be expressed either *in vitro* or in bacteria to provide a far more abundant source of the corresponding protein than could be obtained from cells which naturally express it.

Table 2.1 Purification of transcription factor NF1 from HeLa cells

	Total protein (mg)	Specific binding of ^{32}P DNA (fmol/mg protein) $\times 10^{-3}$	Purification (fold)	Yield (%)
HeLa cell extract*	4590	3.1	1.0	100
Biorex 70 column	550	27.1	8.7	104
E. coli DNA cellulose	65.2	181	58.4	83
NF1 affinity matrix				
1st passage	2.1	4510	1455	67
2nd passage	1.1	7517	2425	57

* Prepared from 6×10^{10} cells or 120 g cells.

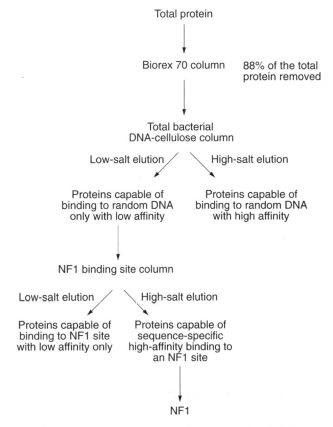

Figure 2.11 Purification of transcription factor NF1 (Rosenfeld and Kelley, 1986). Following removal of most cellular proteins on a Biorex 70 ion-exchange column, proteins which bind to all DNA sequences with high affinity were removed on a bacterial DNA–cellulose column. Subsequent application of the remaining proteins to a column containing the NF1 binding site results in the purification of NF1, since it is the only protein which binds with low affinity to random DNA but with high affinity to an NF1 site.

2.3.2 Gene cloning

In order to isolate the gene encoding a particular transcription factor, a complementary DNA (cDNA) library is prepared from mRNA isolated from a cell type expressing the factor. (For details of the methods used in preparing these libraries, see Sambrook *et al.* (1989).) Some means is then required to identify a clone derived from the mRNA encoding the factor amongst all the other clones in the library. Two methods are normally used for this purpose.

(a) Use of oligonucleotide probes predicted from the protein sequence of the factor

If a particular transcription factor has been purified it is possible to obtain portions of its amino-acid sequence. In turn such sequences can be used to predict oligonucleotides containing a DNA sequence capable of encoding these protein fragments. Owing to the redundancy of the genetic code whereby several different DNA codons can encode a particular amino-acid, there will be multiple different oligonucleotides capable of encoding a particular amino acid sequence. All these possible oligonucleotides are synthesized chemically, made radioactive and used to screen the cDNA library. The oligonucleotide in the mixture which does correspond to the amino-acid sequence of the transcription factor will hybridize to the corresponding sequence in a cDNA clone derived from mRNA encoding the factor. Hence such a clone can be readily identified in the cDNA library (Figure 2.12).

In cases where purified protein is available as in those discussed in the previous section, this approach represents a relatively simple method for isolating cDNA clones. It has therefore been widely used to isolate cDNA clones corresponding to purified factors such as Sp1 (Kadonaga *et al.*, 1987) (Figure 2.12), NF1 (Santoro *et al.*, 1988) and the serum response factor (Norman *et al.*, 1988) (for methodological details, see Nicolas and Goodwin 1998)).

(b) Use of oligonucleotide probes derived from the DNA-binding site of the factor

Although relatively simple, the use of oligonucleotides derived from protein sequence does require purified protein. As we have seen, purification of a transcription factor requires a vast quantity of cells and is technically difficult. Moreover, eventual determination of the

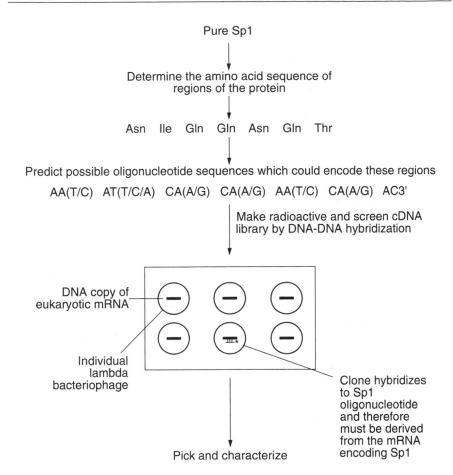

Figure 2.12 Isolation of cDNA clones for the Sp1 transcription factor by screening with short oligonucleotides predicted from the protein sequence of Sp1. Because several different triplets of bases can code for any given amino acid, multiple oligonucleotides that contain every possible coding sequence are made. Positions at which these oligonucleotides differ from one another are indicated by the brackets containing more than one base.

partial amino-acid sequence of the protein requires access to expensive protein-sequencing apparatus.

To bypass these problems, Singh *et al.* (1988) devised a procedure which is based on the fact that information is usually available about the specific DNA sequence to which a particular transcription factor binds. Hence a cDNA clone expressing the factor can be identified in a library by its ability to bind the appropriate DNA sequence. This method relies, therefore, on DNA–protein binding rather than DNA–DNA-binding. Hence the library must be prepared in such a way that the cloned cDNA

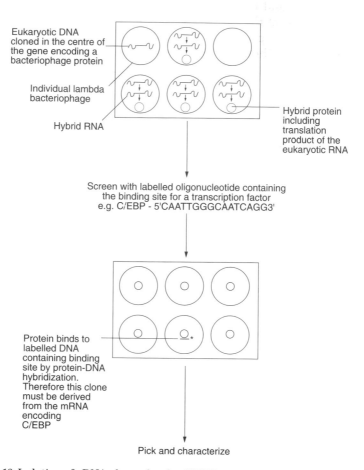

Make cDNA library in such a way that eukaryotic
mRNA will be expressed in the bacteria

Eukaryotic DNA
cloned in the centre of
the gene encoding a
bacteriophage protein

Individual lambda
bacteriophage

Hybrid RNA

Hybrid protein
including
translation
product of the
eukaryotic RNA

Screen with labelled oligonucleotide containing
the binding site for a transcription factor
e.g. C/EBP - 5'CAATTGGGCAATCAGG3'

Protein binds to
labelled DNA
containing binding
site by protein-DNA
hybridization.
Therefore this clone
must be derived
from the mRNA
encoding
C/EBP

Pick and characterize

Figure 2.13 Isolation of cDNA clones for the C/EBP transcription factor by screening an expression library with a DNA probe containing the binding site for the factor.

inserts are translated by the bacteria into their corresponding proteins. This is normally achieved by inserting the cDNA into the coding region of the bacteriophage λ β-galactosidase gene resulting in its translation as part of the bacteriophage protein. The resulting fusion protein binds DNA with the same sequence specificity as the original factor. Hence a cDNA clone encoding a particular factor can be identified in the library by screening with a radioactive oligonucleotide containing the binding site (Figure 2.13). This technique has been used to isolate cDNA clones encoding several transcription factors such as the CCAAT box-binding factor C/EBP (Vinson *et al.*, 1988) and the octamer-binding proteins Oct-1 (Sturm *et al.*, 1988) and Oct-2 (Staudt *et al.*, 1988) (for methodological details, see Cowell and Hurst 1998)).

The development of these two methods of screening with oligonucleotides derived from the protein sequence or oligonucleotides derived from the binding site has, therefore, resulted in the isolation of cDNA clones corresponding to very many transcription factors.

2.3.3 Use of cloned genes

This isolation of cDNA clones has in turn resulted in an explosion of information on these factors. Thus, once a clone has been isolated, its DNA sequence can be obtained allowing prediction of the corresponding protein sequence and comparison with other factors. Similarly, the clone can be used to identify the mRNA encoding the protein and examine its expression in various tissues by Northern blotting, to study the structure of the gene itself within genomic DNA by Southern blotting, and as a probe to search for related genes expressed in other tissues or other organisms.

Most importantly, however, the isolation of cDNA clones provides a means of obtaining large amounts of the corresponding protein for functional study. This can be achieved either by coupled *in vitro* transcription and translation (Figure 2.14a) (see, for example, Sturm *et al.*, 1988), or by expressing the gene in bacteria either in the original expression vector used in the screening procedure (see Section 2.3.2b) or more commonly by sub-cloning the cDNA into a plasmid expression vector (Figure 2.14b) (see, for example, Kadonaga *et al.*, 1987).

The protein produced in this way has similar activity to the natural protein, being capable of binding to DNA in footprinting or mobility shift assays (see, for example, Kadonaga *et al.*, 1987) and of stimulating the transcription of appropriate DNAs containing its binding site when added to a cell-free transcription system (see, for example, Mueller *et al.*, 1990).

Moreover, once a particular activity has been identified in a protein produced in this way, it is possible to analyse the features of the protein which produce this activity in a way that would not be possible using the factor purified from cells which normally express it. Thus, because the cDNA clone of the factor can be readily cut into fragments and each fragment expressed as a protein in isolation, particular features exhibited by the intact protein can readily be mapped to a particular region. Using the approach outlined in Figure 2.15, for example, it has proved possible to map the DNA-binding abilities of specific transcription factors such as the octamer-binding proteins Oct-1 (Sturm *et al.*, 1987) and Oct-2 (Clerc *et al.*, 1988) to a specific short region of the protein. Once

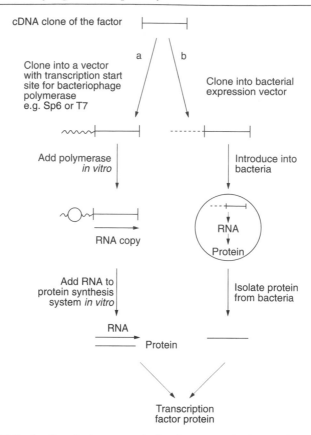

Figure 2.14 Methods of producing transcription factor protein from a cloned transcription factor cDNA. In the coupled *in vitro* transcription and translation method (a), the cDNA is cloned downstream of a promoter recognized by a bacteriophage polymerase and transcribed *in vitro* by addition of the appropriate polymerase. The resulting RNA is translated in an *in vitro* protein synthesis system to produce transcription factor protein. Alternatively (b), the cDNA can be cloned downstream of a prokaryotic promoter in a bacterial expression vector. Following introduction of this vector into bacteria, the bacteria will transcribe the cDNA into RNA and translate the RNA into protein which can be isolated from the bacteria.

this has been done, particular bases in the DNA encoding the DNA-binding domain of the factor can then be mutated so as to alter its amino-acid sequence, and the effect of these mutations on DNA-binding can be assessed as before by expressing the mutant protein and measuring its ability to bind to DNA.

Approaches of this type have proved particularly valuable in defining DNA-binding motifs present in many factors and in analysing how differences in the protein sequence of related factors define which DNA sequence they bind. This is discussed in Chapter 8.

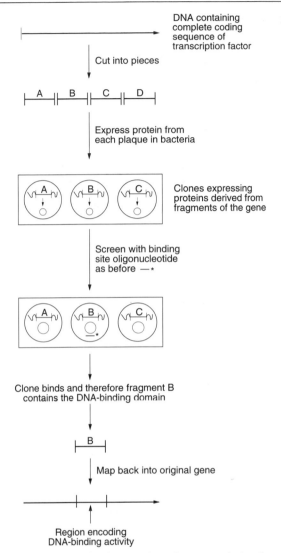

Figure 2.15 Mapping of the DNA-binding region of a transcription factor by testing the ability of different regions to bind to the appropriate DNA sequence when expressed in bacteria.

One other piece of information to emerge from these studies is that the binding to DNA of a small fragment of the factor does not normally result in the activation of transcription. Thus, a 60-amino-acid region of the yeast transcription factor GCN4 can bind to DNA in a sequence-specific manner but does not activate transcription of genes bearing its binding site (Hope and Struhl, 1986). Although DNA-binding is necessary for transcription therefore, it is not sufficient. This indicates that transcription factors have a modular structure in which the DNA-binding domain is distinct from another domain of the protein which mediates transcriptional activation.

The identification of the activation domain in a particular factor is complicated by the fact that DNA-binding is necessary prior to activation. Hence the activation domain cannot be identified simply by expressing fragments of the protein and monitoring their activity. Rather the various regions of the cDNA encoding the factor must each be linked to the region encoding the DNA-binding domain of another factor and the hybrid proteins produced. The ability of the hybrid factor to activate a target gene bearing the DNA-binding site of the factor supplying the DNA-binding domain is then assessed (Figure 2.16). In these so-called 'domain-swap' experiments, binding of the factor to the appropriate DNA-binding site will be followed by gene activation only if the hybrid

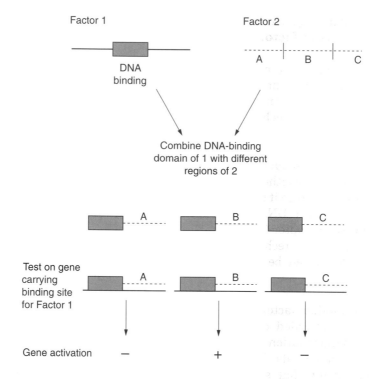

Figure 2.16 Domain-swapping experiment in which the activation domain of factor 2 is mapped by combining different regions of factor 2 with the DNA-binding domain of factor 1 and assaying the hybrid proteins for the ability to activate transcription of a gene containing the DNA-binding site of factor 1.

factor contains the region encoding the activation domain of the factor under test, allowing the activation domain to be identified.

Thus, if another 60-amino-acid region of GCN4 distinct from the DNA-binding domain is linked to the DNA-binding domain of the bacterial Lex A protein, it can activate transcription in yeast from a gene containing a binding site for Lex A. This cannot be achieved by the Lex A DNA-binding domain or this region of GCN4 alone, indicating that this region of GCN4 contains the activation domain of the protein which can activate transcription following DNA-binding and is distinct from the GCN4 protein DNA-binding domain (Hope and Struhl, 1986).

As with DNA-binding domains, the identification of activation domains and comparisons between the domains in different factors has provided considerable information on the nature of activation domains and the manner in which they function. This is discussed in Chapter 9.

2.3.4 Determining the DNA-binding specificity of an uncharacterized factor

As indicated above, it is common for a transcription factor to be identified on the basis of its binding to a known DNA sequence and the gene encoding the factor then cloned. It is also possible, however, for a novel gene to be cloned on the basis, for example, that its expression changes in response to a particular stimulus or that it is mutated in a specific disease (see Section 7.1). On inspection of the DNA sequence and predicted protein sequence, it then appears that this gene encodes a transcription factor either because it is homologous to known transcription factors or because it contains regions with structures similar to those known to mediate DNA-binding (see Chapter 8) or transcriptional activation (see Chapter 9).

Obviously all the techniques for analysing a cloned factor mentioned in Section 2.3.3 can be applied to analysing this factor, for example, examining its expression pattern or determining whether regions within it mediate transcriptional activation when linked to the DNA-binding domain of another factor. Unlike the situation for transcription factors which were identified on the basis of their DNA-binding specificity, however, no information will be available on the DNA sequences to which this novel factor binds. It is evidently essential for the further study of this novel factor that such sequences are identified, so allowing, for example, an analysis of the effect of the factor on artifical promoters carrying its binding site and the identification of its target genes.

To do this, Pollock and Treisman (1990) used a method in which oligonucleotides containing a randomized, central, 26-base-pair sequence flanked by two, defined, 25-nucleotide sequences were prepared (Figure

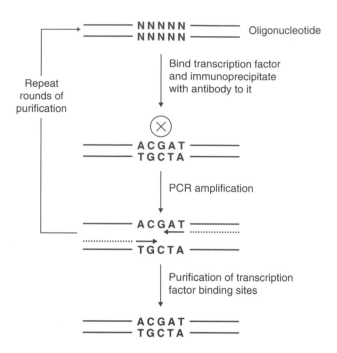

Figure 2.17 Transcription factor binding sites can be cloned using oligonucleotides containing a random central sequence (NNNNN) flanked by defined sequences (solid lines). Repeated cycles of transcription factor binding (X), immunoprecipitation and PCR amplification with primers complementary to the defined end sequences (dotted lines) will eventually result in the purification of oligonucleotides containing the binding site for the factor (ACGAT in this case).

2.17). These sequences were then mixed with transcription factor protein. An antibody to the transcription factor was then used to immunoprecipitate the factor together with the oligonucleotides to which it had bound. This procedure should select from the pool of random oligonucleotides those which contain the binding site for the factor within their central 26-base-pair sequence whilst removing those which contain all other sequences.

However, after a single round of immunoprecipitation, these oligonucelotides will be present in insufficient amounts and purity for further analysis. The immunoprecipitated sequences are therefore amplified by the polymerase chain reaction (PCR) using primers corresponding to the defined 25-base-pair sequences at the ends of each oligonucleotide. Further cycles of transcription factor binding, immunoprecipitation and PCR are then carried out to further purify the binding sequences. Ultimately, the oligonucleotides which bind the factor are cloned and subjected to sequence analysis to identify the common sequence which they contain and which is, therefore, the binding site for the factor.

This method thus allows the identification of specific binding sites for the transcription factor and has been used, for example, to identify the DNA-binding site for the Brn-3 POU family transcription factors (Gruber *et al.*, 1997), which were originally isolated on the basis of homology to other members of the POU family (He *et al.*, 1989) (see Section 6.3.1 for further discussion of POU family transcription factors). Binding sites identified in this way can then, for example, be linked to a gene promoter and introduced into cells with an expression vector encoding the transcription factor itself to determine whether the factor acts as an activator or repressor of gene expression. Similarly, by inspecting the sequences of promoter or enhancer elements of known genes, it may be possible to identify putative target genes for the factor.

A more direct approach to identify target genes for a previously uncharacterized factor was devised by Kinzler and Vogelstein (1989). This method is essentially the same as that of Pollock and Treisman (1990) except that the starting material is not random oligonucleotides but total genomic DNA. This DNA is digested with a restiction enzyme and small defined DNA sequences are added to the ends of the fragments. The transcription factor binding and immunoprecipitation steps are carried out as before, resulting in the purification of pieces of genomic DNA containing the binding site for the transcription factor. These are then PCR amplified as before, using primers corresponding to the defined DNA sequences, which were added at the fragment ends and are then cloned.

Although this method is more technically difficult than the use of oligonucleotides owing to the complexity of genomic DNA, it has the great advantage that the DNA-binding sites are obtained linked to the sequences to which they are normally joined in the genome rather than

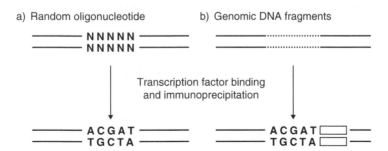

Figure 2.18 Whilst the purification of transcription factor binding sites using random oligonucleotides (as in Figure 2.17) simply isolates the binding site (a), the use of genomic DNA sequences (dotted lines) in the purification results in the isolation of the binding site linked to a fragment of its target gene (boxed), which can then be characterized (b).

in isolation (Figure 2.18). Hence these linked sequences can immediately be characterized and used to identify a target gene for the factor. This method has thus been used, for example, to identify novel target genes for members of the nuclear receptor transcription factor family discussed in Section 4.4, such as the oestrogen receptor (Inoue *et al.*, 1993) and the thyroid hormone receptor (Caubin *et al.*, 1994).

Hence these methods allow the isolation of transcription factor binding sites either alone or linked to their target genes. They thus allow the DNA-binding specificity of cloned transcription factors to be determined just as the cloning techniques described earlier (Section 2.3.2) allow the cloning of factors isolated on the basis of their DNA-sequence specificity. Interestingly, these two sets of techniques have been combined by Nallur *et al.* (1996) who carried out a DNA-binding selection procedure using random oligonucleotides in conjunction with total nuclear extracts from T-lymphocyte cells rather than a single purified transcription factor. This allowed them to isolate a variety of binding sites for transcription factors expressed in these cells. These included some binding sites for known factors and others for previously uncharacterized factors. These uncharacterized factors could then be cloned by using their newly characterized binding sites to screen a cDNA expression library as described in Section 2.3.2b. Hence, in this case, the technique was actually used to clone novel transcription factors on the basis of the prior isolation of their binding sites.

2.4 CONCLUSIONS

This chapter has described a number of methods which allow the investigation of the interaction of a transcription factor with DNA, its purification, gene cloning and dissection of its functional domains as well as the identification of its DNA-binding site. The information obtained by the application of these procedures to particular factors is discussed in subsequent chapters.

REFERENCES

Ares, M. Jr, Chung, J.S., Giglio, L. and Weiner, A.M. (1987). Distinct factors with Sp1 and NF-A specificities bind to adjacent functional elements of the human U2 snRNA gene enhancer. *Genes and Development* **1**, 808–817.

Baumruker, T., Sturm, R. and Herr, W. (1988). OBP 100 binds remarkably degenerate octamer motifs through specific interactions with flanking sequences. *Genes and Development* **2**, 1400–1413.

Caubin, J., Iglesias, T., Bernal, J., Munoz, A., Marquez, G., Barbero, J.L. and Zaballos, A. (1994). Isolation of genomic DNA fragments corresponding to genes modulated *in vivo* by a transcription factor. *Nucleic Acids Research* **22**, 4132–4138.

Clerc, R.G., Corcoran, L.M., LeBowitz, J.H., Baltimore, D. and Sharp, P.A. (1988). The B-cell specific Oct-2 protein contains POU box and homeo box type domains. *Genes and Development* **2**, 1570–1581.

Cowell, I.G. and Hurst, H.C. (1998) Cloning transcription factors from a cDNA expression library. In: *Transcription Factors: a Practical Approach*, 2nd edn (Latchman, D.S. ed.). Oxford, New York: IRL Press (in press).

Dignam, J.D., Lebovitz, R.M. and Roeder, R.G. (1983). Accurate transcription initiation by RNA polymerase II in a soluble extract from isolated mammalian nuclei. *Nucleic Acids Research* **11**, 1575–1489.

Dynan, W.S. and Tjian, R. (1983). The promoter specific transcription factor Sp1 binds to upstream sequences in the SV40 promoter. *Cell* **35**, 79–87.

Fletcher, C., Heintz, N. and Roeder, R.G. (1987). Purification and characterization of OTF-1, a transcription factor regulating cell cycle expression of human H2B gene. *Cell* **51**, 773–781.

Fried, M. and Crothers, D.M. (1981). Equilibria and kinetics of lac repressor-operator interactions by polyacrylamide gel electrophoresis. *Nucleic Acids Research* **9**, 6505–6525.

Galas, D. and Schmitz, A. (1978). DNAse footprinting: a simple method for the detection of protein-DNA-binding specificity. *Nucleic Acids Research* **5**, 3157–3170.

Garner, M.M. and Revzin, A. (1981). A gel electrophoresis method for quantifying the binding of proteins to specific DNA regions: application to components of the *Escherichia coli* lactose operon regulatory system. Nucleic Acids Research **9**, 3047–3060.

Gruber, C.A., Rhee, J.M., Gleiberman, A. and Turner, E.E. (1997). POU domain factors of the Brn-3 class recognize functional DNA elements which are distinctive, symmetrical and highly conserved in evolution. *Molecular and Cellular Biology* **17**, 2391–2400.

He, X., Treacey, M.N., Simmonds, D.M., Ingraham, H.A., Swanson, L.W. and Rosenfeld, M.G. (1989). Expression of a large family of POU-domain genes in mammalian brain development. *Nature* **340**, 35–42.

Herrera, R.E., Shaw, P.E. and Nordheim, A. (1989). Occupation of the c-*fos* serum response element *in vivo* by a multi-protein complex is unaltered by growth factor induction. *Nature* **340**, 68–71.

Hope, I.A. and Struhl, K. (1986). Functional dissection of a eukaryotic transcriptional activator GCN4 of yeast. *Cell* **46**, 885–894.

Inoue, S., Orimo, A., Hosoi, T., Kondo, S., Toyoshima, H., Kondo, T., Ikegami, A., Ouchi, Y., Orimoto, H. and Muramatsu, M. (1993). Genomic binding site cloning reveals an estrogen responsive gene that encodes a RING finger protein. *Proceedings of the National Academy of Sciences USA* **90**, 11117–11121.

Kadonaga, J.T., Carner, K.R., Masiarz, F.R. and Tjian, R. (1987). Isolation of cDNA encoding the transcription factor Sp1 and functional analysis of the DNA-binding domain. *Cell* **51**, 1079–1090.

Kadonaga, J.T. and Tjian, R. (1986). Affinity purification of sequence-specific DNA-binding proteins. *Proceedings of the National Academy of Sciences USA* **83**, 5889–5893.

Kinzler, K.W. and Vogelstein, B. (1989). Whole genome PCR: application to the identification of sequences bound by gene regulatory proteins. *Nucleic Acids Research* **17**, 3645–3653.

Kreale, G.G. (ed) (1994) *DNA–Protein Interactions*. New Jersey: Humana Press.

Lassar, A.B., Buskin, J.N., Lockshun, D., Davis, R.L., Apone, S., Hanaschka, S.D. and Weintraub, H. (1989). Myo D is a sequence-specific DNA-binding protein requiring a region of myc homology to bind to the muscle creatine kinase enhancer. *Cell* **58**, 823–831.

Latchman, D.S. (ed.) (1998). *Transcription Factors: A Practical Approach*. 2nd edn. Oxford, New York: IRL Press (in press).

Lenardo, M., Pierce, J.W. and Baltimore, D. (1987). Protein binding sites in Ig gene enhancers determine transcriptional activity and inducibility. *Science* **236**, 1573–1577.

Manley, J.L., Fire, A., Cano, A., Sharp, P.A. and Gefter, M.L. (1980). DNA-dependent transcription of adenovirus genes in a soluble whole-cell extract. *Proceedings of the National Academy of Sciences USA* **77**, 3855–3859.

Maxam, A.M. and Gilbert, W. (1980). Sequencing end labelled DNA with base-specific chemical cleavages. *Methods in Enzymology* **65**, 499–60.

Mueller, C.R., Macre, P. and Schibler, U. (1990). DBP a liver-enriched transcriptional activator is expressed late in ontogeny and its tissue specificity is determined post-transcriptionally. *Cell* **61**, 279–291.

Mueller, P.R. and Wold, B. (1989). In vivo footprinting of a muscle specific enhancer by ligation mediated PCR. *Science* **246**, 780–786.

Nallur, G.N., Prakash, K., and Weissman, S.M. (1996). Multiplex selection technique (MUST): an approach to clone transcription factor binding sites. *Proceedings of the National Academy of Sciences USA* **93**, 1184–1189.

Nicolas, R.H. and Goodwin G.H. (1998). Purification and cloning of transcription factors. In: *Transcription Factors: A Practical Approach*, 2nd edn (Latchman, D.S., ed,). Oxford, New York: IRL Press (in press).

Norman, C., Runswick, M., Pollock, R. and Treisman, R. (1988). Isolation and properties of cDNA clones encoding SRF, a transcription factor that binds to the c-*fos* serum response element. *Cell* **55**, 989–1003.

Orchard, K., Perkins, N.D., Chapman, C., Harris, J., Emery, V., Goodwin, G., Latchman, D.S. and Collins, M.K.L. (1990). A novel T cell protein recognizes a palindromic element in the negative regulatory element of the HIV-1 LTR. *Journal of Virology* **64**, 3234–3239.

Papavassilou, A.G. (1995). Chemical nucleases as probes for studying DNA–protein interactions. *Biochemical Journal* **305**, 345–357.

Pollock, R. and Treisman, R. (1990). A sensitive method for the determination of protein–DNA-binding specificities. *Nucleic Acids Research* **18**, 6197–6204.

Rosenfeld, P.J. and Kelley, T.J. (1986). Purification of nuclear factor 1 by DNA recognition site affinity chromatography. *Journal of Biological Chemistry* **261**, 1398–1408.

Sambrook, J., Fritsch, E.F. and Maniatis, T. (1989). *Molecular Cloning: A Laboratory Manual*. Cold Spring Harbor, New York: Cold Spring Harbor Laboratory Press.

Santoro, C., Mermod, N., Andrews, P.C. and Tjian, R. (1988). A family of human CCAAT box binding proteins active in transcription and DNA replication: cloning and expression of multiple cDNAs. *Nature* **334**, 218–224.

Siebenlist, U. and Gilbert, W. (1980). Contacts between the RNA polymerase and an early promoter of phage T7. *Proceedings of the National Academy of Sciences USA* **77**, 122–126.

Singh, H., Le Bowitz, J.H., Baldwin, A.S. and Sharp, P.A. (1988). Molecular cloning of an enhancer binding protein: isolation by screening of an expression library with a recognition site DNA. *Cell* **52**, 415–429.

Smith, M.D., Dent, C.L. and Latchman, D.S. (1998). The DNA mobility shift assay. In: *Transcription Factors: A Practical Approach*, 2nd edn (Latchman, D.S., ed.) Oxford, New York: IRL Press (in press).

Spiro, C. and McMurray, C.T. (1998). Footprint analysis of DNA–protein complexes *in vitro* and *in vivo*. In: *Transcription Factors: A Practical Approach*, 2nd edn (Latchman, D.S. ed.). Oxford, New York: IRL Press (in press).

Staudt, L.M., Clerc, R.G., Singh, H., Le Bowitz, J.H., Sharp, P.A. and Baltimore, D. (1988). Cloning of a lymphoid-specific cDNA encoding a protein binding the regulatory octamer DNA motif. *Science* **241**, 577–580.

Sturm, R., Baumruker, T., Franza, R. Jr. and Herr, W. (1987). A 100-kD HeLa cell octamer binding protein (OBP 100) interacts differently with two separate octamer-related sequences within the SV40 enhancer. *Genes and Development* **1**, 1147–1160.

Sturm, R.A., Das, G. and Herr, W. (1988). The ubiquitous octamer-binding protein Oct-1 contains a POU domain with a homeobox subdomain. *Genes and Development* **2**, 1582–1599.

Treisman, R. (1987). Identification and purification of a polypeptide that binds to the c-fos serum response element. *The EMBO Journal* **6**, 2711–2717.

Vinson, C.R., La Marco, K.L., Johnson, P.F., Landschulz, W.H. and McKnight, S.Z. (1988). In situ detection of sequence-specific DNA-binding activity specified by a recombinant bacteriophage. *Genes and Development* **2**, 801–806.

CHAPTER THREE

Transcription factors and constitutive transcription

3.1 RNA POLYMERASES

Transcription involves the polymerisation of ribonucleotide precursors into an RNA molecule using a DNA template. The enzymes which carry out this reaction are known as RNA polymerases. In eukaryotes three different enzymes of this type exist which are active on different sets of genes and can be distinguished on the basis of their different sensitivities to the fungal toxin α-amanitin (Table 3.1) (for a review, see Sentenac, 1985). All the genes which code for proteins as well as those encoding some of the small nuclear RNAs involved in splicing are transcribed by RNA polymerase II. Because of the very wide variety of regulatory processes which these genes exhibit, much of this book is concerned with the interaction of different transcription factors with RNA polymerase II. Information is also available, however, on the interaction of such factors with RNA polymerase I, which transcribes the genes encoding the 28S,

Table 3.1 Eukaryotic RNA polymerases

	Genes transcribed	Sensitivity to α-amanitin
I	Ribosomal RNA (45S precursor of 28S, 18S and 5.8S rRNA)	Insensitive
II	All protein-coding genes, small nuclear RNAs U1, U2, U3, etc.	Very sensitive (inhibited 1 µg/ml)
III	Transfer RNA, 5S ribosomal RNA, small nuclear RNA U6, repeated DNA sequences: Alu, B1, B2 etc., 7SK, 7SL RNA	Moderately sensitive (inhibited 10 µg/ml)

18S and 5.8S ribosomal RNAs (Sommerville, 1984), and with RNA polymerase III, which transcribes the transfer RNA and 5S ribosomal RNA genes (Cilberto *et al.*, 1983). These interactions are therefore discussed where appropriate.

All three RNA polymerases are large multi-subunit enzymes, RNA polymerase II, for example, having 10–14 subunits with sizes ranging from 220 to 10 kDa in size (Sentenac, 1985; Saltzman and Weinmann, 1989), which interact with one another to form a highly complex multimeric molecule (Acker *et al.*, 1997). Interestingly, the cloning of the genes encoding the largest subunits of each of the three polymerases has revealed that they show homology to one another (Memet *et al.*, 1988). Similarly, chemical labelling experiments have indicated that the second largest subunit of each polymerase contains the active site of the enzyme (Riva *et al.*, 1987), whilst at least three smaller, non-catalytic subunits are shared by the three yeast polymerases (Woychik *et al.*, 1990). Such relationships evidently indicate a basic functional similarity between the three eukaryotic RNA polymerases and may also be indicative of a common evolutionary origin.

In addition to the conservation of function between the three eukaryotic enzymes, each individual enzyme exhibits a strong conservation between different organisms. Thus the largest subunit of the mammalian RNA polymerase II enzyme is 75% homologous to that of the fruit fly *Drosophila* (Saltzman and Weinmann, 1989) and also shows homology to the equivalent enzymes in yeast (Memet *et al.*, 1988) and even *E. coli* (Ahearn *et al.*, 1987). All the eukaryotic RNA polymerase II enzymes contain a repeated region at the carboxyl end of the largest subunit which contains multiple copies of the sequence Tyr-Ser-Pro-Thr-Ser-Pro-Ser. This sequence is unique to the largest subunit of RNA polymerase II and is present in multiple copies being repeated 52 times in the mouse protein and 26 times in the yeast protein. As expected from its evolutionary conservation, the repeated region is essential for the proper functioning of the enzyme and hence for cell viability, although its size can be reduced to some extent without affecting the activity of the enzyme (for a review, see Young, 1991).

Interestingly, this repeated region serves as a site for phosphorylation and it is likely that such phosphorylation is critical for functioning of the polymerase (for a review, see Drapkin *et al.*, 1993). Thus it appears that the dephosphorylated form of RNA polymerase II is the form that enters the basal transcriptional complex (see Section 3.2.4), whilst its phosphorylation triggers the start of transcriptional elongation to produce the RNA product. Such phosphorylation appears to be a means of regulating the rate of transcription with specific stimuli such as growth factors resulting in enhanced phosphorylation of the polymerase (Dubois *et al.*, 1994).

In addition, as will be discussed in Chapter 9 (Section 9.2.3), this region may be a target for transcriptional activators either directly or, more probably, indirectly via intermediate proteins. Moreover, recent studies have suggested that factors involved in post-transcriptional processes such as RNA splicing may associate with this region of the polymerase so that the nascent RNA transcript produced by the polymerase can actually be spliced by factors which are bound to the polymerase itself (for review, see Corden and Patturajan, 1997; Steinmetz, 1997). Hence this region appears to represent a critical target for cellular transcriptional and post-transcriptional regulatory processes.

Whether this is the case or not, it is clear that whilst the RNA polymerases possess the enzymatic activity necessary for transcription, they cannot function independently. Rather transcription involves numerous transcription factors which must interact with the polymerase and with each other if transcription is to occur. The role of these factors is to organize a stable transcriptional complex containing the RNA polymerase and which is capable of repeated rounds of transcription.

3.2 THE STABLE TRANSCRIPTIONAL COMPLEX

3.2.1 Characteristics of the stable transcriptional complex

For all three eukaryotic polymerases, the initiation of transcription requires a multi-component complex containing the RNA polymerase and transcription factors. This complex has several characteristics which have led to it being referred to as a stable transcriptional complex (Brown, 1984). These are as follows:

1. The assembled complex is stable to treatment with low concentrations of specific detergents or to the presence of a competing DNA template, both of which would prevent its assembly.
2. The complex contains factors which are necessary for its assembly but not for transcription itself. These factors can, therefore, be dissociated once the complex has formed without affecting transcription.
3. The complex of RNA polymerase and other factors necessary for transcription is stable through many rounds of transcription, resulting in the production of many RNA copies from the gene.

These characteristics are illustrated in Figure 3.1.

Much of the information on these complexes has been obtained by studying the relatively simple systems of RNA polymerases I and III and applying the information obtained to the situation with RNA polymerase II.

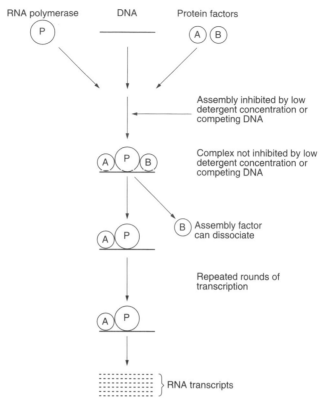

Figure 3.1 Stages in the formation of the stable transcriptional complex. The initial binding of the transcription factor (A) and the assembly factor (B) results in a metastable complex which can be dissociated by low levels of detergent or competing DNA. Following RNA polymerase binding, however, a stable complex is formed. This complex cannot be dissociated by low levels of detergent or competing DNA, is stable through multiple rounds of transcription and retains activity if the assembly factor (B) is removed.

The stable complex formed by each of these enzymes will therefore be discussed in turn.

3.2.2 RNA polymerase I

The simplest complex known is found for the transcription of the ribosomal RNA genes by RNA polymerase I in *Acanthamoeba* (for a review, see Paule, 1990). In this organism only one transcription factor known as TIF-1 is required for transcription by the polymerase. This factor binds to the ribosomal RNA promoter protecting a region from 12 to 70 bases upstream of the transcriptional start site from DNAseI digestion. Subsequently, the polymerase itself binds to the DNA just downstream of TIF-1, protecting a region between 18 and 52 bases upstream of the start site. Interestingly, binding of the polymerase is not dependent on the specific DNA sequence within this region since it can be replaced with a completely random sequence without affecting binding of the polymerase. Hence RNA polymerase is positioned on the promoter by protein–protein interaction with TIF-1, which has previously bound in a sequence-specific manner (Figure 3.2). When the RNA polymerase moves along the DNA transcribing the gene, TIF-1 remains bound at the promoter, allowing subsequent rounds of transcription to occur following binding of another polymerase molecule.

This system therefore represents a simple one in which one single factor is necessary for transcription and is active through multiple rounds

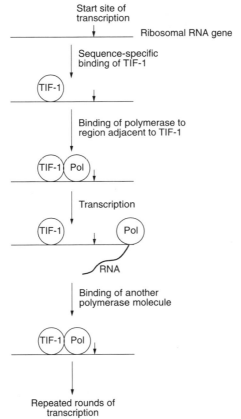

Figure 3.2 Transcription by *Acanthamoeba* RNA polymerase I involves the binding of transcription factor TIF-1 to a specific DNA sequence followed by binding of the polymerase in a non-sequence-specific manner to the DNA region adjacent to TIF-1. When the RNA polymerase moves away as it transcribes the gene, TIF-1 remains bound at the promoter, allowing another RNA polymerase molecule to bind and initiate a new round of transcription.

of transcription. In vertebrate rRNA gene transcription, the situation is more complex, however, with an additional factor UBF (upstream binding factor) also being involved (for reviews, see Geiduschek and Kassavetis, 1995; Jacob, 1995). UBF binds specifically to the promoter and upstream elements of the ribosomal RNA genes and stimulates transcription. This is achieved, however, by interaction with the vertebrate TIF-1 homologue, known as SL1. Thus, although a low basal rate of transcription is observed in the absence of UBF, no transcription is detectable unless SL1 is present. Unlike TIF-1, SL1 does not exhibit sequence-specific binding to the ribosomal RNA promoter. Hence UBF acts by binding to the DNA in a sequence-specific manner and facilitating the binding of SL1. Thus, whilst both SL1 and its homologue TIF-1 act as transcription factors necessary for polymerase I binding, UBF is an additional assembly factor required for binding of SL1 in vertebrates but not of TIF-1 in *Acanthamoeba*. This example, therefore, illustrates the distinction between factors required only for assembly of the complex or for binding of the polymerase and transcription itself (Figure 3.3).

3.2.3 RNA polymerase III

The different roles of transcription factors and assembly factors is also well illustrated by the RNA polymerase III system (for reviews, see Gabrielsen and Sentenac, 1991; Geiduschek and Kassavetis, 1995; White, 1994). Thus three different classes (I–III) of RNA polymerase III transcription unit exist, all of which require the essential factor TFIIIB for transcription (for a review, see Hernandez, 1993). In the case of class I transcription units encoding the 5S ribosomal RNAs, transcription by RNA polymerase III requires the binding of three additional factors TFIIIA, TFIIIB and TFIIIC. Although both TFIIIA and TFIIIC exhibit the ability to bind to 5S DNA in a sequence-specific manner, TFIIIB like SL1 cannot do so unless TFIIIC has already bound. Once the complex of all these factors has formed and the RNA polymerase has bound, TFIIIA and TFIIIC can be removed and transcription continues with only TFIIIB and the polymerase bound to the DNA. Hence like UBF, TFIIIA and TFIIIC are assembly factors which are required for the binding of the transcription factor TFIIIB. In turn, bound TFIIIB is recognized by the polymerase itself and transcription begins (Figure 3.4). As with RNA polymerase I, RNA polymerase III binds to the region of DNA adjacent to that which has bound the transcription factor, binding of the polymerase being independent of the DNA sequence in this region.

Although the transcription of the class II RNA polymerase III transcription units such as those encoding the tRNAs is similar to that

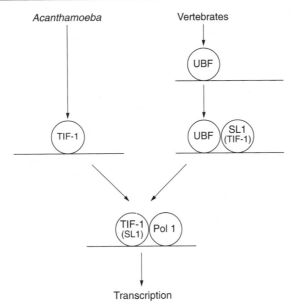

Figure 3.3 Comparison of ribosomal RNA gene transcription in *Acanthamoeba* and vertebrates. Transcription requires both the TIF-1 homologue SL1 and an additional assembly factor UBF whose prior binding is necessary for subsequent binding of SL1.

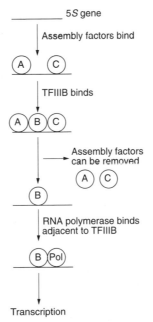

Figure 3.4 Binding of factors to the 5S RNA gene. Transcription requires the initial binding of the assembly factors TFIIIA and TFIIIC with subsequent binding of the transcription factor TFIIIB and of RNA polymerase III itself.

described for the 5S RNA genes, TFIIIA is not required. Rather transcription is dependent only upon TFIIIB and TFIIIC, with binding of TFIIIC being sufficient for subsequent binding of TFIIIB and the polymerase. Similarly, the class III RNA polymerase III transcription units, which have a TATA box in the promoter (for a review, see Sollner-Webb, 1988) that resembles that found in RNA polymerase II promoters (see Section 1.2.2), also require TFIIIB for transcription together with other accessory factors (for a discussion, see Hernandez, 1993).

The process of transcription by RNA polymerases I and III therefore involves the binding of a single transcription factor to the promoter allowing subsequent binding of the RNA polymerase to an adjacent region of DNA. The transcription factor remains bound at the promoter as the polymerase moves down the DNA allowing repeated binding of polymerase molecules and hence repeated rounds of transcription. Binding of the polymerase to the promoter requires prior binding of the transcription factor since the polymerase does not recognize a specific sequence in the promoter but rather makes protein–protein contact with the transcription factor and binds to the adjacent region of the DNA.

In different systems, however, different requirements exist for the binding of the transcription factor itself. Thus in the *Acanthamoeba* system, TIF-1 can bind to DNA in a sequence-specific manner and hence is the only factor required. In most other systems, this is not the case and the transcription factors do not bind to the DNA unless other assembly factors which exhibit sequence-specific DNA binding are present. Once the transcription factor has bound, these assembly factors can be removed, for example, by detergent treatment, without affecting subsequent transcription. It is unclear, however, whether these factors do actually dissociate from the complex under normal conditions *in vivo* once the transcription factor has bound (for a discussion, see Paule, 1990). Whatever the case, the transcription factor itself remains bound at the promoter even after the polymerase has moved down the gene, allowing repeated binding of polymerase molecules and hence repeated rounds of transcription.

Although assembly factors play only an accessory role in transcription itself, they are essential if the complex is to assemble. Hence both assembly factors and transcription factors can be the target for processes which regulate the rate of transcription (for a review, see White, 1994). Thus whilst the high rate of polymerase III transcription in embryonal carcinoma cells is dependent on a high level of transcription factor TFIIIB, the increase in transcription by this polymerase following adenovirus infection is due to an increase in the activity of the assembly factor TFIIIC. Similarly, alterations in the level of TFIIIA during *Xenopus* development control the nature of the 5S rRNA genes which are transcribed at different developmental stages. In addition, as will be discussed in Section 7.3.3, the retinoblastoma anti-oncoprotein inhibits cellular growth by interacting with UBF to inhibit RNA polymerase I activity and with TFIIIB to inhibit RNA polymerase III activity.

3.2.4 RNA polymerase II

Stepwise assembly of the RNA polymerase II basal transcriptional complex

Although some regulation of RNA polymerase I and III activity does occur, therefore, this is much less extensive compared to the very wide variety of regulatory events affecting the activity of genes transcribed by RNA polymerase II. As discussed above, this results in a bewildering array of transcription factors interacting with this enzyme and conferring particular patterns of regulation. These factors will be discussed in Chapters 4–7. Interestingly, however, even the basic transcriptional complex which is essential for any transcription by this enzyme contains far more components than is the case for the other RNA polymerases (for reviews, see Orphanides *et al.*, 1996; Roeder, 1996; Nikolov and Burley, 1997).

One component of this complex which has been intensively studied and plays an essential role in RNA polymerase II mediated transcription is TFIID (for a review, see Burley and Roeder, 1996). In promoters containing a TATA box (Section 1.2.2), TFIID binds to this element, protecting a region from 35 bases to 19 bases upstream of the start site of transcription in the human *hsp70* promoter, for example. The binding of TFIID to the TATA box or equivalent region is the earliest step in the formation of the stable transcriptional complex, such binding being facilitated by another factor TFIIA (Figure 3.5a).

Interestingly, as TFIID is progressively purified, its requirement for TFIIA to aid its activity decreases. This is because in less purified preparations and in the intact cell, TFIID is associated with a number of inhibitory factors such as DR1 and DR2 (for a review, see Drapkin *et al.*, 1993), which act by preventing its binding to the DNA and/or its interaction with other components of the basal complex such as TFIIB (see below). (For further discussion of the role of Dr1, see Section 9.3.2.) One role of TFIIA appears to be to bind to TFIID and overcome this inhibition, thereby stimulating the activity of TFIID. Hence the need for TFIIA decreases as TFIID is purified away from these inhibitory factors, although it is likely to play a critical role in the intact cell. In addition, TFIIA may also play a role in the response to transcriptional activators acting as a co-activator molecule linking DNA-bound activators and the basal transcriptional complex.

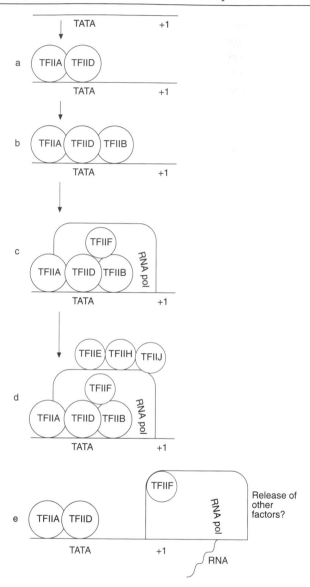

Figure 3.5 Stages in the assembly of the stable transcriptional complex for RNA polymerase II transcription. As the polymerase moves away from the promoter to transcribe the gene, TFIIF remains associated with it whilst TFIIA and TFIID remain bound at the TATA box, allowing the formation of a new stable complex and further rounds of transcription.

Hence, rather than acting as a basal transcription factor essential for all transcription, TFIIA appears to play a key role in the response of the complex to activating and inhibiting molecules. Such a role is of particular importance since the antagonism between positively and negatively acting factors in the assembly of the basal transcriptional complex may play a critical role in regulating the rate of transcription representing a major target for activators and repressors of transcription (For a further discussion of the mechanisms by which specific factors activate or inhibit transcription, see Chapter 9.)

Once TFIID has bound to the DNA, another transcription factor TFIIB, joins the complex by binding to TFIID (Figure 3.5b). This binding of TFIIB is an essential step in initiation complex formation since, as well as binding to TFIID, TFIIB can also bind to the RNA polymerase itself. Hence it acts as a bridging factor allowing the recruitment of RNA polymerase to the complex in association with another factor TFIIF (Figure 3.5c). Following polymerase binding, three other transcription factors TFIIE, TFIIH and TFIIJ, rapidly associate with the complex (Figure 3.5d). At this point, TFIIH which has a DNA helicase activity, unwinds the double-stranded DNA so allowing it to be copied into RNA. Subsequently, the kinase activity of TFIIH, which allows it to phosphorylate other proteins, phosphorylates the C-terminal domain of RNA polymerase (for a review, see Orphanides *et al.*, 1996). This converts it from the non-phosphorylated form which joins the complex to the phosphorylated form which is capable of transcriptional elongation to produce the RNA product (Figure 3.6) (see Section 3.1).

Hence TFIIH via its kinase and helicase activities plays a critical role in allowing the basal transcriptional complex to initiate transcription. Moreover, TFIIH also plays a critical role in the repair of damaged DNA providing a possible link between the processes of DNA repair and transcription (for reviews of TFIIH, see Hoeijmakers *et al.*, 1996; Svejstrup *et al.*, 1996). Interestingly, it has recently been shown that the kinase activity associated with TFIIH can also phosphorylate the retinoic acid receptor, which is a member of the nuclear receptor transcription factor family discussed in Sections 4.4 and 8.3.2. This phosphorylation stimulates the ability of the retinoic acid receptor to activate transcription (Rochette-Egly *et al.*, 1997) indicating that TFIIH may play a role in the regulation of transcription factor activity by phosphorylation (see Section 10.3.4).

The complex of the seven factors (TFIIA, B, D, E, F, H and J) and the polymerase is thus sufficient for transcription to occur. As the polymerase moves down the gene during this process, TFIIF remains associated with it, whilst TFIIA and TFIID remain bound at the promoter and are capable of binding another molecule of polymerase, allowing repeated rounds of transcription as with the other polymerases (Figure 3.5e).

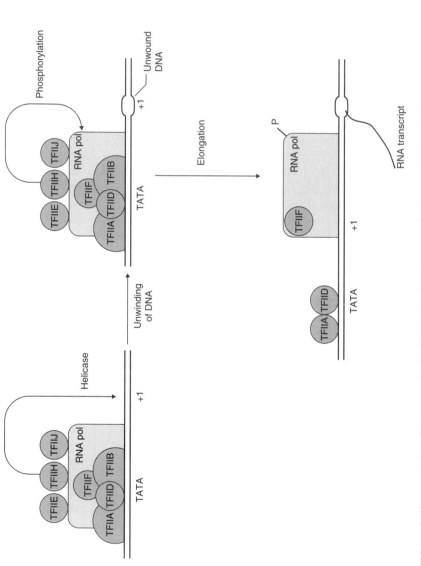

Figure 3.6 TFIIH has a helicase activity which unwinds the DNA, allowing its transcription into RNA and a kinase activity that phosphorylates the C-terminal region of RNA polymerase which allows it to begin transcription.

The RNA polymerase holoenyzme

Although the step-by-step pathway of assembling the basal transcriptional complex described above was proposed on the basis of a number of studies, an alternative pathway has also been proposed based on the finding that some RNA polymerase is found in solution which is already associated with TFIIB, TFIIF and TFIIH in the absence of DNA. This so-called RNA polymerase holoenzyme has now been observed in a wide range of organisms ranging from yeast to man. Therefore, it is probable that, in some cases, following binding of TFIIA and TFIID to the promoter, this complex of RNA polymerase and associated factors may bind, resulting in a reduced number of steps being required for complex formation (Figure 3.7) (for a discussion, see Pugh, 1996; Greenblatt, 1997).

Interestingly, the RNA polymerase holoenzyme also contains a number of other components apart from RNA polymerase itself and the basal transcription factors. Thus, it includes a complex of proteins known

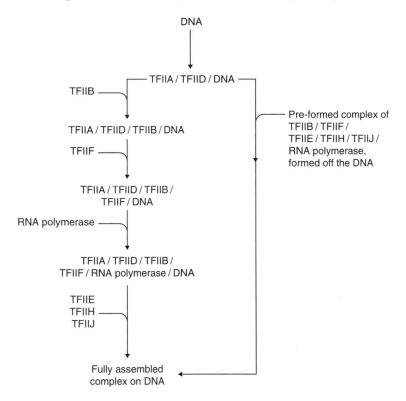

Figure 3.7 Alternative pathways in the assembly of the stable transcriptional complex for RNA polymerase II, involving either the step-by-step pathway (see Figure 3.5), or the binding of a pre-formed complex of RNA polymerase and its associated factors to DNA, which has already bound TFIIA and TFIID.

as the mediator complex which appears to be required, at least in yeast, for the response to transcriptional activators. Hence, the mediator may serve as a link between these activators and the components of the basal transcriptional complex whose activity they stimulate (for a discussion of transcriptional activation, see Section 9.2). In addition, the holoenzyme can also associate with the SWI/SNF complex discussed in Section 1.4.2, whose role is to remodel the chromatin into a form which allows the binding of transcriptional activators and transcription itself. Hence, at least in some cases, this remodelling complex can be recruited to DNA together with the RNA polymerase and its associated proteins.

The RNA polymerase holoenzyme is thus a highly complex structure, which, as well as RNA polymerase itself and basal transcription factors also contains factors, involved in the response to transcriptional activators and others which remodel chromatin structure. Although this holoenzyme represents only one of the two possible methods by which the basal transcription complex assembles on the DNA, it is clear that, regardless of its method of assembly, the basic stable transcriptional complex for RNA polymerase II requires a number of factors in addition to the polymerase itself and is therefore much more complex than that of RNA polymerase I or III.

3.2.5 TBP, The universal transcription factor?

Most of the transcription factors described in the previous sections were isolated by the biochemical fractionation of cellular extracts and were then shown to have a particular functional activity in modulating the rate of transcription when mixed with RNA polymerase and other sub-cellular fractions. When these factors were characterized in more detail by further fractionation and subsequent cloning, however, many of them were shown to consist of several different proteins, which together are responsible for the properties ascribed to the original factor. Thus, although these factors have been dealt with for simplicity in the previous sections as single factors, most of them are in fact complexes of several different proteins, for example, TFIIE and TFIIF both contain two distinct proteins, whilst TFIIH is a multi-protein complex with one of the component proteins having the kinase activity which results in phosphorylation of the RNA polymerase and another has the helicase activity which unwinds the DNA (see Section 3.2.4) (for reviews, see Hoejmakers *et al.*, 1996; Svejstrup *et al.*, 1996).

This responsibility of one component of the complex for an activity formerly ascribed to the whole complex is seen most clearly in TFIID. Thus, TFIID is a multi-protein complex in which only one protein known as TBP (TATA-binding protein) directs the binding to the TATA box, whilst the other components of the complex known as TAFs (TBP-associated factors) do not bind directly to the TATA box and appear to allow TFIID to respond to stimulation by transcriptional activators (see Section 9.2.4) (for a review, see Tansey and Herr, 1997). They thus represent co-activator molecules, linking transcriptional activators and the basal trascriptional complex.

Hence TBP plays a critical role in the transcription of TATA box-containing RNA polymerase II promoters by binding to the TATA box as the first step in the assembly of the basal transcriptional complex. In view of this critical role, it is not surprising that TBP is one of the most highly conserved eukaryotic proteins. The structure of this protein has been defined by X-ray crystallography and shown to have a saddle structure in which the concave underside binds to DNA and the convex outer surface is accesible for interactions with other factors. Most interestingly, binding of TBP to the DNA deforms the DNA so that it follows the concave curve of the saddle (Figure 3.8). Moreover, structural studies of the TFIID complex (consisting of TBP and the TAFs) bound to DNA have indicated that it resembles the complex of the eight histone molecules around which DNA is wound in the nucleosome to form the normal chromatin structure (see Section 1.4.1). Hence the DNA may bend around TFIID at the promoter in a similar manner to the folding of the rest of DNA in the basic nucleosome structure of chromatin (for reviews, see Surridge, 1996; Hoffmann *et al.*, 1997). This role for TFIID in altering nucleosome structure at the promoter is also supported by the

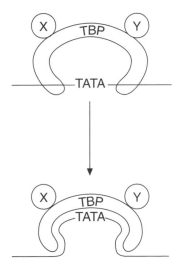

Figure 3.8 The saddle structure of TBP, as determined by X-ray crystallography, allows the concave surface to interact with the TATA box whilst the convex surface associates with other accessory transcription factors (X and Y). The initial binding induces the bending of the DNA so that it follows the concave under surface of the saddle. See also colour Plate 1.

finding that $TAF_{II}250$, one of the subunits of TFIID, has histone acetyltransferase activity (Mizzen *et al.*, 1996), since acetylation of histones appears to play a key role in modulating chromatin structure (see Section 1.4.3).

The bent DNA with TFIID bound to it serves as the central platform on which the basal transcriptional complex assembles. Thus, structural studies have shown that TFIIA binds to the amino-terminal stirrup of the TBP saddle and interacts only with the DNA upstream of the TATA box. This allows it to fulfil its role of protecting TFIID from inhibition by transcriptional repressors and allowing it to respond to activators bound to upstream DNA sequences (see Section 3.2.4). In contrast, TFIIB binds to the carboxyl-terminal stirrup of the TBP saddle and binds to the DNA downstream (as well as upstream) of the TATA box. This allows it to fulfil its role of acting as a bridge between TBP and RNA polymerase II, so positioning the start site of transcription by the polymerase relative to the TATA box (see Plate 1; Geiger *et al.*, 1996) (for reviews, see Burley, 1996; Roeder, 1996; Nikolov and Burley, 1997).

Paradoxically, in view of its TATA box binding ability, TBP also plays a critical role in the transcription of the subset of RNA polymerase II genes which do not contain a TATA box (see Section 1.2.2) (for a review, see Weis and Reinberg, 1992). In this case, however, TBP does not bind to the DNA but is recruited to the promoter by another DNA-binding protein which binds to the initiator element overlapping the transcriptional start site. TBP then binds to this initiator binding protein allowing the recruitment of TFIIB and the RNA polymerase itself as for promoters containing a TATA box. Hence TBP plays a critical role in the assembly of the transcription complex for RNA polymerase II, although it joins the complex by binding to DNA in the case of TATA-box – containing promoters (Figure 3.9a) and is recruited by protein–protein interactions in the case of promoters which lack a TATA box (Figure 3.9b).

These findings have led to the suggestion that TBP represents the basic transcription factor for RNA polymerase II, paralleling the role of SL1 for RNA polymerase I and TFIIIB for RNA polymerase III. This idea was supported by the amazing finding that TBP is actually also a component of both SL1 and TFIIIB (for a review, see White and Jackson, 1992). Thus the SL1 factor is actually a complex of four factors, one of which is TBP. Hence when SL1 is recruited to the promoter by UBF (see Section 3.2.2), TBP is delivered to the DNA exactly as in the RNA polymerase II promoters which do not contain a TATA box, where TBP is recruited by the prior binding of another protein to the initiator element.

Similarly, in the case of RNA polymerase III transcription where TBP is part of the multi-component TFIIIB complex (for a review, see Rigby, 1993), TBP is delivered to class I polymerase III promoters by protein–

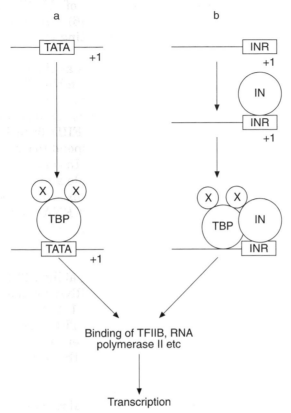

Figure 3.9 Transcription of promoters by RNA polymerase II involves the recruitment of TBP (and associated factors (X) forming the TFIID complex) to the promoter. This may be achieved by direct DNA binding to the TATA box where this is present (a) or by protein–protein interaction with a factor (IN) bound to the initiator element where the TATA box is absent (b).

protein interaction following the prior binding of TFIIIA and TFIIIC, and is delivered to class II polymerase III promoters by the prior binding of TFIIIC (Section 3.2.3) (Figure 3.10a). Interestingly, however, as noted in Section 3.2.3, the class III group of RNA polymerase III promoters contain a TATA box and hence, in this case, TBP can bind directly (Figure 3.10b). As in RNA polymerase II promoters, distinct mechanisms ensure the recruitment of TBP to all RNA polymerase III promoters.

The similarities between the three RNA polymerases discussed in Section 3.1 are therefore paralleled by the involvement of a common factor, TBP, in transcription by all three RNA polymerases. In each case TBP forms a part of the multi-protein complexes which have been shown to be essential for transcription itself, binding via the TATA box or by protein–

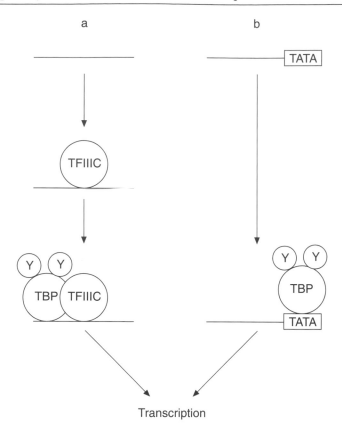

Figure 3.10 Transcription of promoters by RNA polymerase III involves the recruitment of TBP (and associated factors (Y) forming the TFIIIB complex) to the promoter. This may be achieved by protein–protein interactions with TFIIIA and TFIIIC in the case of class I promoters, with TFIIIC alone in the case of class II promoters (a) or by direct DNA binding to the TATA box in class III promoters where the TATA box is present (b).

protein interactions with assembly factors (for a review, see Struhl, 1994). In view of this it has been suggested that TBP represents a evolutionarily ancient transcription factor preceding the division of the three RNA polymerases and having a universal and essential role in eukaryotic transcription (for a review, see Hernandez, 1993). Indeed, a TBP homologue is also found in the archaebacteria which constitutes a separate kingdom distinct from the eukaryotes and the eubacteria. Hence the existence of TBP appears to predate not only the divergence of the three RNA polymerases, but also the divergence of the eukaryotic and archaebacterial kingdoms (for a review of archaebacterial transcription, see Reeve *et al.*, 1997).

3.3 UPSTREAM PROMOTER ELEMENTS

3.3.1 Constitutive transcription factors

The complexity of the RNA polymerase II stable transcriptional complex is paralleled by the existence of numerous factors binding to elements upstream of the TATA box and increasing or decreasing the level of transcription (see Section 1.2). Many of these factors are active only in the presence of an inducing stimulus or in a specific tissue, thereby producing a specific pattern of inducible or tissue-specific gene expression. In addition, however, other factors which are constitutively active bind to specific upstream sequences. The binding of these factors and their interaction with the basal transcriptional complex results in increased levels of transcription in all tissues. This may be dependent on increased stability of the complex in the presence of these factors or to its increased activity. In the absence of such factors, therefore, the basal transcriptional complex can produce only a very low level of transcription, and the binding of one or more of these factors is necessary if significant levels of transcription are to occur. Two factors of this type will be discussed, namely those which bind to the Sp1 and CCAAT box elements found upstream of many promoters.

3.3.2 Sp1

The Sp1 box, which has the consensus sequence GGGCGG, was originally defined in the promoter of the eukaryotic virus SV40, which contains six copies of this motif, each of which binds the transcription factor Sp1 (McKnight and Tjian, 1986). Subsequently, this sequence has been found in a variety of other promoters such as the herpes simplex virus thymidine kinase promoter, the mouse dihydrofolate reductase promoter and the human metallothionein IIA promoter (for reviews, see La Thangue and Rigby, 1988; Jones *et al.*, 1988). As expected, the Sp1 protein is present in all cell types; two closely related Sp1 polypeptides of 105 kDa and 95 kDa having been purified, for example, from HeLa cells (for review of Sp1, see Lania *et al.*, 1997). The addition of these proteins to an *in vitro* transcription reaction specifically stimulates the transcription of genes containing Sp1 binding sites, and the single gene encoding them has been cloned.

Using the techniques described in Section 2.3.3, the DNA binding activity of the protein encoded by the Sp1 gene has been shown to be dependent upon a region at the carboxyl terminus which contains three so-called zinc finger motifs also found in other transcription factors (for further details of this DNA binding motif, see Section 8.3.1). Similarly the ability of the protein to activate transcription has been mapped to two glutamine-rich regions of the protein, which act by interacting with TAFs that are associated with TBP within the TFIID complex (for further details of this activation motif and its mechanism of action, see Section 9.2).

Hence the presence of the Sp1 protein in all cell types, allows it to bind to its binding sites within specific gene promoters and results in their constitutive activation. In agreement with this critical role for Sp1, mice lacking a functional gene encoding Sp1 are not viable and die during embryonic development prior to birth (Marin *et al.*, 1997). Interestingly, however, Sp1 is not the only factor capable of binding to its DNA-binding site. Thus other zinc-finger proteins related to Sp1 such as Sp2, Sp3, Sp4 have been identified and some of these are expressed in a tissue-specific manner, suggesting that Sp family proteins may play a role in regulated as well as constitutive transcription (for a review, see Lania *et al.*, 1997).

3.3.3 CCAAT-box-binding proteins

Like the Sp1 motif, the CCAAT box has been found in a wide variety of promoters such as those of the thymidine kinase and *hsp 70* genes, and plays an essential role in their activity (for reviews, see McKnight and Tjian, 1986; Jones *et al.*, 1988). The CCAAT box binds a number of different proteins, some of which are expressed in all tissues, whilst others are expressed in a tissue-specific manner (for reviews see La Thangue and Rigby, 1988; Johnson and McKnight, 1989).

Initially, two distinct factors binding to this sequence were identified. One of these CTF (CCAAT box transcription factor) consists of a family of polypeptides of 52–66 kDa in size, which are encoded by multiple genes with diversity being increased by alternative splicing of the primary transcripts. As well as their ability to stimulate transcription of genes containing a CCAAT box, these proteins can also stimulate DNA replication, CTF having been shown to be identical to nuclear factor 1, a protein which can stimulate the replication of adenovirus DNA *in vitro*. Hence, like SV40 large T antigen and the octamer binding protein Oct-1 (also referred to as nuclear factor III), this protein is capable of stimulating both transcription and DNA replication and is hence referred to as CTF/NF1.

The second CCAAT-box-binding protein to be defined, C/EBP, differs from CTF/NF1 in a number of properties such as its size (42 kDa), heat stability and the production of a different DNaseI footprint on the CCAAT box (Graves *et al.*, 1987; Johnson *et al.*, 1987). Most importantly, however, this protein differs from CTF/NF1 in its sequence-specificity.

Thus a C to G change in the first base of the CCAAT box impedes binding of CTF/NF1 but actually enhances the binding of C/EBP (Graves *et al.*, 1987). This enhanced binding of C/EBP reflects its ability to bind also to a core motif present in a wide variety of eukaryotic viral enhancers, which has the consensus sequence TGTGGA/TA/TA/TG (Johnson *et al.*, 1987). The ability of C/EBP to bind with high affinity to these two distinct motifs led to its being termed CCAAT/enhancer binding protein (C/EBP) (for a discussion, see Johnson and McKnight, 1989).

The distinction between CTF/NF1 and C/EBP was confirmed when the genes encoding them were cloned and shown to be completely non-homologous to one another. Indeed, although these two proteins can bind the identical DNA sequence in the CCAAT box, they do so using two completely different DNA-binding motifs. Thus C/EBP contains the leucine zipper motif with adjacent basic DNA-binding domain common to a number of transcription factors, whilst CTF/NF1 contains a distinct DNA-binding motif first described in this factor (for discussion of these and other DNA-binding motifs, see Chapter 8). Hence sequence-specific binding to the same DNA sequence can be mediated by two distinct DNA-binding structures.

Following the initial definition of C/EBP as distinct from CTF/NFI in this manner, it became clear that, as with CTF, a family of C/EBP proteins exist including C/EBPα, C/EBPβ and C/EBPδ, each of which are encoded by different genes (for a review, see Hurst, 1996).

The existence of multiple CCAAT box proteins of both the CTF/NFI and C/EBP type clearly suggested that this sequence might play an important role in gene regulation apart from acting simply as an activator of constitutive gene regulation in all tissues and evidence is available that this is the case. Thus, whilst the CTF/NF1 protein, like the Sp1 protein, is present in all tissues, C/EBPα is expressed in the liver at much higher levels than in other tissues (Xanthopoulos *et al.*, 1989). Hence, while binding of CTF/NF1 to the CCAAT box is likely to play a role in constitutive expression of particular genes, C/EBPα may activate the expression of genes expressed specifically in the liver. In agreement with this, the CCAAT box in several genes expressed specifically in the liver has been shown to play a role in their tissue-specific pattern of expression (see, for example, Lichtsteiner *et al.*, 1987; Nerlov and Ziff, 1994). Interestingly, the disruption of C/EBPα binding by a mutation in its binding site within one such liver-specific gene (encoding factor IX) has been shown to be the cause of the human disease, haemophilia B (Crossley and Brownlee, 1990).

In contrast to this role of C/EBPα in constitutive gene expression in the liver, C/EBPβ and C/EBPδ are activated in liver cells and other cell types following exposure to interleukin-6 (IL-6) and other cytokines. They thus play a key role in the response to IL-6 and in its ability to activate a set of genes in the liver, known as the acute-phase response genes (for a review, see Kishimoto *et al.*, 1994). This distinct role of members of the C/EBP family is also seen during the differentiation of fat cells. Thus C/EBPα is specifically expressed during the terminal phase of adipocyte differentiation (for a review, see Vasseur-Cognet and Lane, 1993) and when introduced into fibroblast cells can actually induce them to differentiate into adipocytes (Yeh *et al.*, 1995). In contrast C/EBPβ and δ are expressed in the early phases of differentiation and are down regulated during terminal adipocyte differentiation.

As well as functioning in a positive manner, the tissue-specific CCAAT-box-binding factors can also be involved in tissue-specific gene regulation by acting negatively to prevent the binding of a constitutively expressed, positively acting factor. Thus, in the sea urchin, a factor known as the CCAAT displacement protein is present in embryonic tissues and binds to a sequence overlapping the CCAAT box in the sperm histone H2B gene. This binding prevents the binding of the positively acting CCAAT-box-binding protein and thereby prevents gene transcription (Figure 3.11). Hence the gene is expressed only in sperm when the displacement protein is absent and not in other tissues even though the positively acting CCAAT-box-binding protein is present (Barberis *et al.*, 1987). A negative effect of C/EBPα itself on the expression of human hepatitis B virus has also been reported suggesting that the negative effect of CCAAT-box-binding proteins may be widespread (Pei and Shih, 1990).

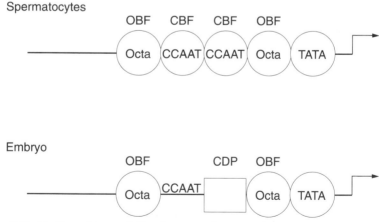

Figure 3.11. Binding of transcription factors to the sea-urchin sperm histone H2B gene in spermatocytes where the gene is expressed and in embryos where it is not expressed. Note that the presence of a CCAAT displacement protein (CDP) in embryos prevents the binding of the positively acting CCAAT-box-binding factor (CBF) to the two CCAAT boxes in the promoter. Expression of the H2B gene is obligately dependent on the binding of CBF and hence the gene is expressed only in sperm where it is present, even though the octamer-binding factor (OBF) is bound to both copies of the octamer motif and TFIID is bound to the TATA box in both cell types.

Therefore, it is clear that, by binding a variety of constitutively expressed and tissue-specific proteins, the CCAAT box can contribute not only to constitutive expression of the genes which contain it but also to a variety of tissue-specific and developmental patterns of gene regulation.

3.4 CONCLUSIONS

The binding of each of the three eukaryotic RNA polymerases to appropriate gene promoters and subsequent transcription is dependent on the prior binding of a specific transcription factor to the promoter. Binding of the polymerase to the DNA adjacent to this factor occurs by recognition of the bound protein rather than by recognition of the specific DNA sequence in this region. In most cases, the binding of the transcription factor itself requires the prior binding of other factors to the DNA. These assembly factors therefore play a critical role in the formation of the stable transcriptional complex but can be dissociated once the complex has formed without affecting its activity. In the case of RNA polymerase II transcription, either the stability of the complex or its activity is greatly affected by the binding of other proteins to sequences upstream of the promoter. Some of these proteins such as Sp1 and CTF/NF1, which have been discussed in this chapter, are present in all tissues and therefore play a role in constitutive gene expression. Others are present or activated only in specific cell types or in response to a specific signal. The role of these proteins in the regulation of inducible or cell-type-specific gene expression is discussed in Chapters 4–6.

REFERENCES

Acker, J., de Graaf, M., Cheynel, I., Khazak, V., Kedinger, C. and Vigneron, M. (1997). Interactions between human RNA polymerase II subunits. *Journal of Biological Chemistry* **272**, 16815–16821.

Ahearn, J.M. Jr, Bartolomei, M.S., West, M.L., Cisek, L.J. and Corden, J.L. (1987). Cloning and sequence analysis of the mouse genomic locus encoding the large subunit of RNA polymerase II. *Journal of Biological Chemistry* **262**, 10695–10705.

Barberis, A., Superti-Furga, G. and Busslinger, M. (1987). Mutually exclusive interaction of the CCAAT binding factor and of a displacement protein with overlapping sequences of a histone gene promoter. *Cell* **50**, 347–359.

Brown, D.D. (1984). The role of stable complexes that repress and activate eukaryotic genes. *Cell* **37**, 359–365.

Burley, S.K. (1996). Picking up the TAB. *Nature* **381**, 112–113.

Burley, S.K. and Roeder, R.G. (1996). Biochemistry and structural biology of transcription factor IID (TFIID). *Annual Review of Biochemistry* **65**, 769–799.

Cilberto, G., Castagnoli, L. and Cortese, R. (1983). Transcription by RNA polymerase III. *Current Topics in Developmental Biology* **18**, 59–88.

Corden, J.L and Patturajan, M. (1997). A CTD function linking transcription to splicing. *Trends in Biochemical Sciences* **22**, 413–416.

Crossley, M. and Brownlee, G.G. (1990). Disruption of a C/EBP binding site in the factor IX promoter is associated with haemophillia B. *Nature* **345**, 444–446.

Drapkin, R., Merino, A. and Reinberg, D. (1993). Regulation of RNA polymerase II transcription. *Current opinion in Cell Biology* **5**, 469–476.

Dubois, M.F., Nguyer, V.T., Dahmus, M.E., Pages, G., Pouyssegur, J. and Bensaudo, O. (1994). Enhanced phosphorylation of the C-terminal domain of RNA polymerase II upon serum stimulation of quiescent cells: possible involvemnet of MAP kinases. *EMBO Journal* **13**, 4787–4797.

Gabrielson, O.S. and Sentenac, A. (1991). RNA polymerase III (C) and its transcription factors. *Trends in Biochemical Sciences* **16**, 412–416.

Geiduschek, E.P. and Kassavetis, G.A. (1995). Comparing transcriptional initiation by RNA polymerase I and III. *Current Opinion in Cell Biology* **7**, 344–351.

Geiger, J.H., Hahn, S., Lee, S. and Sigler, P.B. (1996). Crystal structure of the yeast TFIIA/TBP/DNA complex. *Science* **272**, 830–836.

Graves, B.J., Johnson, P.F. and McKnight, S.L. (1987). Homologous recognition of a promoter domain common to the MSV LTR and the HSV tk gene. *Cell* **44**, 565–576.

Greenblatt, J. (1997). RNA polymerase II holoenzyme and transcriptional regulation. *Current Opinion in Cell Biology* **9**, 310–319.

Hernandez, N. (1993) TBP, a universal eukaryotic transcription factor. *Genes and Development* **7**, 1291–1308.

Hoeijmakers, J.H.J., Egly, J.M. and Vermueler, W. (1996). TFIIH: a key component in multiple DNA transactions. *Current Opionion in Genetics and Development* **6**, 26–33.

Hoffman, A., Oelgeschlage, T. and Roeder, R. (1997). Considerations of transcriptional control mechanisms: Do TFIID-core promoter complexes recapitulate nucleosome-like functions? *Proceedings of the National Academy of Sciences USA* **94**, 8923–8925.

Hurst, (1996). Leucine zippers. *Protein Profile* **3**, 1–72.

Jacob, S.T. (1995). Regulation of ribosomal gene transcription. *Biochemical Journal* **306**, 617–626.

Johnson, P.F., Landschulz, W.H., Graves, B.J. and McKnight, S.L. (1987). Identification of a rat liver nuclear protein that binds to the enhancer core element of three animal viruses. *Genes and Development* **1**, 133–146.

Johnson, P.F. and McKnight, S.L. (1989). Eukaryotic transcriptional regulatory proteins. *Annual Review of Biochemistry* **58**, 799–839.

Jones, N.C., Rigby, P.W.J. and Ziff, E.B. (1988). Trans-acting protein factors and the regulation of eukaryotic transcription. *Genes and Development* **2**, 267–281.

Kishimoto, T., Taga, T. and Akira, S. (1994). Cytokine signal transduction. *Cell* **76**, 253–262.

Lania, L., Majello, B. and de Luca, P. (1997). Transcriptional regulation by the Sp family proteins. International Journal of Biochemistry and Cell Biology. **29**, 1313–1323.

La Thangue, N.B. and Rigby, P.W.J. (1988). Trans-acting protein factors and the regulation of eukaryotic transcription. In: *Transcription and Splicing* (Hanes, B.D. and Glover, D., eds), pp. 3–42. IRL Press Oxford: IRL Press.

Lichtsteiner, S., Wuari, J. and Schibler, U. (1987). The interplay of DNA-binding proteins on the promoter of the mouse albumin gene. *Cell* **51**, 963–973.

McKnight, S. and Tjian, R. (1986). Transcriptional selectivity of viral genes in mammalian cells. *Cell* **46**, 795–805.

Marin, M., Karis, A., Visser, P., Grosveld, F. and Philipsen, S. (1997). Transcription factor Sp1 is essential for early embryonic development but dispensable for cell growth and differentiation. *Cell* **89**, 619–628.

Memet, S., Saurn, W. and Sentenac, A. (1988). RNA polymerases B and C are more closely related to each other than to RNA polymerase A. *Journal of Biological Chemistry* **263**, 10048–10051.

Mizzen, C.A., Yang, X.J., Kokubo, T., Brownell, J.E., Bannister, A.J., Owen-Hughes, T., Workman, J., Wang, L., Berger, S.L., Kouzarides, T., Nakatani, Y. and Allis, C.D. (1996). The TAFII 250 subunit of TFIID has histone acetyltransferase activity. *Cell* **87**, 1261–1270.

Nerlov, C. and Ziff, E.B. (1994). Three levels of functional interaction determine the activity of the CCAAT enhancer binding protein-α on the serum albumin promoter. *Genes and Development* **8**, 350–362.

Nikolov, D.B. and Burley, S.K. (1997). RNA polymerase II transcription initiation: a structural view. *Proceedings of the National Academy of Sciences USA* **94**, 15–22.

Orphanides, G., Lagrange, T. and Reinberg, D. (1996). The general transcription factors of RNA polymerase II. *Genes and Development* **10**, 2657–2683.

Paule, M.R. (1990). In search of the single factor. *Nature* **344**, 819–820.

Pei, D. and Shih, C. (1990). Transcriptional activation and repression by cellular DNA binding protein C/EBP. *Journal of Virology* **64**, 1517–1522.

Pugh, B.F. (1996). Mechanisms of transcripition complex assembly. *Current Opinion in Cell Biology* **8**, 303–311.

Reeve, J.N., Sandman, K. and Daniels, C.J. (1997). Archaeal histones nucleosomes and transcriptional initiation. *Cell* **89**, 999–1002.

Rigby, P.W.J. (1993). Three in one and one in three: it all depends on TBP. *Cell* **72**, 7–10.

Riva, M., Schaffner, A.R., Sentenac, A., Hartmann, G.R., Mustner, A.A., Zaychikov, F. and Grachev, M.A. (1987) Active site labelling of the RNA polymerases A, B and C from yeast. *Journal of Biological Chemistry* **262**, 14377–14380.

Rochette-Egly, C., Adam, S., Rossignol, M., Egly, J.M. and Chambon, P. (1997). Stimulation of RARα activation function AF-1 through binding to the general transcription factor TFIIH and phosphorylation by CDK7. *Cell* **90**, 97–107.

Roeder, R.G. (1996). The role of general initiation factors in transcription by RNA polymerase II. *Trends in Biochemical Sciences* **21**, 327–334.

Saltzman, A.G. and Weinmann, R. (1989). Promoter specificity and modulation of RNA polymerase II transcription. *FASEB Journal* **3**, 1723–1733.

Sentenac, A. (1985). Eukaryotic RNA polymerases. *CRC Critical Reviews in Biochemistry* **1**, 31–90.

Sollner-Webb, B. (1988) Surprises in RNA polymerase III transcription *Cell* **52**,153–154.

Somerville, J. (1984). RNA polymerase I promoters and cellular transcription factors. Nature 310, 189–190.

Steinmetz, E.J. (1997). Pre-mRNA processing and the CTD of RNA polymerase II: the tail that wags the dog? *Cell* **89**, 491–494.

Struhl, K. (1994). Duality of TBP: the universal transcription factor. *Science* **263**, 1103–1104.

Surridge, C. (1996). The core curriculum. *Nature* **380**, 287–288.

Svejstrup, J.Q., Vichi, P. and Egly, J.M. (1996). The multiple roles of transcription/repair factor TFIIH. *Trends in Biochemical Sciences* **21**, 346–350.

Tansey, W.P. and Herr, W. (1997). TAFs: guilt by association. *Cell* **88**, 729–732.

Vasseur-Cognet, M. and Lane, M.B. (1993). Trans-acting factors involved in adipocyte differentiation. *Current Opinion in Genetics and Development* **3**, 238–245.

Weis, L. and Reinberg, D. (1992). Transcription by RNA polymerase II initiator directed formation of transcription-competent complexes. *FASEB Journal* **6**, 3300–3309.

White, R.J. (1994). *RNA Polymerase III Transcription*. Austin, Texas: R.G. Landes Company.

White, R.J. and Jackson, S.P. (1992). The TATA binding protein: a central role in transcription by RNA polymerases I and III. Trends in Genetics 8, 284–288.

Woychik, N.A., Liao, S.M., Koldrieg, P.A. and Young, R.A. (1990). Subunits shared by eukaryotic RNA polymerases. *Genes and Development* **4**, 313–323.

Xanthopoulos, K.G., Mirkovitch, J., Decker, T., Kuo, C.F. and Darnell, J.E. Jr. (1989). Cell-specific transcriptional control of the mouse DNA-binding protein mC/EBP. *Proceedings of the National Academy of Sciences USA* **86**, 4117–4121.

Yeh, W.C., Cao, Z., Classon, M. and McKnight, S.L. (1995). Cascade regulation of terminal adipocyte differentiation by three members of the C/EBP family of leucine zipper proteins. *Genes and Development* **9**, 168–181.

Young, R. (1991). RNA polymerase II. *Annual Review of Biochemistry* **60**, 689–715.

CHAPTER FOUR

Transcription factors and inducible gene expression

4.1 INDUCIBLE GENE EXPRESSION

All cells from bacteria to mammals respond to various treatments by activating or repressing the expression of particular genes. As discussed in Section 1.2.3, genes that are activated in response to a specific treatment share a short DNA sequence in their promoters or enhancers whose transfer to another gene renders that gene inducible by the specific treatment. In turn, such sequences act by binding a specific transcription factor which becomes activated in response to the stimulus. Once activated, this factor interacts with the constitutive transcription factors discussed in Chapter 3 resulting in increased transcription of the gene.

A selection of DNA sequences which enable a gene to respond to a particular stimulus and the transcription factors which they bind is given in Table 4.1 (for a review, see Latchman, 1998). Rather than discuss each of these examples individually, we will focus on three cases of inducible gene expression. These examples, the induction of gene activity by heat shock, by cyclic AMP and by steroid hormones, have been chosen to illustrate the manner in which transcription factors can become active in response to a signal as well as how the active factor induces increased transcription of the target gene.

Table 4.1 Sequences that confer response to a particular stimulus

Consensus sequences	Response to	Protein factor	Gene containing sequences
CTNGAATNTT CTAGA	Heat	Heat-shock transcription factor	*hsp70, hsp83, hsp27*, etc.,
T/G T/A CGTCA	Cyclic AMP	CREB/ATF	Somatostatin fibronectin, α-gonadotrophin *c-fos, hsp70*
TGAGTCAG	Phorbol esters	AP1	Metallothionein IIA, α_1-antitrypsin, collagenase
CC(A/T)$_6$GG	Growth factor in serum	Serum response factor	*c-fos, Xenopus* γ-actin
RGRACNNN TGTYCY	Glucocorticoid	Glucocorticoid receptors	Metallothionein IIA, tryptophan oxygenase, uteroglobin, lysozyme
RGGTCANNN TGACCY	Oestrogen	Oestrogen receptor	Ovalbumin, conalbumin, vitellogenin
RGGTCAT GACCY	Thyroid hormone retinoic acid	Thyroid hormone receptors	Growth hormone, myosin heavy chain
TGCGCCCGCC	Heavy metals	Mep-1	Metallothionein genes
AGTTTCNN TTTCNC/T	Interferon-α	Stat-1 Stat-2	Oligo A synthetase guanylate-binding protein
TTNCNNNAA	Interferon-γ	Stat-1	Guanylate-binding protein, Fc

N indicates that any base can be present at that position, R indicates a purine, i.e. A or G, Y indicates a pyrimidine, i.e. C or T.

4.2 HEAT-INDUCIBLE TRANSCRIPTION

4.2.1 The heat-shock factor

When cells from a variety of species ranging from bacteria to mammals are subjected to elevated temperatures (heat-shock) or other stresses, such as anoxia, they respond by inducing increased transcription of a small number of genes known as the heat-shock genes (for reviews, see Morimoto, 1993; Parsell and Lindquist, 1993). As discussed in Section 1.2.3, these heat-inducible genes share a common DNA sequence which, when transferred to another gene, can render the second gene heat inducible. This sequence is known as the heat-shock element (HSE). The manner in which a *Drosophila* HSE when introduced into mammalian cells functioned at the mammalian rather than the *Drosophila* heat-shock temperature suggested that this sequence acted by binding a protein rather than by acting directly as a thermosensor (see Figure 1.3).

Direct evidence that this was the case, was provided by studying the proteins bound to the promoters of the *hsp* genes before and after heat shock. Thus, prior to heat shock, the TFIID complex (see Sections 3.2.4 and 3.2.5) is bound to the TATA box and another transcription factor known as GAGA is bound upstream (Figure 4.1a) (Wu, 1985; Tsukiyama *et al.*, 1994).

Following heat shock, however, an additional factor is observed which is bound to the HSE (Figure 4.1b). The amount of this factor bound to the HSE increased with the time of exposure to elevated temperature and with the extent of temperature elevation. Moreover, increased protein binding to the HSE was also observed following exposure to other agents that also induce the transcription of the heat-shock genes such as 2, 4-dinitrophenol (Figure 4.2). Thus activation of the heat shock genes, mediated by the HSE is accompanied by the binding of a specific transcription factor to this DNA sequence. This factor which has been variously referred to as the heat-shock transcription factor or heat-shock activator protein is now generally known as the heat-shock factor (HSF).

Hence, prior to heat shock, the heat-shock genes are poised for transcription. Thus, whilst the bulk of cellular DNA is associated with histone proteins to form a tightly packed chromatin structure, the binding of the GAGA factor to the heat-shock gene promoters has resulted in the displacement of the histone-containing nucleosomes from the promoter region (for a review, see Wilkins and Lis, 1997). This opens up the chromatin and renders the promoter region exquisitely sensitive to digestion with the enzyme DNaseI.

a)　Prior to heat shock

b)　After heat shock

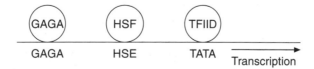

Figure 4.1 Proteins binding to the promoter of the *hsp70* gene before (a) and after (b) heat shock.

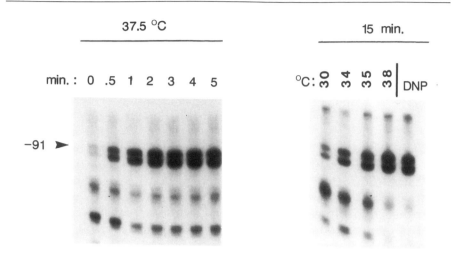

HEAT SHOCK

Figure 4.2 Detection of HSF binding to the HSE 91 bases upstream (−91) of the start site for transcription in the *Drosophila hsp82* gene and protecting this region from digestion with exonuclease III. Note the increased binding of HSF with increasing time of exposure to heat shock or increased severity of heat shock. HSF binding is also induced by exposure to 2,4-dinitrophenol (DNP) which is known to induce transcription of the heat shock genes.

Although such a DNaseI hypersensitive site marks a gene as poised for transcription (for a review, see Latchman, 1998), it is not in itself sufficient for transcription. The binding of the GAGA factor thus opens up the gene and renders it poised for transcription in response to a suitable stimulus. This role for the GAGA factor in chromatin remodelling is not confined to the heat-shock genes. Thus mutations in the gene encoding GAGA result in the *Drosophila* mutant trithorax in which a number of homeobox genes (which control the formation of the correct body plan, see Section 6.2) are not converted from an inactive to an active chromatin state and are hence not transcribed (for a review, see Schumacher and Magnuson, 1997). This mutation thus produces a fly with an abnormal body pattern and thus has a similar effect to the *brahma* mutation in the SWI2 component of the SWI/SNF chromatin remodelling complex which was discussed in Section 1.4.2. Indeed, the GAGA factor has been shown to be associated with a multi-protein complex known as nucleosome remodelling factor (NURF), which, like SWI/SNF, can hydrolyse ATP and alter chromatin structure (Figure 4.3) (for a review, see Tsukiyama and Wu, 1997).

Hence, following binding of GAGA, the gene is in a state poised for the

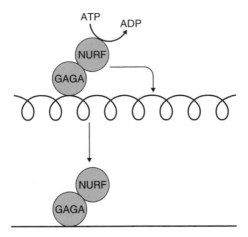

Figure 4.3 The GAGA factor can bind to DNA, which is in a tightly packed chromatin structure (wavy line), and recruit the nucleosome remodelling factor (NURF). NURF then hydrolyses ATP and uses the energy to remodel the chromatin to a more open structure (solid line) to which other activating proteins can bind.

binding of an activating transcription factor which in turn will result in transcription of the gene. In the case of the heat-shock genes, this is achieved following heat shock by the binding of the HSF to the HSE (Figure 4.4). This factor then interacts with TFIID and other components of the basal transcription complex resulting in the activation of transcription (for further discussion see Section 9.2).

The critical role of the HSF in this process obviously begs the question of how this factor is activated in response to heat.

4.2.2 Activation of HSF by heat

If cells are heat treated in the presence of cycloheximide, which is an inhibitor of protein synthesis, increased binding of HSF to the HSE is observed exactly as in cells treated in the absence of the drug (Zimarino and Wu, 1987). This indicates that the observed binding of HSF following heat shock does not require *de novo* protein synthesis. Rather, this factor must pre-exist in non-heat-treated cells in an inactive form whose ability to bind to the HSE sequence in DNA is activated post-translationally by heat. In agreement with this, activation of HSF can also be observed following heat treatment of cell extracts *in vitro* when new protein synthesis would not be possible (Larson *et al.*, 1988).

Analysis of the activation process using *in vitro* systems from human cells (Larson *et al.*, 1988) has indicated that it is a two-stage process

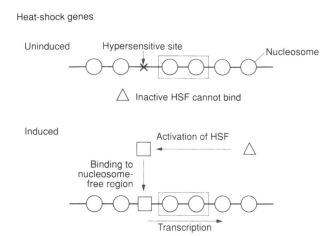

Figure 4.4 Activation of HSF by heat is followed by its binding to a pre-existing nucleosome-free region in the heat-shock gene promoters, which is marked by a DNaseI hypersensitive site and was produced by the prior binding of the GAGA factor. Binding of HSF then results in the activation of heat-shock gene transcription.

(Figure 4.5). In the first stage, the HSF is activated to a form which can bind to DNA by an ATP-independent mechanism which is directly dependent on elevated temperature. Subsequently, this protein is further modified by phosphorylation allowing it to activate transcription. Interestingly, the second of these two stages appears to be disrupted in murine erythro-leukaemia (MEL) cells in which heat-shock results in increased binding of HSF to DNA but transcriptional activation of the heat-shock genes is not observed (Hensold *et al.*, 1990).

The activation of HSF into a form capable of binding DNA involves its conversion from a monomeric to a trimeric form which can bind to the HSE (for a review, see Morimoto, 1993). The maintenance of the monomeric form of HSF prior to heat shock is dependent on a region at the C-terminus of the molecule since when this region is deleted, HSF spontaneously trimerizes and can bind to DNA even in the absence of heat shock (Rabindran *et al.*, 1993). The C-terminal region contains a motif known as the leucine zipper which contains a leucine residue every seven amino acids. As leucine zippers are known to be able to interact with one another (see Section 8.4), it is thought that this region acts by interaction with another leucine zipper located adjacent to the N-terminal DNA-binding domain promoting intra-molecular folding which masks the DNA-binding domain. Following heat shock, HSF unfolds, unmasking the DNA-binding domain and allowing a DNA-binding trimer to form (Figure 4.6).

The two-stage process described above represents a common mechanism for the activation of HSF in higher eukaryotes such as *Drosophila*

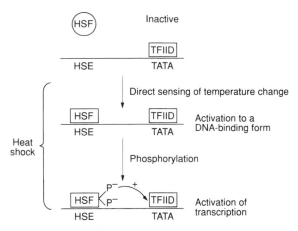

Figure 4.5 Stages in the activation of HSF in mammalian and *Drosophila* cells. Initial activation of HSF to a DNA-binding form following elevated temperature is followed by its phosphorylation, which converts it to a form capable of activating transcription.

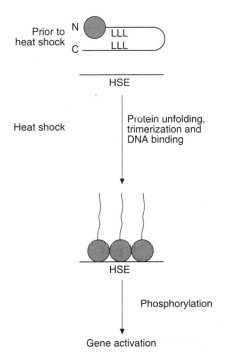

Figure 4.6 Prior to heat shock, HSF is present in a monomeric form in which the leucine zipper motifs (L) at the C-terminus and within the molecule promote intra-molecular folding, which masks the N-terminal DNA-binding domain (shaded) preventing binding to the HSE. Following heat shock, the protein unfolds and forms the DNA-binding trimeric form. This form binds to the HSE and activates transcription following its subsequent phosphorylation.

and mammals. In contrast, however, the *Saccharomyces cerevisiae* (budding yeast) HSF is activated by a much simpler mechanism. Thus, unlike *Drosophila* or mammalian HSF, the budding yeast protein lacks the C-terminal leucine zipper region that promotes monomer formation and therefore exists as a trimer prior to heat-shock. As expected from this, HSF can be observed bound to the HSE even in non-heat-shocked cells (Sorger *et al.*, 1987). HSF can activate transcription, however, only following heat treatment when the protein becomes phosphorylated. Interestingly, in *Schizosaccharomyces pombe* (fission yeast) HSF regulation follows the *Drosophila* and mammalian system with HSF becoming bound to DNA only following heat-shock (Gallo *et al.*, 1991).

Hence in mammals, *Drosophila* and fission yeast, activation of HSF is more complex than in budding yeast, involving an initial stage activating the DNA-binding ability of HSF in response to heat as well as the stage, common to all organisms, in which the ability to activate transcription is stimulated by phosphorylation (Figure 4.7). Such phosphorylation is suggested to result in the creation of a highly negatively charged region on the HSF. This region would be analogous to the negatively charged acidic activation domains of other transcription factors (see Section 9.2.2). It would therefore allow HSF to interact with components of the constitutive transcriptional apparatus such as TFIID and activate transcription of the heat-shock genes. In agreement, with this idea, HSF has been shown to possess a heat-inducible activation domain (Nieto-Sotelo *et al.*, 1990). Interestingly, this domain can be converted into a form which activates transcription at any temperature by deletion of a seven-amino-acid peptide which is located adjacent to the activation

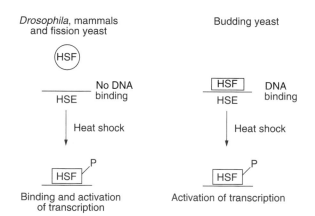

Figure 4.7 HSF activation in *Drosophila*, mammals and fission yeast compared to that in budding yeast. Note that in budding yeast HSF is already bound to DNA prior to heat shock and hence its activation by heat involves only the second of the two stages seen in other organisms, namely its phosphorylation, allowing it to activate transcription.

domain within HSF, indicating that this peptide normally inhibits the activity of the activation domain in a temperature-dependent manner (Jakobsen and Pelham, 1991).

In summary, therefore, heat-shock induces the increased transcription of a small number of cellular genes via the post-translational modification of a pre-existing transcription factor, HSF. This factor binds to a sequence known as the HSE which is located in a region of the heat-shock gene promoters that is free of nucleosomes and contains a DNaseI hypersensitive site even prior to heat shock. The activated form of HSF is capable, following binding, of interacting with components of the constitutive transcriptional apparatus which are bound at other sites in the promoter region and thereby stimulating transcription.

4.3 CYCLIC AMP-INDUCIBLE TRANSCRIPTION

4.3.1 CREB and the CRE

The interaction of a number of different hormones such as adrenaline, glucagon, vasopressin and thyroid-stimulating hormone with their specific cell-surface receptors results in the activation of the membrane-associated enzyme adenylyl cyclase. The activation of this enzyme then leads to a rise in the intracellular levels of cyclic $3',5'$-monophosphate (cyclic AMP). In turn this elevation in cyclic AMP levels results in the induction of many different genes (for reviews, see Montminy, 1997; Sassone-Corsi, 1998a, 1998b).

As with the heat-inducible genes discussed in the previous section, cyclic AMP inducible genes contain a short sequence in their regulatory regions which can confer responsiveness to cyclic AMP when it is transferred to another gene that is not normally cyclic AMP inducible. This sequence, which is known as the cyclic AMP response element (CRE), consists of the eight-base pair palindromic sequence TGACGTCA.

The first transcription factor that was shown to bind to this site was a 43 kDa protein which was named CREB (cyclic AMP response element binding protein). This factor has a basic DNA-binding domain with adjacent leucine zipper dimerization motif (Figure 4.8) (for further discussion of this motif, see Section 8.4) and binds to the palindromic CRE as a dimer with each CREB monomer binding to one half of the palindrome.

The CREB factor plays a key role in the activation of gene expression via the CRE following cyclic AMP treatment. Like HSF, the CREB factor is present in cells in an inactive form prior to exposure to the activating stimulus. Unlike HSF in higher eukaryotes, which only binds to the HSE

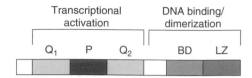

Figure 4.8 Structure of the CREB transcription factor indicating the glutamine-rich activation domains (Q_1 and Q_2), the phosphorylation box (P) containing the serine 133 residue, and the basic DNA-binding domain (BD) with associated leucine zipper (LZ).

following heat shock (see Section 4.2), CREB is actually bound to the CRE prior to exposure to cyclic AMP but this DNA-bound CREB does not activate transcription. Elevated levels of cyclic AMP result in the activation of the protein kinase A enzyme (for a review, see McKnight *et al.*, 1988), which in turn phosphorylates CREB on the serine amino acid at position 133 in the molecule. This serine residue is located in a region of CREB known as the phosphorylation box (P-box) which is flanked on either side by regions rich in glutamine amino acids which act as transcriptional activation domains (see Section 9.2) (Figure 4.8). The phosphorylation of CREB on serine 133 results in a change in the structure of the molecule which now allows it to activate transcription (Figure 4.9). Hence this case parallels the activation of HSF in *S.cerevisiae* and the second stage of its activation in other eukaryotes (see Section 4.2.2), where phosphorylation activates the transcriptional activation ability of a DNA-bound factor. The mechanism by which phosphorylation activates CREB in this manner has been intensively studied and is discussed in the next section.

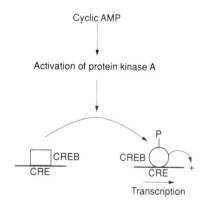

Figure 4.9 Activation of the CREB factor by cyclic AMP-induced phosphorylation. The ability of DNA-bound CREB to activate transcription is produced by the cyclic AMP-dependent activation of protein kinase A, which phosphorylates the CREB protein resulting in its activation.

4.3.2 CREB and CBP

As well as being used to isolate proteins able to bind to specific DNA-binding sites, the cDNA expression libraries described in Section 2.3.2b can also be screened with labelled protein to isolate cDNA clones encoding other proteins with which the labelled protein interacts. The screening of such a cDNA expression library with CREB protein phosphorylated on serine 133 resulted in the isolation of cDNA clones encoding CBP (CREB-binding protein) (Chrivia *et al.*, 1993). CBP is a 265 kDa protein which associates only with phosphorylated CREB and not with the unphosphorylated form (for a review, see Shikama *et al.*, 1997). This pattern of association immediately suggests that CBP plays a critical role in the ability of CREB to activate transcription only after phosphorylation. In agreement with this, injection of cells with antibodies to CBP prevents gene activation in response to cyclic AMP indicating that CBP is essential for this effect (for a review, see Montminy, 1997; Sassone-Corsi, 1998b).

Hence CBP is a co-activator molecule whose binding to phosphorylated CREB is essential for transcriptional activation to occur (Figure 4.10). Two possible mechanisms by which CBP achieves this effect have been described. Thus CBP has been shown to interact via a protein–protein interaction with several components of the basal transcriptional complex such as TFIIB (see Section 3.2.4) and has been identified as part of the RNA polymerase II holoenzyme complex (which also contains RNA polymerase II, components of the basal transcriptional complex and other regulatory proteins) (Nakajima *et al.*, 1997). Thus CBP may serve as a bridge between CREB and the basal transcriptional complex either interacting with components of the complex to enhance their activity or serving to recruit the RNA polymerase holoenzyme to the DNA by the CBP component binding to CREB (Figure 4.11).

As well as this mechanism, however, it is also possible that CBP acts via a mechanism involving alterations in chromatin structure. Thus both CBP and the related co-activator p300 have been shown to have histone

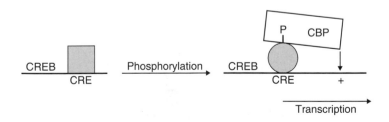

Figure 4.10 The phosphorylation of CREB on serine 133 allows it to bind the CBP *co-activator, which then stimulates transcription.*

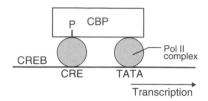

Figure 4.11 CBP can bind to both CREB and the basal transcriptional complex. It may therefore act as a bridge between CREB and the complex, allowing transcriptional activation to occur.

acetyltransferase activity (Ogryzko *et al.*, 1996). As discussed in Section 1.4.2, acetylated histones are associated with the more open chromatin structure which is required for transcription. Hence the binding of CBP to CREB which recruits it to DNA may then result in the acetylation of histones leading to a chromatin structure compatible with transcription (Figure 4.12) (for a review, see Wade *et al.*, 1997).

Evidently these two possibilities are not mutually exclusive and CBP may act both at the level of the basal transcriptional complex and at the level of chromatin structure. Whether one or both of these mechanisms operate, however, it is clear that CBP plays a critical role in the response to cyclic AMP mediated via CREB. Interestingly, however, although CBP was discovered via its association with CREB, its role is not confined to this situation alone. Thus, as will be discussed in subsequent sections, it

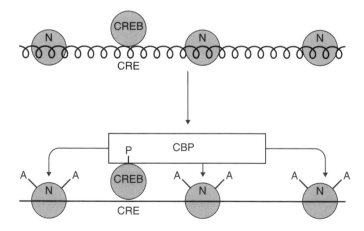

Figure 4.12 CBP has histone acetyltransferase activity, therefore, following binding to phosphorylated CREB, it can acetylate (A) histones within the nucleosome (N), resulting in a more open chromatin structure (wavy versus solid line) compatible with transcription.

plays a key role as a co-activator for transcription factors activated by other signalling pathways including the nuclear receptors (see Section 4.4) and AP1 (see Section 7.2.1) (for review, see Janknecht and Hunter, 1996; Shikama *et al.*, 1997).

4.3.3 Other cyclic AMP regulated transcription factors

Although CREB was the first factor shown to bind to the CRE, other factors which do so have subsequently been defined. They constitute a family of related proteins known as the CREB/ATF (activating transcription factor) family and all have a basic domain/leucine zipper structure (for a review, see Hurst, 1996). Thus the CREM factor resembles CREB in being phosphorylated following cyclic AMP treatment at a site located between two glutamine-rich activation domains. Like CREB, it can therefore bind to the CRE and activate transcription in response to cyclic AMP by binding CBP (for reviews, see Lalli and Sassone-Corsi, 1994; Sassone-Corsi, 1998a, 1998b).

Interestingly, however, alternative splicing produces distinct forms of the CREM factor which lack the activation domains, although they retain the leucine zipper and basic DNA-binding domain (Figure 4.13a). These forms can therefore bind to DNA but cannot activate transcription since they lack an activation domain. They therefore inhibit transcription by competing for binding to the CRE with the activating forms (Figure 4.14) (for a discussion of indirect repression of this type, see Section 9.3.1). Since the proportion of the activating and inhibitory forms of CREM varies in different cell types, the level of transcription directed by a CRE following cyclic AMP treatment will be different in these cells depending on the precise balance between the activating and inhibitory forms.

As well as producing distinct forms with and without the activation domain, the CREM factor also undergoes alternative splicing in another manner. Thus two distinct exons in the CREM gene contain two distinct DNA-binding domains. Alternative splicing results in the proteins which either do or do not contain the activation domains also having one or other of the DNA-binding domains (Figure 4.13b). As the relative usage of the two DNA-binding domains is different in different cell types, this effect is likely to have biological significance but its precise role is at present unclear.

The different forms of the CREM factor which have been discussed so far are all produced by alternative splicing of a single RNA transcript whose rate of production is unaffected by cyclic AMP. Thus the ability of the CREB factor and the activating forms of CREM to switch on gene expression is stimulated post-translationally by their phosphorylation following cyclic AMP treatment, hence allowing them to switch on gene expression in response to cyclic AMP.

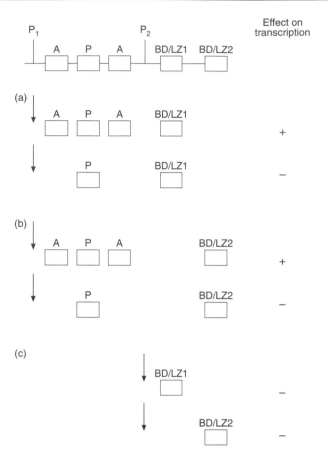

Figure 4.13 The CREM protein contains two transcriptional activation domains (A), a region containing a site for cyclic AMP-induced phosphorylation (P), and two DNA-binding domains containing a basic domain and leucine zipper (BD/LZ). After transcription from the P_1 promoter, alternative splicing can result in forms with or without the activation domains (a) or having either of the DNA-binding domains (b). In addition, cyclic AMP-inducible transcription from the P_2 promoter can produce forms containing only one or other of the DNA-binding domains but lacking the activation domains and the phosphorylated region (c). Arrows indicate the transcriptional start sites used in each case.

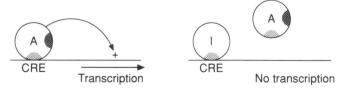

Figure 4.14 Gene activation by the activating forms of the CREM protein (A) can be inhibited by forms (I), which contain the DNA-binding domain (light shading) but lack the activation domain (heavy shading). They therefore bind to the CRE and prevent binding by the activating forms.

In contrast to such post-translational regulation, the CREM gene also contains a promoter which is activated in response to cyclic AMP. This promoter produces transcripts encoding short proteins which contain one or other of the DNA-binding domains and the phosphorylated region but lack the activation domain (Figure 4.13c). These proteins can therefore bind to the cyclic AMP response element and repress transcriptional activation by the activating forms, exactly as described above for the alternatively spliced forms lacking the activation domain. These forms are therefore known as ICERs (inducible cyclic AMP early repressors). As they are inducible by cyclic AMP, these forms are likely to play a key role in making the cyclic AMP response self-limiting. Thus, following cyclic AMP treatment, CREB and CREM will activate the expression of promoters containing a CRE including that which produces the ICERs. The ICERs produced in this manner will then bind to the CRE and switch off the inducible genes by preventing the binding of CREB and CREM (Figure 4.15), thereby making the cyclic AMP response a transient one.

The regulation of cyclic AMP inducible transcription by the CREB and CREM factor is, therefore, extraordinarily complex with both alternative splicing and the use of two different promoters in the CREM gene. Indeed, this may seem a highly complex system to use simply to mediate a relatively simple process like cyclic AMP inducibility. However, the findings that cyclic AMP inducible transcription mediated by these proteins plays a critical role in processes as diverse as pituitary gland development and long-term memory (for reviews, see Lalli and Sassone-Corsi, 1994; Stevens, 1994; Abel *et al.*, 1998) indicate that this complexity may be required in order to regulate these very complex biological phenomena properly.

In summary, therefore, the phosphorylation of CREB and CREM following cyclic AMP treatement allows them to bind the CBP activator. In turn, this results in the activation of transcription mediated by the ability of CBP to interact with the basal transcriptional complex and/or acetylate histones. The production of the ICER forms of CREM, which are cyclic AMP-inducible but have an inhibitory role, renders the response self-limiting and ensures only a transient response to cyclic AMP.

4.4 STEROID-INDUCIBLE TRANSCRIPTION

4.4.1 Steroid receptors

The steroid hormones are a group of substances derived from cholesterol which exert a very wide range of effects on biological processes such as growth, metabolism and sexual differentiation (for a review, see King and

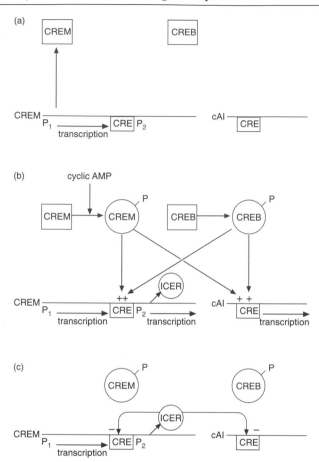

Figure 4.15. (a) In the absence of cyclic AMP, the CREM gene is transcribed from the P_1 promoter. However, neither the CREM produced in this way or the CREB protein can activate transcription until they are activated post-translationally. (b) Following cyclic AMP treatment, the CREB and CREM proteins become activated post-translationally by phosphorylation. They therefore activate the cyclic AMP-inducible genes (CAI), which contain a cyclic AMP response element (CRE) in their promoters. In addition, they also activate the P_2 promoter of CREM, which also contains a CRE. (c) The ICER produced by the CREM P_2 promoter, binds to the CREs and prevents activation by CREB and CREM, thereby repressing transcription.

Mainwaring, 1974). Early studies using radioactively labelled hormones showed that they act by interacting with specific receptor proteins. This binding of hormone to its receptor activates the receptor and allows it to bind to a limited number of specific sites in chromatin. In turn, this DNA-binding activates transcription of genes carrying the receptor-binding site. Hence, as with the heat-shock factor, these receptor proteins are transcription factors becoming activated in response to a specific signal and in turn activating specific genes (for reviews, see Beato *et al.*, 1995; Gronemeyer and Laudet, 1995; Mangelsdorf *et al.*, 1995; Perlmann and Evans, 1997). These receptor proteins were, therefore, amongst the earliest transcription factors to be identified, well before the techniques described in Chapter 2 were in routine use, simply on the basis of their ability to bind radioactively labelled steroid ligand.

As with heat shock, genes which are induced by a particular steroid hormone contain a specific binding site for the receptor–hormone complex. The responses to different steroid hormones such as glucocorticoids and oestrogen are mediated by distinct palindromic sequences that are related to one another. In turn, such sequences are related to one of the sequences that mediates induction by other substances which are related to steroids such as thyroid hormone and retionic acid. Similarly, repeated elements with different spacings between the repeats also mediate responses to these different substances (Table 4.2) (for a review, see Gronemeyer and Moras, 1995).

The basis of this binding-site relationship was revealed when the genes encoding the receptor proteins were cloned. Thus they were found to constitute a family of genes encoding closely related proteins of similar structure with particular regions being involved in DNA-binding, hormone binding and transcriptional activation (Figure 4.16). This has led to the idea that these receptors are encoded by an evolutionarily related gene family which is known as the steroid–thyroid hormone receptor or nuclear receptor gene super family (for reviews, see Gronemeyer and Laudet, 1995; Mangelsdorf *et al.*, 1995; Perlmann and Evans, 1997). The structure of the thyroid hormone receptor bound to its ligand, thyroid hormone, is illustrated in Plate 2 (Wagner *et al.*, 1996).

Table 4.2. Relationship of various hormone response elements

(a) Palindromic repeats	
Glucocorticoid	RGRACANNNTGTYCY
Oestrogen	RGGTCANNNTGACCY
Thyroid	RGGTCA – TGACCY
(b) Direct repeats	
9-*cis* Retinoic acid	$AGGTCAN_1AGGTCA$
All-trans retinoic acid	$AGGTCAN_2AGGTCA$
	$AGGTCAN_5AGGTCA$
Vitamin D_3	$AGGTCAN_3AGGTCA$
Thyroid hormone	$AGGTCAN_4AGGTCA$

N indicates that any base can be present at that position, R indicates a purine, i.e. A or G, Y indicates a pyrimidine, i.e. C or T indicates A or T. A dash indicates that no base is present, the gap having been introduced to align the sequence with the other sequences.

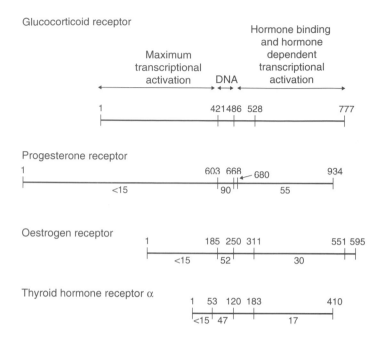

Figure 4.16 Domain structure of individual members of the steroid–thyroid hormone receptor super-family. The proteins are aligned on the DNA binding domain, which shows the most conservation between different receptors. The percentage homologies in each domain of the receptors to that of the glucocorticoid receptor are indicated.

As shown in Figure 4.16, the most conserved region between the different receptors is the DNA-binding domain explaining the ability of the receptors to bind to similar DNA sequences. Clearly, the differences in amino-acid sequences between the DNA-binding motifs in the different receptors will determine the precise DNA sequence which each receptor recognizes. The replacement of particular amino acids in the DNA-binding domain with their equivalents in other receptors has therefore provided considerable information on the features which determine the precise sequence recognized or the optimal spacing between the two halves of the sequence. These experiments and the 'so-called' zinc finger motifs in the different receptors which mediate their DNA-binding are discussed in Section 8.3.2. Interestingly, both DNaseI protection and methylation studies support the idea that the receptor binds to DNA as a dimer, each receptor molecule binding to one half of the recognition sequence.

Having established therefore that steroid hormones exert their effect on gene expression via specific receptor proteins, two questions remain.

These are, firstly, how does binding of hormone to the receptor activate its ability to bind to specific DNA sequences and, secondly, how does such binding activate transcription? These questions will be considered in turn.

4.4.2 Activation of the receptor

Following identification of the hormone receptors, it was very rapidly shown that the receptors were only found associated with DNA after hormone treatment. These early studies were subsequently confirmed by using DNaseI footprinting on whole chromatin to show that the receptor was only bound to the hormone response sequence following hormone treatment (Becker *et al.*, 1986). These studies were therefore consistent with a model in which the hormone induces a conformational change in the receptor activating its ability to bind to DNA and thereby activate transcription.

Subsequent studies have suggested that the situation is more complex however. Thus, although in the intact cell the receptor binds to DNA only in the presence of the hormone, purified receptor can bind to DNA *in vitro* in a band shift or footprinting assay regardless of whether hormone is present or not (Wilmann and Beato, 1986) (Figures 4.17 and 4.18).

This discrepancy led to the suggestion that the receptor is inherently capable of binding to DNA but is prevented from doing so in the absence of steroid because it is anchored to another protein. The hormone acts to release it from this association and allow it to fulfil its inherent ability to bind to DNA. In agreement with this possibility, in the absence of hormone, the glucocorticoid receptor protein is found in the cytoplasm complexed to a 90 000 molecular weight heat-inducible protein (hsp90) in an *8S* complex. This complex is dissociated upon steroid treatment releasing the *4S* receptor protein (for reviews, see Pratt, 1997; Pratt and Toft, 1997). The released receptor is free to dimerize and move into the nucleus. Since these processes have been shown to be essential for DNA-binding and transcriptional activation by steroid hormone receptors, dissociation of the receptor from hsp90 is essential if gene activation is to occur. In agreement with this, antiglucocorticoids which inhibit the positive action of glucocorticoids have been shown to stabilize the *8S* complex of hsp90 and the receptor. Similar complexes with hsp90 have also been reported for the other steroid hormone receptors. Thus the activation of the different steroid receptors such as the glucocorticoid and oestrogen receptors by their specific hormones is likely to involve disruption of the protein–protein interaction with hsp90 (Figure 4.19). The manner in which the steroid achieves this effect is discussed further in Section 10.3.2.

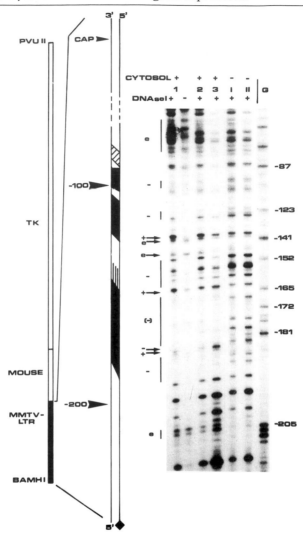

Figure 4.17 DNAseI footprint analysis of the binding of the glucocorticoid receptor to the glucocorticoid-inducible mouse mammary tumour virus long terminal repeat promoter (MMTV-LTR). In tracks I and II the DNAseI digestion has been carried out without any added receptor. In tracks 1–3, glucocorticoid receptor has been added prior to DNAseI digestion either alone (track 1+), with the glucocorticoid hormone corticosterone (track 2) or with the anti-hormone RU486, which inhibits steroid-induced activation of the receptor (track 3). Track 1– shows the result of adding receptor to the DNA in the absence of DNAseI addition in which some cleavage by endogenous nucleases (e) occurs, whilst track G is a marker track produced by cleaving the same DNA at each guanine residue. Minus signs indicate footprinted regions protected by receptor, plus signs are hypersensitive sites at which cleavage is increased by the presence of the receptor. The DNA fragment used and position of the radioactive label (diamond) are shown together with the distances upstream from the initiation site for transcription. Note that the identical footprint is produced by the receptor either alone or in the presence of hormone or anti-hormone. Hence, *in vitro*, the receptor can bind to DNA in the absence of hormone.

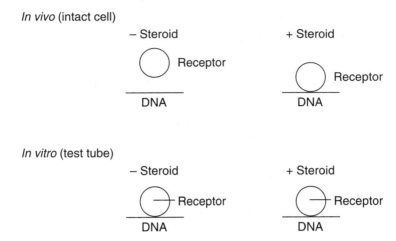

Figure 4.18 Comparison of steroid receptor binding to DNA in the presence or absence of hormone *in vivo* and *in vitro*. Note that, whilst *in vivo* DNA binding can occur only in the presence of hormone, *in vitro* it can occur in the presence or absence of hormone.

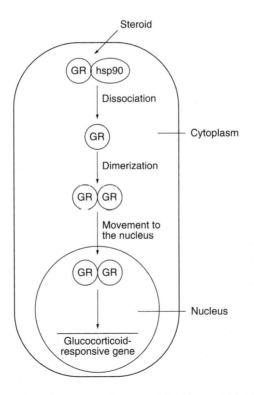

Figure 4.19 Activation of the glucocorticoid receptor (GR) by steroid involves dissociation of hsp90, allowing dimerization and movement to the nucleus.

In addition to the steroid-induced dissociation of the receptors from hsp90, it is clear that a second step following dissociation from hsp90 is also required for receptor activation. Thus, in a cell-free system in which the progesterone receptor exists in a *4S* form, free of bound hsp90, the addition of progesterone is still required for the activation of progesterone-responsive genes. This indicates that the hormone has an additional effect on the receptor apart from dissociating it from hsp90. This effect involves the unmasking of a previously inactive transcriptional activation domain in the receptor, allowing it to activate gene expression in a hormone-dependent manner following DNA-binding. Thus, domain-swopping experiments (see Section 2.3.3) have identified C-terminal regions in both the glucocorticoid and oestrogen receptors which, when linked to the DNA-binding domain of another factor, can activate transcription only following hormone addition (see Figure 4.16). These regions hence constitute hormone-dependent activation domains. Moreover, in the case of the oestrogen receptor, it has been shown that the oestrogen antagonist 4-hydroxytamoxifen induces the receptor to bind to DNA (presumably by promoting dissociation from hsp90 and dimerization) but does not induce gene activation, suggesting that it fails to activate the oestrogen-responsive trans-activation domain. Hence the mechanism by which the steroid receptors are activated is now thought to involve both dissociation from hsp90 and a change in their transcriptional activation ability (Figure 4.20a).

Interestingly, other members of the nuclear receptor family which bind to substances that are related to steroids such as retinoic acid or thyroid hormone do not associate with hsp90 and are bound to DNA prior to exposure to ligand. Their activation by their appropriate ligand thus involves only the second stage discussed above, namely a ligand-induced structural change in their C-terminal activation domain which is adjacent to the ligand-binding domain, allowing it to activate transcription (for a review, see Mangelsdorf and Evans, 1995) (Figure 4.20b). Indeed, crystallographic studies of the ligand-binding domain and the C-terminal activation domain of the retinoic acid receptors both in the presence or absence of hormone have provided direct evidence for this change. Thus, as illustrated in Plate 3, the activation domain is not closely associated with the ligand-binding domain in the absence of ligand, but is much more closely associated with it following ligand-binding and forms a lid covering the ligand-binding region (Renaud *et al.*, 1996). Although first defined in the retinoic acid receptors, a similar structural change occurs upon ligand binding in other members of the nuclear receptor family including the glucocorticoid and oestrogen receptors (Wurtz *et al.*, 1996). Indeed, it has recently been shown that, whilst oestrogen induces this realignment of the oestrogen receptor activation domain, the oestrogen antagonist raloxifene does not do so,

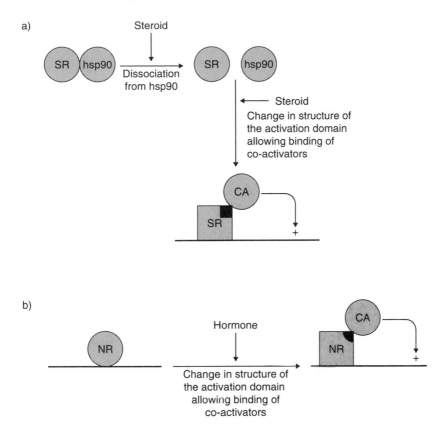

Figure 4.20 (a) Activation of the steroid receptors (SR) by treatment with steroid. As well as inducing dissociation of the receptor from hsp90, steroid treatment also increases the ability of the receptor to activate transcription following DNA binding by changing the structure of the activation domain (solid area), allowing it to bind co-activator proteins (CA) which stimulate transcription. (b) Activation of other members of the nuclear receptor family, which bind non-steroids such as retinoic acid or thyroid hormone, involves only the second of these stages.

thereby explaining its antagonistic action (Brzozosowski *et al.*, 1997) (Figure 4.21). In turn this ligand-induced structural change allows the activation domain to bind co-activator proteins which bind to the receptors only after exposure to hormone and appear to play a key role in the ability to activate transcription (Figure 4.20). Several such co-activator molecules have been defined (for a review, see Glass and Rosenfeld, 1997). Interestingly, these co-activators include the CBP protein which was discussed earlier (see Section 4.3.2) and which binds to several nuclear receptors including the oestrogen, thyroid hormone and retinoic acid receptors only following addition of ligand.

Figure 4.21 (a) The binding of the ligand (L) induces the realignment of the C-terminal activation domain of the nuclear receptors (light shading) so that it forms a lid over the ligand-binding domain and the activation domain then stimulates transcription. (b) This realignment is not induced by binding of antagonists (A), which therefore do not stimulate transcriptional activation.

Hence, in all the nuclear reeptors, activation by ligand involves a structural change in the C-terminal activation domain which allows it to bind co-activators. In the steroid hormone receptors, this is preceded by an earlier step which involves the disruption of the receptor hsp90 association. The manner in which the activated receptor stimulates transcription is discussed in the next section.

4.4.3 Activation of transcription

Having considered the activation of the receptor by hormone, it is necessary to discuss how binding of the activated receptor in turn activates transcription of the genes bearing its target site (for a review, see Beato *et al.*, 1995).

As discussed in Section 4.2.2, the heat-shock factor binds to a region of chromatin which is free of histone-containing nucleosomes and is hypersensitive to DNaseI digestion. This hypersensitive site exists prior to heat shock, marking the potential binding site for the transcription factor. In contrast, in a number of cases, steroid hormone treatment has been shown to cause the induction of a DNaseI hypersensitive site located at the DNA sequence to which the receptor binds. Hence the binding of the receptor may activate transcription by displacing or altering the structure of a nucleosome within the promoter of the gene creating the hypersensitive site. In turn, this would facilitate the binding of other transcription factors necessary for gene activation whose binding sites would be exposed by the change in the position or structure of the nucleosome. These factors would be present in the cell in an active form prior to steroid treatment but could not bind to the gene because their binding sites were masked by a nucleosome (Figure 4.22) (for a review, see Beato and Eisfeld, 1997). In agreement with this idea, the binding sites for TFIID and CTF/NFI (see Sections 3.2.3 and 3.3.3) in the glucocorticoid-responsive mouse mammary tumour virus promoter are occupied only following hormone treatment, although these factors are present in an active DNA-binding form at a similar level in treated and untreated cells.

Interestingly, the nuclear receptors may alter chromatin structure by both the mechanisms described in Section 1.4. Thus, as discussed in the preceding section, following ligand binding, the receptors bind co-activator molecules such as CBP, PCAF, SRC-1 and ACTR, which are known to have histone acetyltransferase activity (see Section 4.3.2) (Chen *et al.*, 1997; Spencer *et al.*, 1997). Hence the receptor-induced change in chromatin structure may be brought about by histone acetylation as discussed in Section 1.4.3. In addition however, it appears that the glucocorticoid receptor can stimulate the activity of the SWI/SNF complex (Ostlund Farrants *et al.*, 1997), allowing it to fulfil its role of hydrolysing ATP and unwinding chromatin (see Section 1.4.2) (Figure 4.23).

This mechanism, in which the receptor acts by altering chromatin structure, allowing constitutive factors access to their binding sites, is

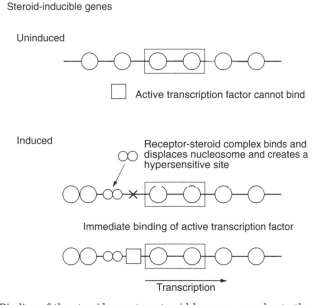

Figure 4.22 Binding of the steroid receptor–steroid hormone complex to the promoter of a steroid-inducible gene results in a change in chromatin structure, creating a hypersensitive site and allowing pre-existing constitutively expressed transcription factors to bind and activate transcription. Note that the binding of these constitutive factors may occur because the receptor totally displaces a nucleosome from the DNA as shown in the diagram or because it alters the structure of the nucleosome so as to expose specific binding sites in the DNA.

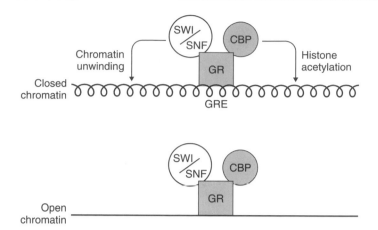

Figure 4.23 By binding histone acetyltransferases such as CBP and chromatin remodelling factors such as SWI/SNF, the steroid receptors can alter chromatin structure from a tightly packed (wavy line) to a more open (solid line) configuration.

clearly in contrast to the binding of HSF to a promoter which already lacks a nucleosome and contains bound GAGA factor and TFIID. In this latter case, activation of transcription must occur not via alteration in chromatin structure but via interaction with the components of the constitutive transcriptional apparatus. It should be noted, however, that these two mechanisms are not exclusive. Thus, as discussed above (Section 4.3.2), the CBP co-activator can also interact with components of the basal transcriptional complex to increase transcription. This finding indicates therefore that the steroid receptors and their associated co-activators promote transcription both by altering chromatin structure to allow constitutive factors to bind and also by interacting directly with other transcription factors such as components of the basal transcriptional complex (Figure 4.24). Activation by steroid hormones would therefore be a two-stage process involving, firstly, alteration of chromatin structure and, secondly, stimulation of the basal transcriptional complex (Jenster *et al.*, 1997). In agreement with this idea, chromatin disruption following binding of the thyroid hormone receptor to DNA is necessary but not sufficient for transcriptional activation to occur (Wong *et al.*, 1997). The activation of transcription by factors which alter chromatin structure and/or interact with the basal transcriptional complex is discussed further in Section 9.2.

Hence, in the case of the steroid receptors, a single factor and its associated co-factors can alter the chromatin structure and then activate transcription. In contrast, in heat-shock gene activation, these functions are performed by separate factors with the GAGA factor displacing a

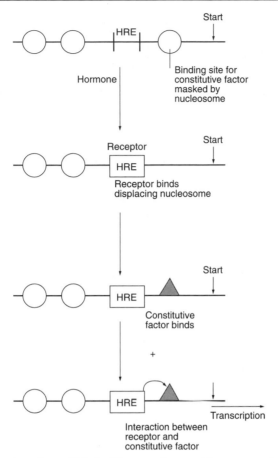

Figure 4.24 Activation of steroid-inducible genes by steroid receptors. As well as altering chromatin structure and allowing constitutive factors to bind, the hormone receptor, and its associated co-factors is also able to interact directly with constitutive factors such as the basal transcriptional complex and directly activate transcription.

nucleosome, allowing HSE to bind and activate transcription following a subsequent heat shock.

4.4.4 Inhibition of gene expression by steroid hormones

In the preceding sections we have discussed gene activation mediated by steroid hormones complexed to their corresponding receptors. In addition, however, these hormones can also inhibit the expression of specific genes, glucocorticoid, for example, inhibiting expression of the genes encoding bovine prolactin and human pro-opiomelanocortin. The

Binding site for positive regulation   **R G R A C A N N N T G T Y C Y**
Binding site for negative regulation   **A T Y A C A N N N T G A T C W**

Figure 4.25 Relationship of the sites in DNA which mediate gene activation or repression by binding the glucocorticoid receptor. Note that the sites are related but distinct.

inhibitory effect observed in these cases is mediated by binding to DNA of the identical receptor/hormone complex which activates glucocorticoid-inducible genes. However, the DNA sequence element to which the complex binds when mediating its negative effect (nGRE) is distinct from the glucocorticoid response element (GRE) to which it binds when inducing gene expression, although the two are related (Figure 4.25).

This has led to the suggestion that the sequence difference causes the receptor–hormone complex to bind to the nGRE in a configuration in which its activation domain cannot interact with other transcription factors to activate transcription as occurs following binding to the positive element (Figure 4.26). In agreement with this idea, the glucocorticoid receptor has been shown to bind to the nGRE in the POMC gene as a trimer rather than the dimer form, which binds to the GRE and stimulates transcription (Drouin *et al.*, 1993). The receptor bound in this configuration to the negative element may exert a direct negative influence resulting in decreased transcription of the gene. More probably, however, it simply acts by preventing binding of a positive-acting factor to this or an adjacent site, thereby preventing gene induction. In agreement with this idea, the nGRE in the human glycoprotein hormone α-subunit gene which overlaps a cyclic AMP response element is only able to inhibit gene expression when the CRE is left intact. Hence, it is likely that receptor bound at the negative element prevents binding of a transcriptional activator to the CRE and thereby inhibits gene expression (Figure 4.27).

In the case of the glucocorticoid receptor, inhibition of gene expression by binding to the nGRE only occurs after glucocorticoid treatment, which allows it to dissociate from hsp90 and bind to DNA. However, the thyroid hormone receptor, which does not complex to hsp90, can bind to its response element (TRE) in the absence of thyroid hormone and inhibit gene expression. This effect is not due to the receptor preventing other positive factors from binding but actually involves a direct inhibitory effect of the receptor on transcription which requires a specific inhibitory domain at the C-terminus of the molecule. In the presence of thyroid hormone, the receptor undergoes a conformational change which exposes its activation domain and converts it from a repressor to an activator (Figure 4.28). Hence, in this case, gene activation or repression can be

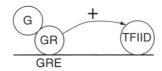

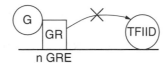

Figure 4.26 Consequences of glucocorticoid receptor binding to the DNA-binding sites which mediate gene activation (GRE) or repression (nGRE). Note that the receptor is likely to bind in a different configuration to the two different sequences, resulting in its ability to activate transcription only following binding to the GRE.

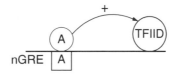

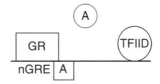

Figure 4.27 Inhibition of gene expression by glucocorticoid receptor binding to an nGRE is likely to be mediated by preventing the binding of a positively acting activator protein (A) to a site adjacent to or overlapping the nGRE.

mediated from the same DNA-binding site with the effect depending on the presence or absence of the hormone.

As described in Section 4.4.3, the ability of the nuclear receptors to activate transcription following ligand binding is dependent upon the fact that the conformational change affecting the activation domain allows it to bind co-activator molecules such as CBP. In turn, these co-activator molecules activate transcription by interacting with the basal

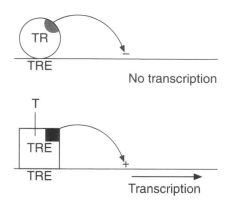

Figure 4.28. In the absence of thyroid hormone (T) the thyroid hormone receptor inhibits gene expression via a discrete inhibitory domain (hatched area). Binding of thyroid hormone (T) exposes the activation domain of the receptor (solid box) and allows it to activate transcription.

transcriptional complex and/or altering chromatin structure via histone acetylation. It has been shown that the inhibitory domains of the thyroid hormone receptor and the related retinoic acid receptors act indirectly by binding a co-repressor molecule known as N-CoR (nuclear receptor co-receptor) (for a review, see Perlmann and Venstrom, 1995). Moreover, it has been shown that N-CoR binds to the receptors together with two other proteins, mSIN3 and mRPD3. Interestingly, mRPD3 has histone deacetylase activity and is likely, therefore, to induce a more tightly packed chromatin structure (for a review, see Wolffe, 1997). Hence the binding of ligand to the receptor induces it to release a co-repressor complex with histone deacetylase activity and bind co-activators with histone acetylase activity (Figure 4.29).

In addition to the thyroid-binding form of the thyroid hormone receptor, alternative splicing generates another form (α2), lacking a part of the hormone-binding domain and therefore unable to bind hormone (Koenig *et al.*, 1989) (Figure 4.30a). Both the α2 form and the hormone-binding α1 form can bind to DNA, however. Binding of α2 to the TRE sequence prevents binding of α1 and thereby prevents gene induction in response to thyroid hormone (Figure 4.30b). As discussed in Section 7.2.2, a similar non-hormone-binding form of the thyroid hormone receptor is encoded by the v-*erbA* oncogene which produces cancer by inhibiting the expression of thyroid-hormone-responsive genes involved in erythroid differentiation.

As well as transcriptional inhibition by binding of the glucocorticoid receptor to the nGRE or by the thyroid receptor in the absence of hormone, it is also possible for the nuclear receptors to interfere with

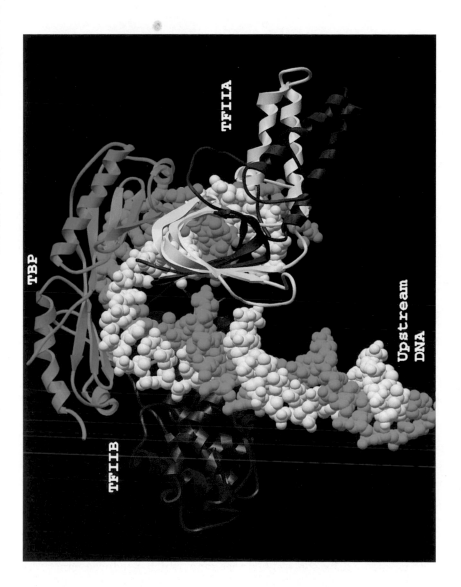

Plate 1 Schematic diagram illustrating the structure of the TFIIB/TBP/TFIIA complex bound to DNA. Note the bending of the DNA induced by TBP binding and the positions of TFIIB and TFIIA relative to TBP.

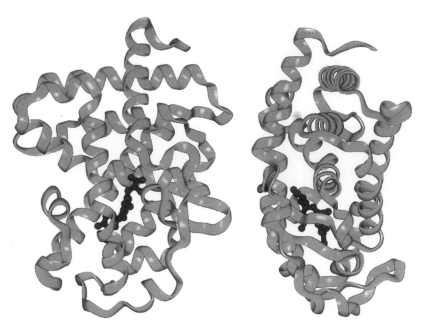

Plate 2 (Above) Two views of the structure of the thyroid hormone receptor ligand-binding domain (blue) with bound thyroid hormone ligand (red). Note that the ligand is completely buried in the interior of the protein.

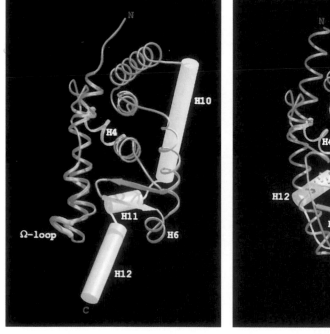
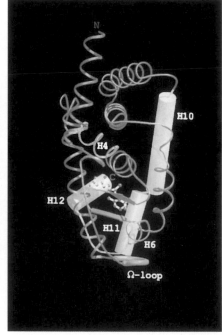

3(a)   3(b)

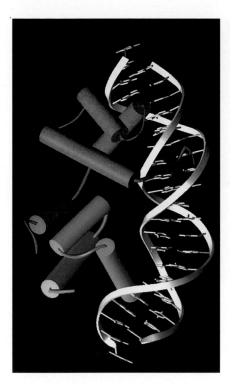

Plate 4 (Left) Binding of the al (blue) / α2 (red) homeodomain heterodimer to DNA. α-helices are shown as cylinders. Note the three-helical structure of the homedomains of al and α2. The C-terminal region of α2, forms an additional α-helix in the presence of al and packs against the al homeodomain, forming the dimerization interface.

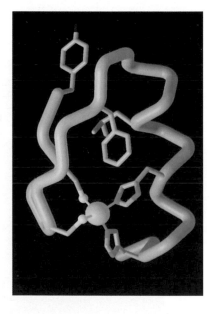

Plate 5 (Right) Structure of the Cys$_4$–His$_2$ zinc finger from Xfin. The Cys residues are shown in yellow and the His residues in dark blue.

Plate 3 (Facing Page) Structure of (a) the RXRα receptor in the absence of ligand (b) the closely related RARα receptor following binding of ligand (light blue atoms joined by white bonds). Note the structural change induced by the binding of ligand involving the movement of the H12 helix towards the ligand-binding core, so creating a sealed pocket in which the ligand is trapped.

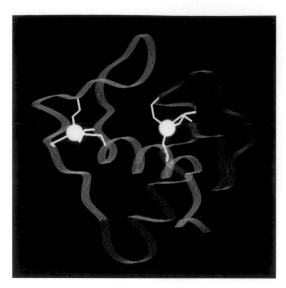

Plate 6 Structure of the two Cys₄ zinc fingers in a single molecule of the glucocorticoid receptor. The first finger is shown in red and the second finger in green with the zinc atoms shown white.

Plate 7 Structure of the oestrogen receptor dimer consisting of two receptor molecules bound to DNA. The two molecules of the receptor are shown in green and blue, respectively, and the DNA is shown in purple.

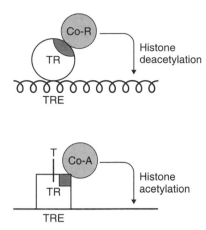

Figure 4.29 The inhibitory domain (hatched semi-circle) of the thyroid hormone receptor binds a co-repressor (Co-R) complex which deacetylates histones, inducing a closed chromatin structure (wavy line). Binding of ligand results in the release of the Co-R and binding of co-activator (Co-A) molecules to the exposed activation domain (hatched square). The co-activators have histone aceytltransferase activity and produce a more open chromatin structure (solid line) compatible with transcription.

transcriptional activation by entirely unrelated transcription factors. Thus glucocorticoid hormones have been known for some time to be a potent inhibitor of the induction of the collagenase gene by phorbol esters resulting in their having an anti-inflammatory effect. This inhibition is mediated by the glucocorticoid receptor which inhibits the activity of the Jun and Fos proteins that normally activate the collagenase gene via the AP-1 sites in its promoter (for a discussion of Fos, Jun and AP1, see Section 7.2.1). This effectively inhibits collagenase gene activation. Unlike the examples of repression by the glucocorticoid receptor discussed above, however, the collagenase promoter does not contain any binding sites for the receptor adjacent to the AP1 sites, nor does the receptor apparently bind to the collagenase promoter.

Interestingly, however, like the glucocorticoid receptor, the Fos/Jun complex requires the CBP protein as a co-activator to activate transcription. Hence the glucocorticoid receptor may compete with Fos/Jun for limited quantitites of the CBP co-activator, which are present in the cell, resulting in a failure of Fos/Jun to activate in the presence of activated glucocorticoid receptor (Kamei *et al.*, 1996) (Figure 4.31). Clearly such competition between Fos/Jun and the glucocorticoid receptor for limited quantities of CBP will also result in inhibition of glucocorticoid-dependent genes in response to hormone in the presence of high concentrations of Fos and Jun, and this is indeed observed (Figure 4.31).

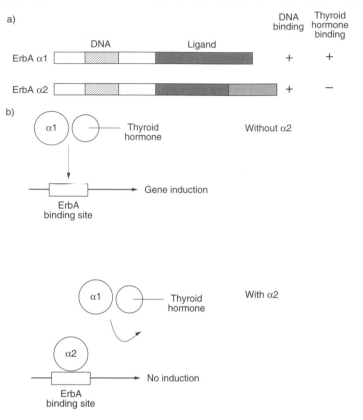

Figure 4.30 (a) Relationship of the ErbA α1 and α2 proteins. Note that only the α1 protein has a functional thyroid hormone binding domain. (b) Inhibition of ErbA α1 binding and of gene activation in the presence of the α2 protein.

Hence mutual transrepression of two different activating proteins can be achieved by competition for a co-activator (Figure 4.31). Moreover, this mutual repression illustrates how different cellular signalling pathways, which are activated, respectively, by phorbol esters and glucocortocoid hormones can interact with one another resulting in cross talk between the pathways (for a review, see Janknecht and Hunter, 1996).

Hence, as well as being able to activate gene expression, the members of the steroid–thyroid hormone receptor family also illustrate three mechanisms by which repression of gene expression can be achieved, namely the direct inhibition of transcription or the neutralization of a positive factor either by preventing its binding to DNA by masking of its site or by competing for a co-activator (for further discussion of the mechanisms of transcriptional repression, see Section 9.3). It should be noted, however, that these three examples differ in that the glucocorti-

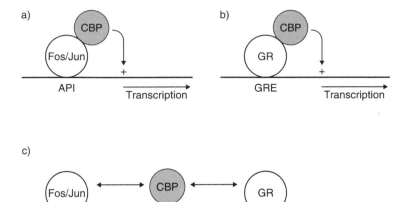

Figure 4.31 Mutual transrepression by Fos/Jun and the glucocorticoid receptor. Competition between Fos/Jun and the glucocorticoid receptor for the CBP co-activator inhibits the expression of genes containing binding sites for either Fos/Jun (AP1 sites) or for the glucocorticoid receptor (GRE).

coid receptor needs to be activated by steroid before it can inhibit gene expression by binding to an nGRE or competing for CBP, whereas the thyroid hormone receptor directly inhibits by binding to its response element in the absence of hormone.

4.5 CONCLUSIONS

This chapter has discussed in detail three systems in which specific genes are switched on in response to a particular inducer. As indicated in Table 4.1, many other inducing agents can activate particular genes containing specific DNA sequences. The examples described here, however, illustrate the general principles by which such induction takes place.

Thus the induction event normally involves the activation of a particular transcription factor which is present in the cell in an inactive form prior to induction. This activation of the transcription factor may take place via post-translational modifications such as phosphorylation, a change in the conformation of the factor and/or via disruption of an inhibitory protein–protein interaction.

In turn, such modifications may produce activation of the factor by allowing it to move to the nucleus and fulfil its inherent DNA-binding ability as in the case of the steroid receptors. Alternatively, they may convert the protein from a form which cannot bind DNA into a DNA-binding form as in the case of the *Drosophila*, mammalian and fission yeast heat-shock factors. Lastly, the conversion to an active form may not involve any alteration in DNA-binding at all but may enhance the ability of bound factor to enhance transcription as in the case of the budding yeast HSF or the CREB factor. These activating effects of inducing agents on transcription factors will be discussed further in Chapter 10 as part of a general discussion on the activation of transcription factor in both tissue specific and inducible gene expression and in development.

Once the transcription factor has been activated, it induces gene expression by interacting with the constitutive components of the transcriptional apparatus discussed in Chapter 3 to enhance the stability of the stable transcription complex and hence increase transcription. This process, which is characteristic of virtually all factors mediating inducible gene expression, is observed for both the heat-shock factor and the steroid receptors that contain activation domains capable of interacting with other transcription factors. In addition however, the steroid receptors illustrate another facet of gene activation not seen in the heat-shock factor. Thus the binding of the receptor to DNA results in an alteration of chromatin structure. This exposes the DNA-binding sites for constitutive factors such as CTF/NF1, allowing these factors to bind to their exposed binding sites in DNA and activate transcription. In the case of the heat-shock genes, this opening of the chromatin has already occurred owing to the binding of the GAGA factor, allowing HSF to bind and activate transcription following heat shock.

The activation of the various transcription factors in response to the inducing stimulus and the consequent effect of the activated factors on the constitutive transcriptional apparatus therefore results in the observed inducible pattern of gene expression.

REFERENCES

Abel, T., Martin, K.C., Bartsch, D. and Kandel, E.R. (1998). Memory supressor genes: inhibitory constraints on the storage of long term memory. *Science* **279**, 338–341.

Beato, M. and Eisfeld, K. (1997). Transcription factor access to chromatin. *Nucleic Acids Research*, **25**, 3559–2563.

Beato, M., Herrlich, P. and Schutz, G. (1995). Steroid hormone receptors: many actors in search of a plot. *Cell* **83**, 851–857.

Becker, P.B., Gloss, B., Schmid, W., Strahle, U. and Schutz, G. (1986). In vivo protein–DNA interactions in a glucocorticoid response element require the presence of the hormone. *Nature* **324**, 686–688.

Brzozowki, A.M., Pike, A.C.W., Dauter, Z., Hubbard, R.E., Bonn, T., Engstrom, O., Ohman, L., Greene, G.L., Gustafsson, J.A. and Carlquist, M. (1997). Molecular basis of agonism and antagonism in the oestrogen receptor. *Nature* **389**, 753–758.

Chen, H., Lin, R.J., Schiltz, R.L., Chakravati, D., Nash, A., Nagy, L., Privalsky, M.L., Nakatani, Y. and Evans, R.M. (1997). Nuclear receptor coactivator ACTR is a novel histone acetyltransferase and forms a multimeric activation complex with P/CAF and CBP/p300. *Cell* **90**, 569–580.

Chrivia, J.C., Kwok, R.P., Lamb, N., Hagiwara, M., Montiminy, M.R. and Goodman, R.H. (1993). Phosphorylated CREB binds specifically to the nuclear protein CBP. *Nature* **365**, 855–859.

Drouin, J., Sun, Y.L., Chamberland, M., Ganthier, Y., De Lean, A., Nemer, M. and Schmidt, T.J. (1993). Novel glucocorticoid receptor complex with DNA element of the hormone repressed POMC gene. *EMBO Journal* **12**, 145–156.

Gallo, G.J., Schuetz, T.J. and Kingston, R.E. (1991). Regulation of heat shock factor in *Schizosaccharomyces pombe* more closely resembles regulation in mammals than in Saccharomyces cerevisiae. *Molecular and Cellular Biology* **11**, 281–288.

Glass, C.K. and Rosenfeld, M.G. (1997). Nuclear receptor co-activators. *Current Opinion in Cell Biology* **9**, 222–232.

Gronemeyer, H. and Laudet, V. (1995). Nuclear receptors. *Protein Profile* **2**, 1173–1308.

Gronemeyer, H. and Moras, D. (1995). Nuclear receptors: how to finger DNA. *Nature* **375**, 190–191.

Hensold, J.O., Hunt, C.R., Calderwood, S.K., Housman, D.E. and Kingston, R.E. (1990). DNA-binding of heat shock factor to the heat shock element is insufficient for transcriptional activation in murine erythroleukemia cells. *Molecular and Cellular Biology* **10**, 1600–1608.

Hurst, H. (1996). Leucine zippers. *Protein Profile* **3**, 1–72.

Jakobsen, B.K. and Pelham, H.R.B. (1991). A conserved heptapeptide restrains the activity of the yeast heat shock transcription factor. *EMBO Journal* **10**, 369–375.

Janknecht, R. and Hunter, T. (1996). A growing co-activator network. *Nature* **383**, 22–23.

Jenster, G., Spencer, T.H., Burcin, M., Tsai, S.Y., Tsai, M.J. and O'Malley, B.W. (1997). Steroid receptor induction of gene transcription: a two step model. *Proceedings of the National Academy of Sciences USA* **94**, 7879–7884.

Kamei, Y., Xu, L., Heinzel, T., Torchia, J., Kurokawa, R., Gloss, B., Lin,

S.C., Heyman, R.A., Rose, D.W., Glass, C.K. and Rosenfeld, M.G. (1996). A CBP integrator complex mediates transcriptional activation and AP-1 inhibition by nuclear receptors. *Cell* **85**, 403–414.

King, R.J.B. and Manwaring, W.I.P. (1974). *Steroid Cell Interactions.* London: Butterworths.

Koenig, R.G., Lazar, M.A., Hodin, R.A., Brent, G.A., Larsen, P.R., Chin, W.W. and Moore, D.D. (1989). Inhibition of thyroid hormone action by a non-hormone binding c-erbA protein generated by alternative RNA splicing. *Nature* **337**, 659–661.

Lalli, E. and Sassone-Corsi, P. (1994). Signal transduction and gene regulation: the nuclear response to cyclic AMP. *Journal of Biological Chemistry* **269**, 17359–17362.

Larson, J.S., Schuetz, T.J. and Kingston, R.E. (1988). Activation *in vitro* of sequence-specific DNA-binding by a human regulatory factor. *Nature* **335**, 372–375.

Latchman, D.S. (1998) *Gene Regulation: A Eukaryotic Perspective*, 3rd edn. London, New York: Chapman and Hall.

McKnight, G.S., Clegg, C.H., Uhler, M., Chrivia, J.C., Cudd, G.G., Correll, L.A. and Otten, A.D. (1988). Analysis of the cAMP-dependent protein kinase system using molecular genetic approaches. *Recent Progress in Hormone Research* **44**, 307–331.

Mangelsdorf, D.J. and Evans, R.M. (1995). The RXR heterodimers and orphan receptors. *Cell* **83**, 841–850.

Mangelsdorf, D.J., Thummel, C., Beato, M., Herrlich, F., Schutz, G., Umesono, K., Blumberg, B., Kustner, P., Mark, M., Chambon, P. and Evans, R.M. (1995). The nuclear receptor superfamily: the second decade. *Cell* **83**, 835–839.

Montminy, M. (1997). Transcriptional regulation by cyclic AMP. *Annual Review of Biochemistry* **66**, 807–822.

Morimoto, R. (1993) Cells in stress: transcriptional activation of heat shock genes. *Science* **259**, 1409–1410.

Nakajima, T., Uchida, C., Anderson, S.F., Parvin, J.D. and Montiminy, M. (1997). Analysis of a cAMP-responsive activator reveals a two component system for transcriptional activation via signal dependent factors. *Genes and Development* **11**, 738–747.

Nieto-Sotelo, J., Wiederrecht, G., Okuda, A. and Parker, C.S. (1990). The yeast heat shock transcription factor contains a transcriptional activation domain whose activity is repressed under non shock conditions. *Cell* **62**, 807–817.

Ogryzko, V.V., Schiltz, L., Russanova, V., Howard, B.H. and Nakatani, Y. (1996). The transcriptional coactivators p300 and CBP are histone acetyltransferases. *Cell* **87**, 953–959.

Ostlund Forrants, A.K., Blomquist, P., Kwon, H. and Wrange, O. (1997). Glucocorticoid receptor–glucocorticoid response element binding stimulates nucleosome disruption by the SWI/SNF complex. *Molecular and Cellular Biology* **17**, 895–905.

Parsell, D.A. and Lindquist, S. (1993). The function of heat shock proteins in stress tolerance: degradation and reactivation of damaged proteins. *Annual Review of Genetics* **27**,437–496.

Perlmann, T. and Evans, R.M. (1997). Nuclear receptors in Sicily: all in the famiglia. *Cell* **90**, 391–397.

Perlmann, T. and Vennstrom, T. (1995) Nuclear receptors: the sound of silence. *Nature* **377**, 387–388.

Pratt, W.B. (1997). The role of the hsp90 based chaperone system in signal transduction by nuclear receptors and receptor signalling via MAP kinase. *Annual Review of Pharmacology and Toxicity* **37**, 297–326.

Pratt, W.B. and Toft, D.O. (1997). Steroid receptor interactions with heat shock proteins and immunophilin chaperons. *Endocrinology Reviews* **18**, 306–360.

Rabindran, S.K., Haroun, R.I., Clos, J., Wisniewski, J. and Wu, C. (1993) Regulation of heat shock factor trimer formation: role of a conserved Leucine zipper. *Science* **259**, 230–234.

Renaud, J.P., Rochel, N., Ruff, M., Vivat, V., Chambon, P., Gronemeyer, H. and Moras, D. (1996). Crystal structure of the RAR-λ ligand binding domain bound to all trans-retinoic acid. *Nature* **378**, 681–689.

Sassone-Corsi, P. (1998a). Transcription factors responsive to cyclic AMP. *Annual Review of Cell and Developmental Biology* **11**, 355–377.

Sassone-Corsi, P. (1998b). Coupling gene expression to cAMP signalling: role of CREB and CREM. *International Journal of Biochemistry and Cell Biology* **30**, 27–38.

Schumacher, A. and Magnuson, T. (1997). Murine polycomb and trithorax group genes regulate homeotic pathways and beyond. *Trends in Genetics* **13**, 167–170.

Shikama, N., Lyon, J. and La Thangue, N.B. (1997). The p300/CBP family: integrating signals with transcription factors and chromatin. *Trends in Cell Biology* **7**, 230–236.

Sorger, P.K., Lewis, M.J. and Pelham, H.R.B. (1987). Heat shock factor is regulated differently in yeast and HeLa cell. *Nature* **329**, 81–84.

Spencer, T.E., Jenster, G., Burcin, M.M., Allis, C.D., Zhou, J., Mizzen, C.A., McKenna, N.J., Onate, S.A., Tsai, S.Y., Tsai, M.J. and O'Malley, B.W. (1997). Steroid receptor co-activator–1 is a histoneacetylase transferase. *Nature* **389**, 194–198.

Stevens, C.F. (1994). CREB and memory consolidation. *Neuron* **13**, 769–770.

Tsukiyama, T., Becker, P.B. and Wu, C. (1994). ATP-dependant nucleosome disruption at a heat shock promotor mediated by DNA-binding of GAGA transcription factor. *Nature* **367**, 525–532.

Tsukiyama, T. and Wu, C. (1997). Chromatin remodelling and transcription. *Current Opinion in Genetics and Development* **7**, 182–191.

Wade, P.A., Pruss, D. and Wolffe, A.P. (1997). Histone acetylation: chromatin in action. *Trends in Biochemical Sciences* **22**, 128–132.

Wagner, R.L., Apriletti, J.W., McGarth, M.E., West, B.L., Baxter, J.D. and Fletterick, R.J. (1996). A structural role for hormone in the thyroid hormone receptor. *Nature* **378**, 690–697.

Wilkins, R.C. and Lis, J.T. (1997). Dynamics of potentiation and activation. GAGA factor and its role in heat shock gene regulation. *Nucleic Acids Research* **25**, 3963–3968.

Wilmann, T. and Beato, M. (1986). Steroid-free glucocorticoid receptor binds specifically to mouse mammary tumour DNA. *Nature* **324**, 688–691.

Wolffe, A. (1997). Sinful repression. *Nature* **387**, 16–17.

Wong, J., Shi, Y.B. and Wolffe, A.P. (1997). Determinants of chromatin disruption and transcriptional regulation instigated by the thyroid hormone receptor: hormone regulated chromatin disruption is not sufficient for transcriptional activation. *EMBO Journal* **16**, 3158–3171.

Wu, C. (1985). An exonuclease protection assay reveals heat shock element and TATA-box binding proteins in crude nuclear extracts. *Nature* **317**, 84–87.

Wurtz, J.M., Bourget, W., Renaud, J.P., Vivat, V., Chambon, P., Moras, D. and Gronemeyer, H. (1996). A canonical structure for the ligand binding domain of nuclear receptors. *Nature Structural Biology* **3**, 87–94.

Zimarino, V. and Wu, C. (1987). Induction of sequence-specific binding of *Drosophila* heat shock activator protein without protein synthesis. *Nature* **327**, 727–730.

CHAPTER FIVE

Transcription factors and cell type-specific transcription

5.1 CELL TYPE-SPECIFIC GENE EXPRESSION

As discussed in Chapter 4, both prokaryotes and eukaryotes respond to specific stimuli by inducing the expression of particular genes, this process being mediated by transcription factors which are activated in response to the stimulus. In eukaryotes, however, transcription factors also play a critical role in processes which have no parallel in prokaryotes. Thus, the higher eukaryote contains a vast range of different cell types, each of which expresses specific genes encoding particular products necessary for the specialized function of that cell type. The role of transcription factors in controlling the cell type-specific expression of particular genes is the subject of this chapter.

A number of different factors that are involved in mediating gene expression in specific cell types have been defined and a selection of these is listed in Table 5.1. These factors are normally synthesized or activated only in one specific tissue, resulting in the cell type-specific transcription of genes whose expression is dependent upon them.

In general, such factors have been defined by studying the regulation of a specific well-characterized gene, which is expressed only in a particular cell type and identifying the cell type-specific transcription factor responsible for this pattern of expression. Subsequently, once this factor has been identified, its role in the regulation of other, less well-characterized genes expressed in the particular cell type can be assessed. For example, the transcription factor GATA-1 (NF-E1), which is expressed at very high levels in erythroid cells, was originally identified

Table 5.1 Transcription factors regulating cell type-specific gene expression in mammals

Factor	Cell type	Gene regulated
GATA-1	Erythroid	Haemoglobin, porphobillinogen deaminase
Is1-1	Islet cells of the pancreas	Insulin
LFB1	Liver	Albumin, α1-antitrypsin, fibrinogen
MyoD1	Skeletal muscle	Myosin, creatine kinase
NFκB	B cells, activated T cells	Immunoglobulin κ light chain, interleukin-2, α-receptor
NFAT	Activated cells	Interleukin-2
Pit-1	Anterior pituitary	Growth hormone

on the basis of its binding to the promoter or enhancer regions of chicken and mammalian globin genes, and was shown to be essential for their erythroid-specific pattern of gene expression, deletion or mutation of GATA-1 binding sites resulting in the abolition of such expression (Reitman and Felsenfeld, 1988). Subsequently, this protein has also been implicated in the erythroid-specific gene expression of the porphobillinogen deaminase gene which also contains a binding site for the factor within its promoter (Mignotte *et al.*, 1989). It was then shown that the introduction of a mutation into the GATA-1 gene, which inactivated it, led to a failure of red blood cell formation, although all other cell types, including white blood cells, were still produced (Pevny *et al.*, 1991). This factor therefore plays a critical role in the regulation of a number of different erythroid-specific genes and thereby in erythroid cell differentiation. Interestingly, it has recently been shown that GATA-1 is also essential for the production of circulating platelets, indicating that it plays a key role in the generation of more than one haematopoetic cell type (Shivadasani *et al.*, 1997).

In Section 5.2 we will consider one example of the role of transcription factors in tissue-specific gene expression which has been defined in this way. This involves the B cell-specific expression of the genes encoding the heavy and light chains of the immunoglobulin molecule which is regulated by transcription factors that are specifically active in B cells.

However, not all tissue-specific transcription factors have been identified in this manner. Thus, the factor MyoD, which activates muscle-specific genes, was not identified on the basis of studying genes of this type. Rather it was isolated on the basis of the finding that its artificial expression within undifferentiated fibroblast-like cells was sufficient to transform them into muscle precursor cells (Davis *et al.*, 1987). Hence this factor is not only capable of switching on muscle-specific genes but, by doing so, actually plays a central role in the production of the differentiated cell type itself. This role of MyoD in determining the muscle cell type is discussed in Section 5.3.

The role of transcription factors in producing specific cell types has been particularly well characterized in homothallic yeasts. In this case, whether a cell is a or α in mating type is determined by which transcription factor gene is expressed at the mating-type locus. This system, which is discussed in Section 5.4, not only offers insights into cell type-specific gene expression but also provides a possible model system for studies on the developmental regulation of gene expression by transcription factors which will be discussed in Chapter 6.

5.2 TRANSCRIPTION FACTORS AND IMMUNOGLOBULIN GENE EXPRESSION

5.2.1 B cell-specific expression of the immunoglobulin genes

The B lymphocytes which circulate in the blood of higher vertebrates have as their primary role the production of antibodies to defend the body against foreign organisms, etc. These antibodies are produced by the association of two immunoglobulin heavy chains and two immunoglobulin light chains to produce a functional antibody molecule (for a recent review of the immune system emphasizing molecular aspects, see Roitt *et al.* (1996)). Consistent with the need to produce antibodies only in B cells, the separate genes encoding the heavy and light chains are expressed in a B cell-specific manner.

This B cell specificity is produced by the combined action of a promoter element located upstream of the transcribed region and an enhancer element located within the intervening sequence (intron) separating the exons encoding the variable and constant regions of the molecule (Figure 5.1). Both these elements, if tested in isolation by linkage to a marker gene, are capable of driving high-level expression in B cells and are hence both B cell specific. In combination, however, the promoter and the enhancer which acts on it constitute a very powerful B cell-specific regulatory element and produce the high-level expression of the immunoglobulin genes observed in B cells.

Interestingly, the promoter and enhancer are separated by a large distance in gene line DNA and are brought together by a DNA rearrangement event which allows the enhancer to act on the promoter producing maximal gene expression (Figure 5.2). The role of this rearrangement in gene regulation is entirely secondary, however, to its primary role in producing antibody diversity by the combination of different DNA sequences encoding different variable, joining and constant regions to produce different antibody molecules (for review see Gellert, 1992).

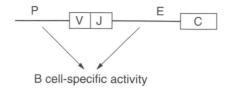

B cell-specific activity

Figure 5.1 Structure of the rearranged immunoglobulin genes in a B cell. The B cell-specific promoter (P) is located adjacent to the DNA segments encoding the variable (V) and joining (J) regions of the molecule. These DNA segments are separated from the DNA segment encoding the constant (C) region of the immunoglobulin molecule by an intervening sequence (intron) which contains the B cell-specific enhancer (E).

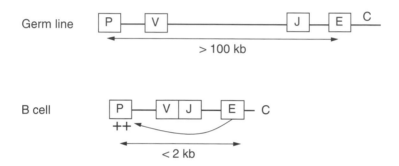

Figure 5.2 The rearrangement of the heavy-chain gene brings an enhancer (E) in the intervening sequence between J and C close to the promoter (P) adjacent to V, and results in the activation of the promoter and gene transcription.

Hence the B cell-specific expression of the immunoglobulin genes is controlled by the inherent B cell specificity of both their promoters and enhancers. An understanding of this B cell specificity requires an understanding therefore of the transcription factors binding to these promoters and enhancers and their role in gene expression.

5.2.2 Structure and activity of immunoglobulin promoter and enhancer elements

The promoters of the immunoglobulin heavy and light chain genes have a relatively simple structure containing only one identifiable transcription factor binding site in addition to the TATA box (Figure 5.3). This octamer motif, which has the consensus sequence ATGCAAAT, is found approximately 70 bases upstream of the transcription initiation site in both light- and heavy-chain genes. Interestingly, the octamer is oriented

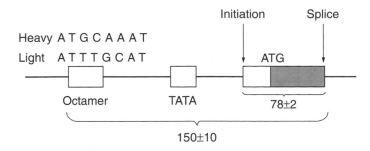

Figure 5.3 The immunoglobulin heavy- and light-chain gene promoter and adjacent transcribed region. The promoter contains a TATA box and an octamer motif. Note the opposite orientation of the octamer motif in the heavy- and light-chain genes relative to the start site of transcription. The initiation site of transcription, the start site (ATG) of the translated region (shaded) and the splice site at the end of the first exon are illustrated. Numbers indicate the average distance (in base pairs) between the different elements in the different immunoglobulin genes.

in the opposite direction relative to the start site of transcription in the heavy- and light-chain genes.

In both these genes the octamer element plays a critical role in the B cell-specific activity of the promoter. Thus its deletion or mutation, leaving the rest of the promoter intact, abolishes B cell-specific expression. Similarly linkage of the octamer motif to a non-immuno-globulin promoter results in a B cell-specific expression pattern of the heterologous promoter (Wirth *et al.*, 1987). Interestingly, the octamer motif is also found within the enhancer elements of the immunoglobulin genes. In this situation also, deletion of the octamer diminishes the activity of the enhancer in B cells. Hence the B cell specificity of both immunoglobulin promoters and enhancers involves the octamer motif.

In DNA mobility shift assays, the octamer motif binds two major proteins in B cell extracts (see Figure 2.2), which are known as Oct-1 and Oct-2 (Singh *et al.*, 1986). One of these proteins, Oct-1, is present in virtually all cell types, whereas the smaller Oct-2 protein is found in B cells which express the immunoglobulin genes and not in most other cell types (Figure 2.2). This tissue-specific expression of Oct-2 is paralleled by the presence of its corresponding mRNA at high levels in B cells, whilst it is absent in most other cell types (Clerc *et al.*, 1988). Hence, unlike the heat-shock factor or the steroid receptors, Oct-2 is not activated from a pre-existing inactive form when it is required, rather the Oct-2 mRNA and protein are synthesized in B cells but not in most other cell types.

This expression pattern immediately suggests that it is the binding of Oct-2 to the octamer motif in the immunoglobulin genes which determines their B cell-specific expression. In fact, however, the situation is not as simple as this. Thus, the activity of the immunoglobulin promoter is not diminished in B cells from mice which lack Oct-2 and such Oct-2 knock-out mice continue to synthesize immunoglobulin (Corcoran *et al.*, 1993). These findings create a paradox in that (as noted above) deletion of the octamer motif from the promoter abolishes its B cell-specific activity. Hence the octamer motif in the promoter can evidently be activated by Oct-1 as well as by Oct-2, but this occurs only in B cells even though Oct-1 is present in all cell types. The solution to this paradox is provided by the finding that a second B cell-specific factor (OCA-B), which does not itself bind to the octamer motif, can bind to DNA-bound Oct-1 or Oct-2 and activate transcription from an octamer-containing promoter (Gstaiger *et al.*, 1995; Strubin *et al.*, 1995).

Hence the B cell-specific activity of the immunoglobulin genes is produced by the tissue-specific expression of this co-activator, which can interact with either Oct-2 or the constitutively expressed Oct-1 to activate transcription from the immunoglobulin promoter (Figure 5.4a). In agreement with this idea, mice in which the gene encoding OCA-B has been inactivated show severe defects in immunoglobulin gene expression (Kim *et al.*, 1996; Schubart *et al.*, 1996). Interestingly, recent data indicate that OCA-B is also synthesized in T lymphocytes following exposure to stimuli which activate these cells, indicating that it may also play a key role in stimulating the expression of specific octamer-containing genes following T-cell activation (Zwilling *et al.*, 1997; for a review, see Graef and Crabtree, 1997).

Although Oct-2 does not therefore play an essential role in immunoglobulin gene expression, it does play an essential role in B lymphocyte maturation which is abnormal in Oct-2 knock-out mice (Corcoran *et al.*, 1993). This indicates that the B-cell specific expression of Oct 2 is likely to be involved in the B cell-specific expression of genes other than those encoding immunoglobulin. In agreement with this idea, the expression of the genes encoding CD36 and cysteine-rich secretory protein-3 (CRISP-3) in B lymphocytes has been shown to be dependent upon Oct-2 (Konig *et al.*, 1995; Pfisterer *et al.*, 1996).

These examples illustrate therefore how the cell type-specific transcription of individual genes can be controlled by cell type-specific expression of DNA binding factors such as Oct-2 or non-DNA-binding co-activators such as OCA-B. Thus some B cell-specific genes such as those encoding immunoglobulin are activated by the B cell-specific expression of OCA-B which can bind to Oct-1 or Oct-2 (Figure 5.4a), whilst others such as those encoding CD36 or CRISP-3 are activated by the B-cell specific expression of Oct-2 itself (Figure 5.4b).

Although the immunoglobulin promoter contains only an octamer motif, the immunoglobulin heavy-chain enhancer also contains binding

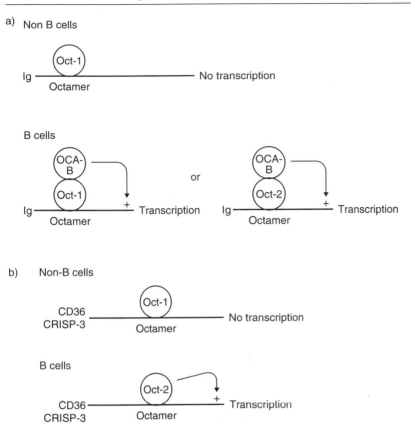

Figure 5.4 Immunoglobulin (Ig) genes are activated in B lymphocytes by the B-cell specific co-activator OCA-B, which can bind to either Oct-1 or Oct-2 (a), whereas other octamer-containing genes such as those encoding CD36 or CRISP-3 are activated in B cells by Oct-2 (b).

sites for several other factors in addition to the octamer motif (Figure 5.5). One of these (μB) binds a factor which is found only in B cells, whilst the others (E1–E4) bind ubiquitous factors present in all cell types. However, deletion or mutation of the binding sites for these ubiquitous factors reduces the activity of the enhancer in B cells even though the octamer motif is intact. Hence, in the heavy-chain gene enhancer B cell-specific activity is achieved by the interaction of B cell-specific factors and ubiquitous factors which bind to adjacent sites in the enhancer.

A similar mixture of binding sites for ubiquitous and B cell-specific proteins is found in the enhancer element of the immunoglobulin κ light-chain genes (Figure 5.6) and is critical for its activity. Thus, for example, the enhancer contains a motif known as E2 which serves as a binding site

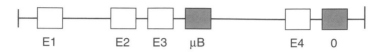

Figure 5.5 Structure of the immunoglobulin heavy-chain gene enhancer. Note the binding sites (shaded) for the B cell-specific factors µB and Oct-2 (0), and the E1–E4 sites (unshaded) which bind constitutive factors.

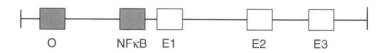

Figure 5.6 Structure of the immunoglobulin light-chain gene enhancer. Note the binding sites (shaded) for the B cell-specific factors NFκB and Oct-2, and the E1–E3 sites (unshaded) which bind constitutive factors.

for two ubiquitously expressed proteins E12 and E47, which are both encoded by the E2A gene, as well as two other binding sites (E1 and E3) for constitutively expressed factors. The E12 and E47 factors play a critical role in B cell development since inactivation of the E2A gene, which encodes both these factors, results in the arrest of B cell development at an early stage (for a review, see Dorshkind, 1994).

As well as binding sites for such constitutively expressed factors, the light-chain enhancer also contains a binding site for Oct-2 and a binding site for another B cell-specific protein, NFκB, which is a heterodimer of two subunits p50 and p65 (for reviews, see Thanos and Maniatis, 1995; Baeuerle and Baltimore, 1996). The binding of NFkB to its binding site in the enhancer plays an especially critical role in the expression of the light-chain gene. Thus mutation or deletion of the NFκB-binding site abolishes the B cell-specific activity of the light-chain enhancer whilst, like the octamer motif, it can render a heterologous promoter B cell specific.

This suggests therefore that binding of NFκB to its site in the enhancer controls the B cell-specific activity of the enhancer. In agreement with this, binding activity can be detected only in B cell lines which express the immunoglobulin κ gene and not in pre-B cells or other cell types which do not. Moreover, treatment of pre-B cells with lipopolysaccharide results in the appearance of NFκB-binding activity and concomitantly activates κ gene expression.

This activation of NFκB by lipopolysaccharide treatment can occur in the presence of the protein synthesis inhibitor cycloheximide. Hence NFκB is present in B cells in an inactive form and can be activated post-translationally without new protein synthesis. Interestingly, this inactive form of NFκB is widely distributed in different cell types and can be activated in both T cells and HeLa cells by treatment with phorbol ester. Although in these cases NFκB activation does not result in immunoglobulin light0-chain gene expression, since the gene has not rearranged and is tightly packed within inactive chromatin, it does play a role in gene regulation. Thus the activation of NFκB by agents which activate T cells, results in the active transcription factor inducing increased expression of cellular genes such as that encoding the interleukin-2 α receptor and is also responsible for the increased activity of the human immunodeficiency virus promoter in activated T cells. Like OCA-B, NFκB therefore plays a role not only in B cell-specific gene activity but also in gene activity specific to activated T cells. Indeed, recent work has suggested an additional role for NFκB in bone development indicating that it plays a key role in a number of different cell types (for a review, see Abu-Amer and Tondravi, 1997).

Thus, unlike Oct-2 and OCA-B, NFκB is present in all tissues and is activated post-translationally. This occurs as with HSF by increased phosphorylation. However, the primary target for phosphorylation is not NFκB but an inhibitory protein associated with it known as IκB (for reviews, see Verma *et al.*, 1995; Baldwin, 1996). Phosphorylation of IκB results in its dissociation from NFκB and its rapid degradation. This allows NFκB to move to the nucleus in a manner similar to the nuclear movement of hsp90-free glucocorticoid receptor following steroid treatment (Figure 5.7). (For further discussion of the mechanisms activating NFκB, see Section 10.3.3.)

In summary, therefore, B cell-specific expression of the immunoglobulin genes is controlled by the B cell-specific activity of their promoters and enhancers, which is in turn controlled primarily by the B cell-specific synthesis of OCA-B and the B cell-specific activation of NFκB.

5.3 MYOD AND THE CONTROL OF MUSCLE-SPECIFIC GENE EXPRESSION

5.3.1 Identification of MyoD

Probably the most novel approach to the cloning of the gene encoding a transcription factor was taken by Davis *et al.* (1987), who isolated cDNA clones encoding MyoD, a factor which plays a critical role in skeletal muscle-specific gene regulation. They used an embryonic muscle fibroblast cell line known as C3H 10T1/2. Although these cells do not exhibit any differentiated characteristics, they can be induced to

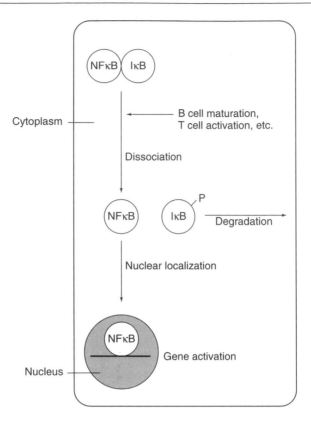

Figure 5.7 Activation of NFκB by dissociation of the inhibitory protein IκB, allowing NFκB to move to the nucleus and switch on gene expression. Note that dissociation of IκB from NFκB is caused by its phosphorylation (P) and degradation. NFκB is shown as a single factor for simplicity, although it normally exists as a heterodimer of two subunits p50 and p65.

differentiate into myoblast cells expressing a number of muscle-lineage genes upon treatment with 5-azacytidine (Constantinides *et al.*, 1977). This agent is a cytidine analogue having a nitrogen instead of a carbon atom at position 5 on the pyrimidine ring and is incorporated into DNA instead of cytidine. Unlike cytidine, however, it cannot be methylated at this position and hence its incorporation results in demethylation of DNA. As methylation of DNA at C residues is thought to play a critical role in transcriptional silencing of gene expression (for reviews, see Razin and Cedar, 1991; Latchman, 1998), this artificial demethylation can result in the expression of particular genes which were previously silent.

In the case of 10T1/2 cells, therefore, this demethylation was thought to result in the expression of previously silent, regulatory loci which are necessary for differentiation into muscle myoblasts. Several experiments also suggested that the activation of only one key regulatory locus might be involved. Thus 5-azacytidine induces myoblasts at very high frequency, consistent with only the demethylation of one gene being required, whilst DNA prepared from differentiated cells can also induce differentiation in untreated cells at a frequency consistent with the transfer of only one activated locus.

Hence differentiation is thought to occur via the activation of one regulatory locus (gene X in Figure 5.8) whose expression in turn switches on the expression of genes encoding muscle lineage markers, which is observed in the differentiated 10T1/2 cells and thereby induces their differentiation. This suggested that the regulatory locus might encode a transcription factor which switched on muscle-specific gene expression.

To isolate the gene encoding this factor, Davis *et al.* (1987) reasoned that it would continue to be expressed in the myoblast cells but would evidently not be expressed in the undifferentiated cells. They therefore prepared RNA from the differentiated cells and removed from it, by subtractive hybridization, all the RNAs which were also expressed in the undifferentiated cells. After various further manipulations to exclude RNAs characteristic of terminal muscle differentiation such as myosin and others induced non-specifically in all cells by 5-azacytidine, the enriched probe was used to screen a cDNA library prepared from differentiated 10T1/2 cells.

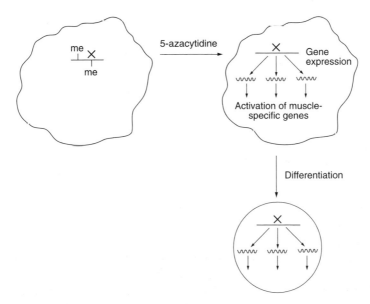

Figure 5.8 Model for differentiation of 10T1/2 cells in response to 5-azacytidine. Activation of a master locus (X) by demethylation allows its product to activate the expression of muscle-specific genes thereby producing differentiation.

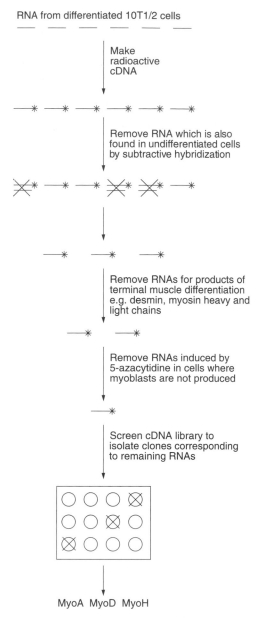

Figure 5.9 Strategy for isolating the master regulatory locus expressed in 10T1/2 cells after but not before treatment with 5-azacytidine. Subtractive hybridization was used to isolate all RNA molecules which are present in 10T1/2 cells only following treatment with 5-azacytidine. After removal of RNAs for terminal differentiation products of muscle and RNAs induced in non-muscle-producing cells by 5-azacytidine, the remaining RNAs were used to screen a cDNA library. Three candidates for the master regulatory locus MyoA, MyoD and MyoH were isolated in this way.

This procedure (Figure 5.9) resulted in the isolation of three clones, MyoA, MyoD and MyoH whose expression was specifically activated when 10T1/2 cells were induced to form myoblasts with 5-azacytidine. When each of these genes was artifically expressed in 10T1/2 cells, MyoA and MyoH had no effect. However, artificial expression of MyoD was able to convert undifferentiated 10T1/2 cells into myoblasts (Figure 5.10). Hence expression of MyoD alone can induce differentiation of 10T1/2 cells.

The differentiated 10T1/2 cells produced in this manner, like those induced by 5-azacytidine, express a variety of muscle lineage markers and, indeed, also switch on both MyoA and MyoH as well as the endogenous MyoD gene itself. This suggests that MyoD is a transcription factor which switches on genes expressed in muscle cells. In agreement with this, MyoD was shown to bind to a region of the creatine kinase gene upstream enhancer which was known to be necessary for its muscle-specific gene activity.

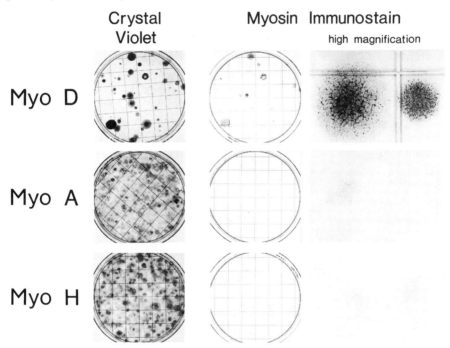

Figure 5.10 Test of each of the putative master regulatory loci *MyoA, MyoD* and *MyoH*. Each of the genes was introduced into 10T1/2 cells and tested for the ability to induce the cells to differentiate into muscle cells. Note that whilst *MyoA* and *MyoH* have no effect, introduction of *MyoD* results in the production of muscle cells which contain the muscle protein myosin. The differentiated muscle cells induced by *MyoD* cease to divide on differentiation, resulting in less cells being detectable by staining with crystal violet compared to the *MyoA*- and *MyoH*-treated cells which continue to proliferate. Hence only *MyoD* has the capacity to cause 10T1/2 cells to differentiate into non-proliferating muscle cells producing myosin, identifying it as a master regulatory locus for muscle differentiation.

Moreover, a recent study (Gerber *et al.*, 1997) has indicated that MyoD can actually bind to its binding sites within target genes when they are in the tightly packed chromatin structure characteristic of genes which are inactive in a particular lineage. This binding results in the remodelling of the chromatin to a more open form and is then followed by enhanced transcription stimulated by MyoD (Figure 5.11). This alteration in chromatin structure is likely to be dependent on the ability of MyoD to interact with the p300 co-activator protein (Puri *et al.*, 1997), which is a close relative of the CBP factor discussed in Section 4.3.2. Like CBP, p300 has histone acetyltransferase activity and is therefore likely to be able to alter chromatin to the more open structure associated with acetylated histones (see Section 1.4.3).

Hence MyoD is capable of activating transcription by two distinct means, namely the remodelling of chromatin and the direct stimulation of enhanced transcription (see Section 9.2 for a discussion of the mechanisms of transcriptional activation). This is particularly important since it allows MyoD to induce the development of myogenic cells from non differentiated precursors in which the genes, which must be switched on, are in an inactive closed chromatin structure which is inaccesible to many transcriptional activators.

Interestingly, as well as stimulating muscle-specific genes, MyoD also promotes differentiation by modulating gene expression so as to inhibit cellular proliferation, thereby producing the non-dividing phenotype characteristic of muscle cells. Thus MyoD has recently been shown to activate the gene encoding the p21 inhibitor of cyclin-dependent kinases (Halevy *et al.*, 1995). This results in the inhibition of these kinases whose activity is necessary for cell division (see Section 7.3.2). In addition, MyoD can also repress the promoter of the c-*fos* gene whose protein product is important for cellular proliferation (see Section 7.2.1), indicating that MyoD can also act by repressing genes whose products are not required in non-dividing muscle cells (Trouche *et al.*, 1993).

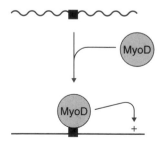

Figure 5.11 MyoD binding to its binding site (solid box) both converts the chromatin structure from a closed (wavy line) to a more open (solid line) structure compatible with transcription and also directly enhances the rate of transcription (arrow).

Like gene activation by MyoD, repression of the c-*fos* promoter is dependent on DNA binding which in this case prevents the binding of a positively acting factor to a site known as the serum response element which overlaps the MyoD-binding site in the c-*fos* promoter (Figure 5.12). Obviously, in contrast to its binding to the creatine kinase enhancer, MyoD must bind to its binding site in the c-*fos* promoter in a form which cannot activate transcription. Hence like the glucocorticoid receptor (see Section 4.4.4), MyoD can have different effects on gene expression depending on the nature of its binding site.

In both cases, however, DNA binding by MyoD is dependent upon a basic region of the protein which binds directly to the DNA, and an adjacent region which can form a helix–loop–helix structure and is essential for dimerization of MyoD. These elements which are also found in other transcription factors such as the E12 and E47 proteins that bind to the E2 site in immunoglobulin enhancers (see Section 5.2.2) are discussed further in Section 8.4.

Hence MyoD is a transcription factor whose activation by 5-azacytidine treatment results in the activation of muscle-specific gene expression. Interestingly, the observation that the introduction of MyoD into cells switches on the endogenous MyoD gene (see above) suggests that a positive feedback loop normally regulates MyoD expression so

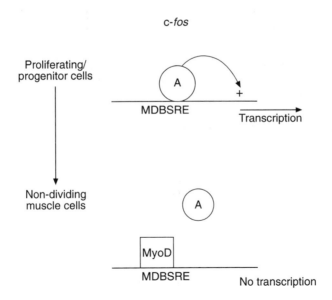

Figure 5.12. MyoD binds to its binding site (MDBS) in the c-*fos* promoter in a configuration which does not activate transcription and prevents binding of an activating factor (A) to the overlapping serum response element (SRE). This therefore results in the repression of c-*fos* transcription in MyoD-expressing muscle cells.

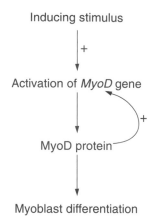

Inducing stimulus

Activation of *MyoD* gene

MyoD protein

Myoblast differentiation

Figure 5.13 Ability of MyoD protein to activate expression of its own gene, creating a positive feedback loop which ensures that, following an initial stimulus, the MyoD protein is continuously produced and hence maintains myoblast differentiation.

that, once the gene is initially expressed, expression is maintained producing commitment to the myogenic lineage (Figure 5.13). This is of importance since MyoD appears to be essential for the repair of damaged muscle in adult animals, indicating that its expression must be maintained throughout life (Megene *et al.*, 1996).

Despite its ability to induce cells in culture to differentiate into muscle cells, however, mice which lack any functional MyoD develop normally and produce skeletal muscle. However, if a second gene encoding the Myf-5 protein is also deleted, no skeletal muscle is formed even though inactivation of Myf-5 alone does not prevent skeletal muscle formation (Rudnicki *et al.*, 1993). Hence either MyoD or Myf-5 alone can induce skeletal muscle formation *in vivo* in the absence of the other, resulting in the lack of skeletal muscle only when both proteins are inactive. As expected from this, both Myf-5 and another related protein myogenin are also able to induce the differentiation of 10T1/2 cells into myoblasts and, together with MyoD, these proteins form a family of related transcription factors containing the helix–loop–helix motif which play a key role in controlling muscle differentiation (for reviews, see Rudnicki and Jaenisch, 1995; Molkentin and Olson, 1996; Firulli and Olson, 1997).

5.3.2 Regulation of MyoD

As expected in view of the critical role which MyoD plays in the development of muscle cells, the MyoD mRNA is present in skeletal muscle tissue taken from a variety of different sites in the body but is absent in all other tissues including cardiac muscle (Davis *et al.*, 1987) (Figure 5.14). Hence, like Oct-2, the MyoD mRNA and protein accumulate only in a specific cell type where it is required and the activation of the MyoD gene during myogenesis is likely to be of central importance in switching on the expression of muscle specific genes. In turn, this suggests that other developmentally regulated transcription factors of the type discussed in Chapter 6 will be involved in switching on MyoD expression during myogenesis. In agreement with this, the paired-type homeobox factor Pax 3 (see Section 6.3.2) has been shown to activate MyoD expression and myogenic differentiation in a variety of non muscle cell types (for a review, see Rawls and Olson, 1997), whilst the classical homeobox factor MSX1 can repress the transcription of the MyoD gene (Woloshin *et al.*, 1995).

However, in addition to its control by regulation of its synthesis, MyoD activity also appears to be regulated in another manner. Thus, in 10T1/2 cells, increased expression of MyoD, whether produced by 5-azacytidine treatment or by introduction of the MyoD gene, results in the formation of myoblasts. However, this increased expression of MyoD does not result

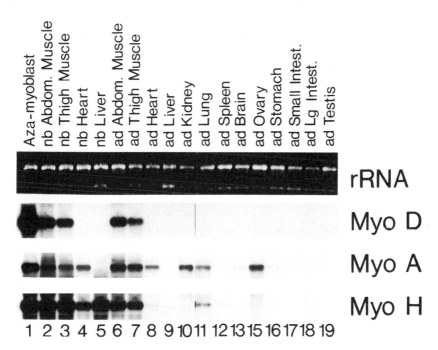

Figure 5.14 Northern blotting experiment to detect the mRNAs encoding MyoA, MyoD and MyoH in different muscle and non-muscle tissues. Note that the MyoD mRNA is present only in skeletal muscle, as expected in view of its ability to produce muscle differentiation, whereas the MyoA and MyoH mRNAs are more widely distributed. nb, newborn; ad, adult; rRNA, the ribosomal RNA control used to show that all samples contain intact RNA.

in the formation of fully differentiated muscle cells or in the activation of several genes which can be activated by MyoD *in vitro* such as creatine kinase. The activation of these products of terminal muscle cell differentiation requires the removal of growth factors from the culture medium of the myoblasts resulting in cell fusion and the formation of fully differentiated myotubes. Although this process results in switching on of MyoD-dependent genes such as creatine kinase, no further increase in the level of MyoD is observed during this differentiation (Figure 5.15).

Hence a paradox exists, with the MyoD-dependent genes encoding muscle differentiation markers only becoming activated when myotubes

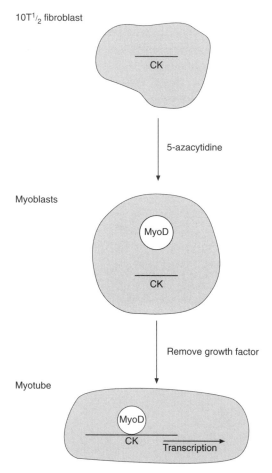

Figure 5.15 Differentiation of 10T1/2 cells into myoblasts by 5-azacytidine and then into myotubes by removal of growth factors. Note that the MyoD-dependent induction of genes encoding terminal differentiation markers such as creatine kinase (CK), which occurs in myotubes, takes place without an increase in MyoD concentration.

are formed even though the MyoD level is the same as that in myoblasts. The explanation of this paradox was provided by the identification of the Id protein (Benezra *et al.*, 1990), which like MyoD contains a helix–loop–helix motif but lacks the basic domain mediating DNA binding. Because the helix–loop–helix motif mediates dimerization of proteins containing it, Id can dimerize with other helix–loop–helix proteins such as MyoD and inhibit their DNA binding since the resulting heterodimer lacks the necessary pair of DNA binding motifs (Figure 5.16). Increasingly, when 10T1/2-derived myoblasts are induced to form myotubes, Id levels decline, indicating that this second stage of myogenesis is mediated by a decline in the inhibitory protein rather than an increase in the activator, MyoD.

The role of inhibitory helix–loop–helix proteins is not confined to myogenesis. Thus the level of Id also declines during differentiation of early embryonic cells, and of erythroid cells whilst the product of the *emc* gene which regulates neurogenesis in *Drosophila* also contains a helix–loop–helix motif and lacks a basic DNA-binding domain (for a review, see Jones, 1990).

In summary, therefore, regulation of MyoD appears to combine the types of regulation we have noted in Oct-2 and NFκB, being controlled

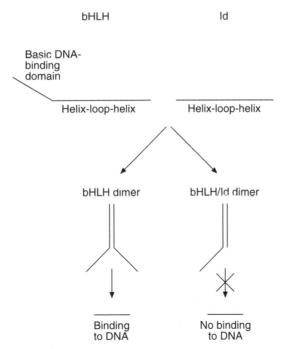

Figure 5.16 Dimerization of functional basic helix–loop–helix proteins (bHLH) with Id. Note that, whilst Id can dimerize with other proteins via the helix–loop–helix domain, it lacks the basic DNA domain and hence the Id-containing heterodimer cannot bind to DNA.

both by its increased synthesis in myoblasts and by its increased activity mediated by changes in protein–protein interaction during the formation of myotubes.

5.4 REGULATION OF YEAST MATING TYPE

5.4.1 Yeast mating type

In the preceding sections, we have discussed some of the control mechanisms which regulate the expression of different genes in the vast array of differentiated cells found in higher eukaryotes. Clearly such expression of different genes within different cells of the same organism does not occur in single-celled eukaryotes such as yeast, where the organism consists of only a single cell. None the less, similar regulatory processes operate in these organisms also. Thus, in yeast, two distinct mating types exist known as a and α, which must fuse together to create a diploid cell. These two mating types each express distinct genes encoding products such as pheromones or pheromone receptors which are involved in mating. Individual yeasts of different mating type can thus be regarded as analogous to the different types of differentiated cell in a higher eukaryote, although they are in fact distinct individual organisms of different phenotype.

In some yeast strains, known as heterothallic strains, the two mating types are entirely separate as in higher organisms. We shall discuss, however, the situation in homothallic yeast strains where each cell is capable of switching its mating type from a to α or vice versa, and behaves following switching exactly as does any other cell with its new mating type (reviewed by Herskowitz, 1989; Dolan and Fields, 1991).

This switching process is a very precise one (Figure 5.17), occurring only in the G1 phase of the cell cycle prior to cell division and only in a mother cell which has already produced a daughter by budding. The daughter itself does not switch until it has grown and produced a daughter itself. This process is thought to be a consequence of the need to produce diploid progeny rapidly, switching from a to α or vice versa, allowing this to occur amongst the progeny of a single cell without the need for contact with another strain of different mating type. This process is of interest in terms of gene regulation from two points of view. Firstly, how does a cell switch from a to α or vice versa and, secondly, how is the expression of genes expressed only in a or α cells regulated. These two aspects will be discussed in turn.

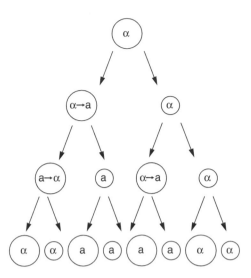

Figure 5.17 Mating-type switching in yeast. In every generation the larger mother cell which has produced a smaller daughter, switches its genotype from a to α or vice versa.

5.4.2 Control of mating-type switching

Genetic analysis of homothallic yeast strains allowed the demonstration that whether a cell was a or α in phenotype was controlled by a single gene locus on chromosome 3, known as MAT (mating type). If this transcriptionally active locus contained an a gene, the cell was of a mating type; if it contained an α gene, then the cell was of α mating type. In addition, however, yeast cells also contain transcriptionally silent copies of both the a and α genes elsewhere on chromosome 3 at the HML and HMR loci. The change in mating type occurs via a cassette mechanism in which one of the silent copies replaces the active gene at the MAT locus changing the mating type (Figure 5.18). This process is controlled by an endonuclease which is the product of the HO (homothallism) gene and which makes a double-stranded cut at the MAT locus initiating switching.

At first sight this process involving DNA cutting by an endonuclease appears to have little relevance to the study of gene regulation by transcription factors. A more detailed consideration of this process reveals, however, a central role for transcription factors. Thus the activity of the HO endonuclease is controlled by transcriptional regulation of its corresponding gene. A number of regulatory processes operate on this gene ensuring that it is transcribed only in the G1 phase of the cell cycle, and only in mother cells and not in daughter cells,

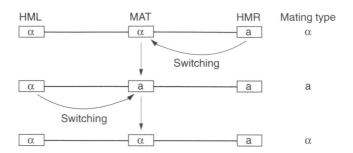

Figure 5.18 Mechanism of mating-type switching involving the movement of an a or α gene from the inactive HML or HMR loci to the active MAT locus.

resulting in the observed restriction of the switching process to the G1 phase in mother cells.

Genetic analysis has allowed the identification of a number of different loci which regulate HO expression. Thus five SWI genes necessary for the expression of HO have been defined whilst six SIN loci which inhibit HO expression have also been identified. It is likely that these genes encode transcription factors which activate or repress HO gene expression in particular situations. Thus the Swi4 and Swi6 gene products have been shown to be necessary for the formation of a protein complex on DNA sequences upstream of the HO gene (for a review, see Andrews and Mason, 1993). This complex binds to a DNA sequence in HO which is necessary for its expression in G1 and which can confer this expression pattern on an unrelated gene. It is likely therefore that the Swi4 and Swi6 gene products play a critical role in the activation of HO expression and therefore of switching only in the G1 phase of the cell cycle.

Similarly, the activation of the HO gene only in mother cells and not in daughter cells is controlled by the Swi5 factor which contains zinc finger motifs characteristic of DNA binding transcription factors (see Section 8.3.1) and binds specifically to sequences in the HO promoter involved in mother–daughter control (Stillman *et al.*, 1988). Although Swi5 is present in both mother and daughter cells, it activates HO gene transcription only in mother cells. This is because daughter cells, but not mother cells, contain the Ash-1 protein which specifically antagonises the action of Swi5 and prevents it activating HO gene transcription (Figure 5.19) (for a review, see Amon, 1996). In turn, the presence of Ash-1 only in daughter cells is controlled by the preferential localization of its mRNA, which is made in both cells but is specifically transported so that it accumulates only in daughter cells (Long *et al.*, 1997). Hence a post-transcriptional regulatory mechanism acting on the Ash-1 mRNA results in HO gene transcription being activated by Swi5 only in mother cells where Ash-1 protein is absent.

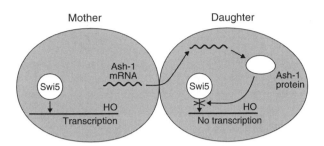

Figure 5.19 The preferential localization of the Ash-1 mRNA in daughter but not mother cells results in the Ash-1 protein accumulating only in daughter cells. It therefore inhibits the action of Swi5, so preventing HO gene transcription in daughter cells. In contrast, in mother cells where Ash-1 is absent, Swi5 stimulates HO gene transcription.

Therefore, even though mating-type switching is actually brought about by the action of an endonuclease, the activity of the endonuclease and hence switching is regulated at the level of gene transcription by several interacting control mechanisms.

5.4.3 Gene regulation by the a and α gene products

Once mating-type switching has determined whether the mating-type locus contains an a or α gene, the expression of this gene in turn regulates the activation or repression of genes specific to a or α cells. This is achieved at the level of transcription, the a or α genes encoding transcription factors which can activate or repress gene expression. Interestingly, DNA sequence analysis of these genes (Shepherd *et al.*, 1984) has revealed that they contain a motif with strong homology to the homeobox which is shared by a number of different transcription factors involved in regulating embryonic development in *Drosophila* and other higher eukaryotes (Figure 5.20) (see Section 6.2), and which mediates DNA binding by the factors which contain it (see Section 8.2).

Hence the a and α products control mating type by modulating the transcription of genes whose products are required for the a or α phenotype. This is not achieved, however, by the a product switching on a-specific genes and the α product switching on α-specific genes. Rather the products of the a and α loci interact together to regulate the a- and α-specific genes as well as the expression of genes whose products are required in both mating types but not in diploid cells (Figure 5.21) (reviewed by Dolan and Fields, 1991).

Genetic analysis of this process has indicated that the a-specific genes are constitutively active but are repressed by one of the products of the α

Ftz	Ser	Leu	Ser	Glu	Arg	Gln	Ile	Lys	Ile	Trp	Phe	Gln	Asn	Arg	Arg	Met	Lys	Ser	Lys
α 2	Ser	Leu	Ser	Arg	Ile	Gln	Ile	Lys	Asn	Trp	Val	Ser	Asn	Arg	Arg	Arg	Lys	Glu	Lys

Figure 5.20 Relationship of the yeast mating-type protein α2 and the homeobox of the *Drosophila* Fushi-tarazu protein. Boxes indicate identical amino acids.

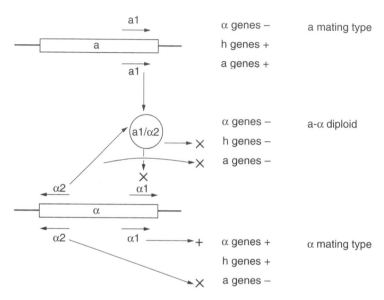

Figure 5.21 Regulatory actions of the a and α gene products. Note that whilst the α1 product is required for the activation of the α-specific genes, the α2 product inhibits the constitutively active a-specific genes. The interaction of the a1 and α2 proteins, which occurs in a/α diploids where both are present, inhibits the expression of the haploid-specific genes (h) as well as that of α1, thereby indirectly inhibiting the α-specific genes.

gene, α2. Hence they are active in a cells where α2 is absent but not in α cells where it is present. Similarly they are repressed in the presence of α2 in a/α diploid cells where the mating genes must be repressed since the cells have already mated and fused to give a diploid.

In contrast to the constitutive activity of the a-specific genes, the α-specific genes are inactive in the absence of the other product of the α gene, α1, and are hence inactive in a cells where it is absent but are active in α cells where it is present. These genes are also inactive, however, in diploids where α1 is present. This paradox is explained by the fact that when both α2 and the product of the a gene, a1, are present, they co-operate to repress both the α1 gene (hence indirectly preventing α gene expression) and also the genes whose expression is required in both mating types but not in non-mating diploids. Interestingly, one of the genes repressed by the a1/α2 complex is that encoding HO which, although required for switching in both a and α cells, is not required in diploids. This gene is therefore repressed by the a1/α2 complex, and its G1-specific expression is hence subject to multiple controls, preventing expression in both haploid daughter cells and in diploids (Figure 5.22). A summary of the regulatory processes acting on the HO gene is given in Figure 5.23.

Hence genetic studies have indicated that gene expression is regulated by the action of the α1 activator protein and the a1 and α2 repressor proteins. These studies have now been supplemented by a more detailed molecular analysis of the a and α proteins and their interaction with the promoters of the genes which they regulate (for a review, see Dolan and Fields, 1991). Thus it has been shown that the constitutive activity of the a-specific genes is controlled by the binding of a constitutively expressed transcription factor MCM1 (also known as GRM or PRTF) to a sequence upstream of the a-specific genes known as the P box. Interestingly, MCM1 is also required for transcription of the α-specific genes and is capable of binding to a sequence upstream of these genes known as the P′ box, which is related to but distinct from the P box. Following binding to this P′ box, however, transcriptional activation does not occur unless the α1 protein is present. This protein forms a complex with MCM1 and transcriptional activation occurs (Figure 5.24).

Therefore the distinction between the a- and α-specific genes is controlled by sequences in their promoters which allow transcriptional activation of the a-specific genes following binding of MCM1 alone, with transcriptional activation of the α-specific genes occurring only following binding of the MCM1/α1 complex and not following binding of MCM1 alone. A possible mechanism for this effect has been suggested by the finding that binding of MCM1 to the P box produces a conformational change in the structure of the protein which does not occur following binding to the P′ box. A similar conformational change in MCM1 is observed however following binding to the P′ box in the presence of the α1 protein. This conformational change is postulated to expose a transcriptional activation domain on MCM1 allowing it to activate transcription. Hence the difference in gene activation following binding to the P and P′ boxes is controlled by the fact that MCM1 alone can undergo the necessary conformational change following binding to the P box but requires α1 to do so following binding to the P′ box.

This model therefore explains the dependence of the a-specific genes on the α1 factor for their expression. A further aspect of the model is required to explain how the constitutively active α-specific genes are repressed in α mating-type cells. This occurs because the P box in these genes is flanked by binding sites for the α2 protein. One molecule of the α2 protein binds on either side of the MCM1 protein bound to the P box

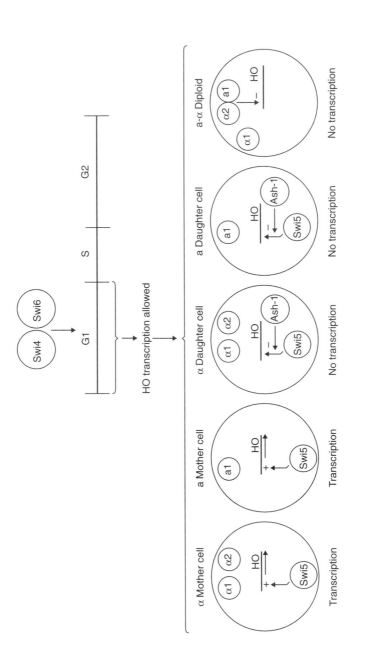

Figure 5.22 Transcription of the HO gene. Note that transcription is only allowed in the G1 phase of the cell cycle when the Swi4 and Swi6 gene products are present and in mother cells where Swi5 is not inhibited by Ash-1. Transcription of HO is also inhibited in a-α diploids by the action of the a1 and α2 proteins.

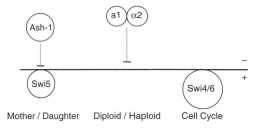

Figure 5.23 Summary of the mechanisms regulating HO transcription involving the negative actions of Ash-1 and the a1/α2 complex and the positive actions of Swi5 and the Swi4/6 complex.

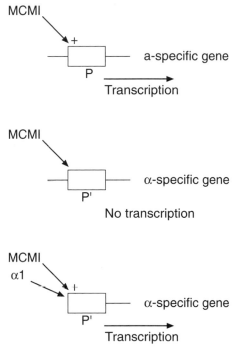

Figure 5.24 Stimulation of the a-specific genes by binding of MCM1 to the P box, and of the α-specific genes by binding of MCM1 and α1 to the P′ box. Note that MCM1 cannot activate the α-specific genes via P′ in the absence of α1.

and prevents the transcription of the a-specific genes (Figure 5.25). This inhibition of the activity of MCM1 does not take place by α2 preventing the binding of MCM1 to the P box. Thus the binding of MCM1 is actually enhanced by the presence of α2 with the two proteins binding co-operatively to the DNA. Hence α2 does not appear to act by interfering with the action of positively acting factors. Rather, it appears to be able to inhibit transcription directly by interacting with the basal transcriptional complex (Herschbach *et al.*, 1994) as occurs for the thyroid hormone receptor in the absence of thyroid hormone (see Section 4.4.4). This latter possibility is supported by the finding that the presence of the α2/MCM1 complex prevents transcription from the promoter directed by any other unrelated activation sequences placed upstream of the α2/MCM1 binding site as would be expected if it directly inhibits the activity of the basal transcriptional complex, rendering it non-responsive to upstream activating proteins (for a discussion of the mechanisms by which transcription factors repress gene expression, see Section 9.3).

The differential activity of the a- and α-specific genes is therefore controlled by the constitutive factor MCM1 and its interaction with the cell type-specific factors α1 and α2 (Figure 5.26).

In addition to its role as a repressor of the a-specific genes, α2 also plays a key role in the repression in diploid cells of genes such as HO whose activity is required for mating in general, and which are therefore expressed in both α and a cells but not in non-mating a/α diploids. This repression is dependent upon the association of the α2 protein with the a1 protein. This association requires the DNA-binding domain of the a1 protein and both the DNA binding domain and a C-terminal region of the α2 protein (Stark and Johnson, 1994), and results in high-affinity co-operative binding of the two proteins upstream of the haploid-specific gene promoters. The inhibitory activity of the α2 protein (see above) then results in the repression of transcription (Figure 5.27). Detailed structural analysis of the a1/α2 complex bound to DNA has shown that, in the presence of a1, the C-terminal region of α2 alters from being

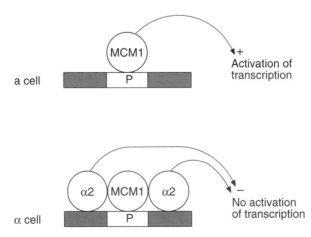

Figure 5.25 Inhibition of the a-specific genes in α cells by binding of α2 to sites flanking the binding site for MCM1.

a cell

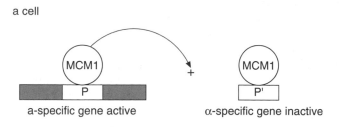

a-specific gene active   α-specific gene inactive

α cell

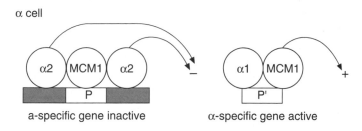

a-specific gene inactive   α-specific gene active

Figure 5.26 Regulation of a and α-specific genes by the constitutive factor MCM1 and the mating-type factors α1 and α2.

relatively unstructured to being α-helical in structure (Figure 5.28 and Plate 4) (Li *et al.*, 1995, for a review, see Andrews and Donoviel, 1995).

Interestingly, a distinct region at the N-terminus of the α2 protein mediates its interaction with the MCM1 protein (see above) producing similar co-operative binding of this complex to the promoters of the a specific genes. Thus the α2 protein, which in isolation binds DNA non-specifically and with low affinity, is directed to distinct DNA-binding sites by interaction with either the MCM1 or a1 proteins. Interaction with MCM1 results in dimerization of α2 which then binds to sites in the a-specific genes. In contrast, interaction with a1 results in the formation of an a1/α2 heterodimer which has a different DNA-binding specificity leading to binding to sites in the haploid-specific genes (Figure 5.27). Hence the α2 protein can be regarded as a generalized transcriptional repressor which is targeted to different genes in different situations depending on the other proteins that are present (Smith and Johnson, 1992; Goutte and Johnson, 1993).

It is clear, therefore, that the products of the a and α mating-type genes act to regulate gene expression via complex interactions both with themselves and with constitutive factors. A detailed analysis of this

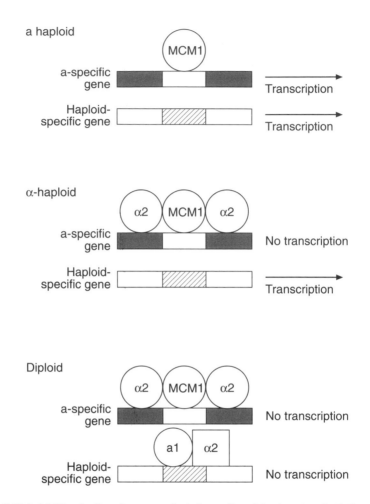

Figure 5.27 Inhibition by the α2 gene product. In a cells, α2 is absent, so both the a-specific genes and the haploid-specific genes are transcribed. In α haploid cells, α2 can repress the a-specific genes but cannot bind to haploid-specific genes, which are therefore transcribed. In diploid cells, however, α2 forms a heterodimer with the a1 factor, which can bind to the haploid-specific genes and repress their transcription.

process and the control of switching itself has been made possible by the combination of the genetic techniques available in yeast and the molecular techniques described in Chapter 2. The possibility that these processes, which are understood in such detail, may serve as a model for similar regulatory processes in higher organisms is discussed in the next section.

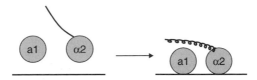

Figure 5.28 Interaction of the α2 protein with the a1 protein results in the disorganized structure of the C-terminal region of α2 becoming α helical. This region interacts with a1, resulting in high-affinity binding of the heterodimer to appropriate target sites in haploid-specific genes.

5.4.4 Relevance of the yeast mating-type system to higher organisms

At first sight, a process such as mating-type switching which involves DNA rearrangement is of apparently little relevance to higher organisms where, with very few exceptions, such as the immunoglobulin genes (see Section 5.2.1), DNA rearrangements have not been shown to be involved in developmental processes. The more detailed discussion of this process given above illustrates however, a number of possible events in this process which are applicable to higher organisms.

Thus both the action of the a and α gene products and the regulation of HO transcription involve the complex interaction of different antagonistic or synergistic transcription factors. Many of these interactions, such as the difference in binding specificity of homo- and heterodimers or the inhibition of gene activation by one factor which is mediated via a second factor, both of which have been extensively studied in this system, have now been shown to have counterparts in higher organisms (see Section 10.3.3). Similarly, the yeast transcription factors themselves show considerable similarity in structure to factors present in higher organisms. Thus MCM1 is a member of the MADS (MCM1-Agamous-Deficens-SRF) family of transcription factors to which the mammalian serum response transcription factor (SRF) also belongs and can bind to sites within mammalian genes which bind this factor (for a review, see Shore and Sharrocks, 1995). Similarly, as discussed above (Section 5.4.3), the a and α gene products contain the homeobox also found in many developmentally regulated transcription factors in higher eukaryotes. Moreover, the interaction between MCM1 and the α2 protein in yeast is paralleled by the interaction between the SRF factor and proteins of the homeobox family in mammalian and *Drosophilia* cells, with homologous regions of MCM1 and SRF being involved in mediating the interaction in each case (for a review, see Treisman, 1995).

Hence the ability to study transcription factors and their interactions both genetically and biochemically in yeast has led to the accumulation of considerable information of potential relevance to gene control in higher eukaryotes.

Interestingly, Herskowitz (1985) has drawn the analogy between mating-type switching and lineage in higher eukaryotes. Thus in the switching system illustrated in Figure 5.16, if α represents a stem cell and a is a differentiated cell derived from it, then if it is assumed that, unlike the yeast situation, switching from α to a is irreversible, the model lineage illustrated in Figure 5.29 is obtained. In this model, the stem cell is continually dividing with one daughter maintaining the stem cell lineage whilst the other differentiates. This type of system is commonly found in higher organisms, being used in the development of a number of different cell types and organs.

Even though such a lineage is unlikely to have a DNA rearrangement as its basis, the simple model of change in expression from one state to another provided by the yeast system can be used therefore to generate models of how differentiation might occur in higher organisms.

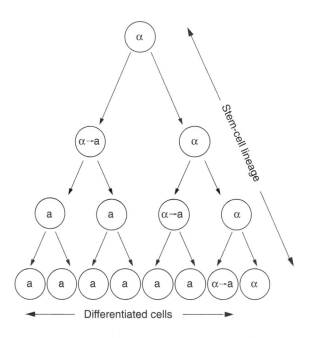

Figure 5.29 Model for the generation of a stem-cell lineage, producing differentiated cells by a system based on α to a mating-type switching. In this model, α represents the stem cell and a the differentiated cell, and it is assumed that, unlike the yeast mating-type system, the α to a switch is irreversible.

5.5 CONCLUSIONS

In this chapter we have discussed the regulation of cell type-specific transcription via the activity of specific transcription factors which are synthesized or are present in an active form only in the particular cell type where the gene(s) they regulate are active. These factors may bind directly to specific DNA sequences in their target genes as occurs for Oct-2 or NFκB or, like OCA-B, may act as co-activators which are recruited to the gene by DNA-bound factors which, like Oct-1, may be expressed in all cell types. By interacting both with each other and constitutively expressed factors, these cell type-specific factors therefore control the specific transcription pattern of the genes which are dependent upon them.

In some cases, such as OCA-B, Oct-2 or NFκB, these factors simply regulate one or more genes which are active in the cell types where they are present. In other cases such as MyoD or the yeast mating-type proteins, however, the factor is capable of directly or indirectly switching on all the genes whose expression is characteristic of the differentiated cell and its expression is therefore sufficient to produce the differentiated cell phenotype.

In lower eukaryotes such as yeast, therefore, the switching on of such a transcription factor in a particular cell will result in the appropriate differentiated phenotype. Although this is the case in higher eukaryotes also, such activation must occur at a particular stage of development and in a particular group of adjacent cells so that a tissue or organ forms at the appropriate time and in the appropriate place. The role of transcription factors in this regulation of gene expression during development is discussed in the next chapter.

REFERENCES

Abu-Amer, Y. and Tondravi, M.M. (1997). NFκB and bone – the breaking point. *Nature Medicine* **3**, 1189–1190.

Amon, A. (1996). Mother and daughter are doing fine: asymmetric cell division in yeast. *Cell* **84**, 651–654.

Andrews, B. and Donoviel, M.S. (1995). A heterodimeric transcriptional repressor becomes clear. *Science* **270**, 251–253.

Andrews, B.J. and Mason, S.W. (1993). Gene expression and the cell cycle; a family affair. *Science* **261**, 1543–1544.

Baeuerle, P.A. and Baltimore, D. (1996) NFκB: ten years after. *Cell* **87**, 13–20.

Baldwin, A.S. (1996). The NFκB and IκB proteins: new discoveries and insights. *Annual Review of Immunology* **14**, 649–681.

Benezra, R., Davis, R.L., Lockshon, D., Turner, D.L. and Weintraub, H. (1990). The protein Id: a negative regulator of helix–loop–helix DNA binding proteins. *Cell* **61**, 49–59.

Clerc, R.G., Corcoran, L.M., LeBowitz, J.H., Baltimore, D. and Sharp, P.A. (1988). The B cell specific Oct-2 protein contains POU box and homeo box type domains. *Genes and Development* **2**, 1570–1581.

Constantinides, P.G., Jones, P.A. and Gevers, W. (1977). Functional striated muscle cells from non-myoblast precursors following 5-azacytidine treatment. *Nature* **267**, 364–366.

Corcoran, L.M., Karvelase, M., Nossal, G.J.V., Ye, Z., Jacks, T. and Baltimore, D. (1993). Oct-2, although not required for early B cell development, is critical for B cell maturation and for post-natal survival. *Genes and Development* **7**, 750–582.

Davis, H.L., Weintraub, H. and Lassar, A.B. (1987). Expression of a single transfected cDNA converts fibroblasts to myoblasts. *Cell* **51**, 987–1000.

Dolan, J.K. and Fields, S. (1991). Cell type-specific transcription in yeast. *Biochimica et Biophysica Acta* **1088**, 155–169.

Dorshkind, K. (1994). Transcriptional control points during lympho-poiesis. *Cell* **79**, 751–753.

Firulli, A.B. and Olson, E.N. (1997). Modular regulation of muscle gene transcription, a mechanism for muscle cell diversity. *Trends in Genetics* **13**, 364–369.

Gellert, M. (1992). Molecular analysis of V(D)J recombination. *Annual Review of Genetics* **22**, 425–446.

Gerber, A.N., Klesert, T.R., Bergstrom, D.A. and Tapscott, S.J. (1997). Two domains of MyoD mediate transcriptional activation of genes in repressive chromatin: a mechanism for lineage determination in myogenesis. *Genes and Development* **11**, 436–450.

Goutte, C. and Johnson, A.D. (1993). Yeast a1 and α2 homeodomain proteins form a DNA binding activity with properties distinct from that of either protein. *Journal of Molecular Biology* **233**, 359–371.

Graef, I.A. and Crabtree, G.R. (1997). The transcriptional paradox: octamer factors and B and T cells. *Science* **277**, 193–194.

Gstaiger, M., Knoepfel, L., Georgiev, O., Schaffner, W. and Hovens, C.M. (1995). A B-cell co-activator of octamer-binding transcription factors. *Nature* **373**, 360–362.

Halevy, O., Novitch, B.G., Spicer, D.B., Skapek, S.X., Rhee, J., Hannon, G.J., Beach, D. and Lassar, A.B. (1995). Correlation of terminal cell cycle arrest of skeletal muscle with induction of p21 by MyoD. *Science* **267**, 1018–1021.

Herschbach, B.M., Arnaud, M.B. and Johnson, A.D. (1994). Transcriptional repression directed by the yeast α2 protein *in vitro*. *Nature* **370**, 309–311.

Herskowitz, L. (1985). Master regulatory loci in yeast and lambda. *Cold Spring Harbor Symposia* **50**, 565–574.

Herskowitz, L. (1989). A regulatory hierachy for cell specialization in yeast. *Nature* **342**, 749–757.

Jones, N. (1990). Transcriptional regulation by dimerization: two sides to an incestuous relationship. *Cell* **61**, 9–11.

Kim, U., Qin, X.F., Gong, S., Stevens, S., Luo, Y., Nussenzweg, M. and Roeder, R.G. (1996). The B-cell specific transcription factor OCA-B/OBF-1/Bob-1 is essential for normal production of immunoglobulin isotypes. *Nature* **383**, 542–547.

Konig, H., Pfisterer, P., Corcoran, L. and Wirth, T. (1995). Identification of CD36 as the first gene dependent on the B cell differentiation factor Oct-2. *Genes and Development* **9**, 1598–1607.

Latchman, D.S. (1998). *Gene Regulation: A Eukaryotic Perspective*. 3rd edn. London, New York: Chapman and Hall.

Li, T., Stark, M.R., Johnson, A.D. and Wolberger, C. (1995). Crystal structure of the MAT a1/MATα2 homeodomain heterodimer bound to DNA. *Science* **270**, 262–269.

Long, R.M., Singer, R.H., Meng, X., Gonzalez, I., Nasmyth, K. and Jansen, R.P. (1997). Mating type switching in yeast controlled by asymmetric localisation of ASH1 mRNA. *Science* **277**, 383–387.

Megene, L.A., Kabler, B., Garrett, K., Anderson, J.E. and Rudnicki, M.A. (1996). MyoD is required for myogenic stem cell function in adult skeletal muscle. *Genes and Development* **10**, 1173–1183.

Mignotte, V., Wall, L., de Boer, E., Grosveld, F. and Romeo, P.H. (1989). Two tissue-specific factors bind the erythroid promoter of the human porphobillinogen deaminase gene. *Nucleic Acids Research* **17**, 37–54.

Molkentin, J.D. and Olson, E.N. (1996). Defining the regulatory networks for muscle development. *Current Opinion in Genetics and Development* **6**, 445–453.

Pevny, L., Simon, M.C., Robertson, E., Klein, W.H., Tsai, S.F., D'Agati, V., Orkin, S.H. and Constantini, F. (1991). Erythroid differentiation in chimaeric mice blocked by a targetted mutation in the gene for transcription factor GATA-1. *Nature* **349**, 257–260.

Pfisterer, P., Konig, H., Hess, J., Lipowsky, G., Haedler, B., Schleuning, W.D. and Wirth, T. (1996). CRISP-3: a protein with homology to plant defence proteins is expressed in mouse B cells under the control of Oct-2. *Molecular and Cellular Biology* **16**, 6160–6168.

Puri, P.L., Avantaggiati, M.L., Balsone, C., Sang, N., Graessmann, A., Giordano, A. and Leureo, M. (1997) p300 is required for MyoD-dependent cell cycle arrest and muscle-specific gene transcription. *EMBO Journal* **6**, 369–383.

Rawls, A. and Olson, E.N. (1997). MyoD meets its maker. *Cell* **89**, 5–8.

Razin, A. and Cedar, H. (1991). DNA methylation and gene expression. *Microbiological Reviews* **55**, 451–458.

Reitman, M. and Felsenfeld, G. (1988). Mutational analysis of the chicken beta globin enhancer reveals two positive acting domains. *Proceedings of the National Academy of Sciences USA* **85**, 6267–6271.

Roitt, I.M., Brostoff, J. and Male, D.K. (1996). *Immunology*, 4th edn. St Louis, MO: Mosby.

Rudnicki, M.A. and Jaenisch, R. (1995). The MyoD family of transcription factors and skeletal myogenesis. *BioEssays* **17**, 203–209.

Rudnicki, M.A., Schnegelsberg, P.N.J., Stead, R.H., Brown, T., Arnold, H.H. and Jaenisch, R. (1993). MyoD or Myf-5 is required for the formation of skeletal muscle. *Cell* **75**, 1351–1359.

Schubart, D.B., Rolink, A., Kosco-Vilbois, M.H., Botteri, F. and Matthias, P. (1996). B-cell specific co-activator OBF-1/OCA-B/Bob-1 required for immune response and germinal centre formation. *Nature* **383**, 538–542.

Shepherd, J.C.W., McGinnis, W., Carrasco, A.E., De Robertis, E.M. and Gehring, W.J. (1984). Fly and frog homeodomains show homologies with yeast mating type regulatory loci. *Nature* **310**, 70–71.

Shivadasami, R.A., Fujiwara, Y., McDevitt, M.A. and Orkin, S.H. (1997). A lineage-selective knockout establishes the critical role of transcription factor GATA-1 in megakaryocyte growth and platelet development. *EMBO Journal* **16**, 3965–3973.

Shore, P. and Sharrocks, A.D. (1995). The MADS-box family of transcription factors. *European Journal of Biochemistry* **229**, 1–13.

Singh, H., Sen, R., Baltimore, D. and Sharp, P.A. (1986). A nuclear factor that binds to a conserved sequence motif in transcriptional control elements of immunoglobulin genes. *Nature* **319**, 154–158.

Smith, O.L. and Johnson, A.D. (1992). A molecular mechanism for combinational control in yeast: MCM1 protein sets the spacing and orientation of the homeodomains of an α2 dimer. *Cell* **68**,133–142.

Stark, M.R. and Johnson, A.D. (1994). Interaction between two homeodomain proteins is specified by a short C-terminal trail. *Nature* **371**,429–432.

Stillman, D.J., Bonkier, A.T., Seddan, A., Groenheat, G. and Nasmylh, K.A. (1988). Characterization of a mother cell transcription factor involved in mother cell specific transcription of the yeast HO gene. *EMBO Journal* **7**, 485–494.

Strubin, M., Newell, J.W. and Matthias, P. (1995). OBF-1, a novel B cell-specific coactivator that stimulates immunoglobulin promoter activity through association with octamer-binding proteins. *Cell* **80**, 497–506.

Thanos, D. and Maniatis, T. (1995). NFκB: a lesson in family values. *Cell* **80**, 529–532.

Treisman, R. (1995). Inside the MADS box. *Nature* **376**, 468–469.

Trouche, D., Grigoriev, M., Lenormard, J.C., Robin, P., Leibovitch, S.A., Sassone-Corsi, P. and Havel-Bellar (1993). Repression of c-*fos* promoter by MyoD on muscle cell differentiation. *Nature* **363**, 79–82.

Verma, I.M., Stevenson, J.K., Schwartz, E.M., Van Antwerp, D. and Miyamoto, S. (1995). Rel/NFκB/IκB family: intimate tales of association and dissociation. *Genes and Development* **9**, 2723–2735.

Vershon, A.K. and Johnson, A.D. (1993). A short disordered protein region mediates interactions between the homeodomain of the yeast α2 protein and the MCM1 protein. *Cell* **72**, 105–112.

Wirth, T., Staudt, L. and Baltimore, D. (1987). An octamer oligonucleotide upstream of a TATA motif is sufficient for lymphoid specific promoter activity. *Nature* **329**, 174–178.

Woloshin, P., Song, K., Degnin, C., Killary, A.M., Goldhamer, D.J., Sasson, D. and Thayer, M.J. (1995). MSXI inhibits MyoD expression in fibroblast X 10T½ cell hybrids. *Cell* **82**, 611–620.

Zwilling, S., Dieckmann, A., Pfisterer, P., Angel, P. and Wirth, T. (1997). Inducible expression and phosphorylation of co-activator BOB.1/OBF.1 in T cells. *Science* **277**, 221–225.

CHAPTER SIX

Transcription factors and developmentally regulated gene expression

6.1 DEVELOPMENTALLY REGULATED GENE EXPRESSION

As discussed in the previous chapter, the ability to carry out genetic experiments in yeast resulted in the isolation of genes whose mutagenesis affects cell type-specific gene regulation, allowing the role of their protein products in this process to be elucidated. Clearly such an approach could be equally valuable in understanding the regulation of gene expression during development in which the processes of cell type-specific gene expression discussed in the previous chapter are integrated and co-ordinated so that each cell type and tissue arises in the right place at the right time. It is evident, however, that this cannot be done in the single-celled yeast. Rather, these studies have focused on the fruit fly *Drosophila melanogaster*, which is well characterized genetically and has a short generation time facilitating its study.

A very large number of mutations which affect the development of this organism have been isolated and their corresponding genes named on the basis of the observed phenotype of the mutant fly (for reviews, see Ingham, 1988; Lawrence and Morata, 1994). Thus mutations in the so-called homeotic genes result in the transformation of one particular segment of the body into another; mutations in the *Antennapedia* gene, for example, causing the transformation of the segment which normally produces the antenna into one which produces a middle leg (Figure 6.1). Similarly, mutations in genes of the gap class result in the total absence of particular segments; mutations in the *Knirps* gene, for example, resulting in the absence of most of the abdominal segments, although the head and thorax develop normally.

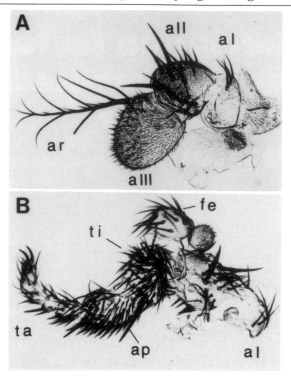

Figure 6.1 Effect of the homeotic mutation *Antennapedia*, which produces a middle leg (B) in the region that would contain the antenna of a normal fly (A). aI, aII, aIII, 1st, 2nd and 3rd antennal segments; ar, arista; ta, tarsus; ti, tibia; fe, femur; ap, apical bristle.

The products of genes of this type therefore play critical roles in *Drosophila* development. The products of the gap genes, for example, are necessary for the production of particular segments, whilst the homeotic gene products specify the identify of these segments. Given that these processes are likely to require the activation of genes whose protein products are required in the particular segment, it is not surprising that many of these genes have been shown to encode transcription factors. Thus the *Knirps* gene product and that of another gap gene, *Kruppel*, contain multiple zinc finger motifs characteristic of DNA-binding transcription factors and can bind to DNA in a sequence-specific manner. Similarly, the tailless gene, whose product plays a key role in defining the anterior and posterior regions of the *Drosophila* embryo, has been shown to be a member of the nuclear receptor super gene family discussed in Section 4.3.

It is clear therefore that the genes identified by mutation as playing a role in *Drosophila* development can encode several different types of transcription factors. However, of the first 25 such genes which were cloned, allowing a study of their protein products, well over half (15) contain a motif known as the homeobox (Gehring *et al.*, 1994) which was originally identified in the homeotic genes of *Drosophila*. The evidence that these homeobox proteins modulate transcription in *Drosophila* and the ways in which they do so are discussed in Sections 6.2.1–6.2.3.

Following the successful use of genetic analysis to isolate the *Drosophila* homeobox genes, attempts have been made to isolate homologues of these genes in other organisms, less amenable to genetic analysis, by using hybridization with the DNA of the *Drosophila* genes. Accordingly, Section 6.2.4 discusses the isolation of homeobox genes in other organisms including mammals, by this and other means as well as the evidence that their protein products are involved in developmental gene regulation in these organisms. Subsequently, other trancription factors have been identified which contain the homeobox as part of a more complex structure. Two classes of such factors, the POU proteins and the PAX proteins, both of which play a key role in developmentally regulated gene expression, are discussed in Section 6.3.

6.2 HOMEOBOX TRANSCRIPTION FACTORS

6.2.1 The homeobox

When the first homeotic genes were cloned, it was found that they shared a region of homology approximately 180 base pairs long and therefore capable of encoding 60 amino acids, which was flanked on either side by regions which differed dramatically between the different genes. This region was named the homeobox (for a review, see Gehring *et al.*, 1994). Subsequently, the homeobox was shown to be present in many other *Drosophila* regulatory genes. These include the *Fushi-tarazu* gene (Ftz), which is a member of the pair-rule class of regulatory loci whose mutation causes alternate segments to be absent, and the *engrailed* gene (eng), which is a member of the class of genes whose products regulate segment polarity. The close similarity of the homeoboxes encoded by the homeotic genes *Antennapedia* and *Ultrabithorax* and that encoded by the Ftz gene is shown in Figure 6.2.

The presence of this motif in a large number of different regulatory genes of different classes strongly suggested that it was of importance in their activity. The evidence that the homeobox-containing proteins are transcription factors whose DNA-binding activity is mediated by the homeobox is discussed in the next section (for reviews, see Hayashi and Scott, 1990; Gehring *et al.*, 1994).

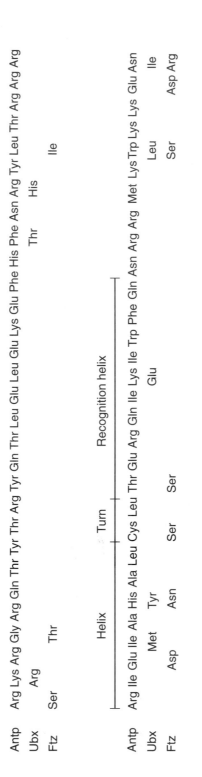

Figure 6.2 Amino-acid sequences of several *Drosophila* homeodomains, showing the conserved helical motifs. Differences between the sequences of the Ubx and Ftz homeodomains from that of Antp are indicated; a blank denotes identity in the sequence. The helix–turn–helix region is indicated.

6.2.2 The homeobox proteins as transcription factors

The first indication that the homeobox proteins were indeed transcription factors came from the finding that the homeobox was also present in the yeast mating type a and α gene products. Thus, as discussed in Section 5.4, these proteins are known to be transcription factors which regulate the activity of a and α-specific genes, hence suggesting, by analogy, that the *Drosophila* proteins also fulfilled such a role.

Direct evidence that this is the case is now available from a number of different approaches. Thus it has been shown that many of these proteins bind to DNA in a sequence-specific manner as expected for transcription factors (Hoey and Levine, 1988). Moreover, binding of a specific homeobox protein to the promoter of a particular gene correlates with the genetic evidence that the protein regulates expression of that particular gene. For example, the Ultrabithorax (Ubx) protein has been shown to bind to specific DNA sequences within its own promoter and in the promoter of the *Antennapedia* (Antp) gene, in agreement with the genetic evidence that *Ultrabithorax* (Ubx) represses *Antennapedia* expression (Figure 6.3).

The ability of the homeobox-containing proteins to bind to DNA is directly mediated by the homeobox itself. Thus, if the homeobox of the Antennapedia protein is synthesized in isolation either in bacteria or by chemical synthesis, it is capable of binding to DNA in the identical sequence-specific manner characteristic of the intact protein. The structural features of the homeodomain which allow it to do this are discussed in Section 8.2.

Although DNA binding is a prerequisite for the modulation of transcription, it is necessary to demonstrate that the homeobox proteins do actually affect transcription following such binding. In the case of the Ubx protein, this was achieved by showing that co-transfection of a plasmid expressing Ubx with a plasmid in which the *Antennapedia* promoter drives a marker gene resulted in the repression of gene expression driven by the *Antennapedia* promoter. Hence the observed binding of Ubx to the Antp promoter (see above) results in down regulation of its activity in agreement with the results of genetic experiments.

Most interestingly, the Ubx expression plasmid was able to up regulate activity of its own promoter in co-transfection experiments, this ability being dependent on the previously defined binding sites for Ubx within its own promoter. Similarly, although Ubx normally has no effect on expression of the alcohol dehydrogenase (Adh) gene, it can stimulate the Adh promotor following linkage of the promoter to a DNA sequence containing multiple binding sites for Ubx. Hence a homeobox protein can produce distinct effects following binding, Ubx activating its

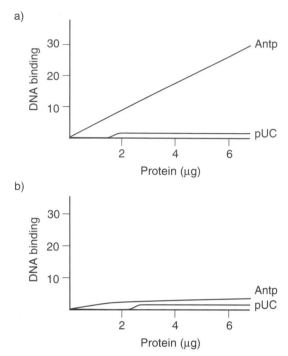

Figure 6.3 Assay of protein binding to a DNA fragment from the *Antennapedia* gene promoter (Antp) or a control fragment of plasmid DNA (pUC) using protein extracts from *E. coli* which have been genetically engineered to express the *Drosophila* Ubx protein (a) or protein extracts from control *E. coli* not expressing Ubx (b). Note the specific binding of Ubx protein to the *Antennapedia* DNA fragment.

own promoter and a hybrid promoter containing Ubx binding sites but repressing the activity of the Antp promoter (Figure 6.4).

A similar transcriptional activation effect of DNA binding has been demonstrated for the Fushi-tarazu (Ftz) protein. This protein binds specifically to the sequence TCAATTAAATGA. As with Ubx, linkage of this sequence to a marker gene confers responsivity to activation by Ftz, such activation being dependent upon binding of Ftz to its target sequence, a one-base-pair change which abolishes binding, also abolishing the induction of transcription (Figure 6.5).

These co-transfection studies carried out by introducing target DNA into cultured cells along with another DNA molecule expressing the homeobox protein have been supplemented by studies using *in vitro* transcription assays. Thus, if the Ubx promoter linked to a marker gene is added to a suitable cell-free extract, transcription of the marker gene driven by the promoter can be observed. Addition of the purified homeodomain protein even-skipped (eve) to this extract inhibits Ubx

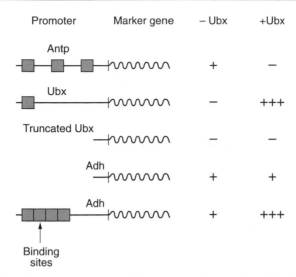

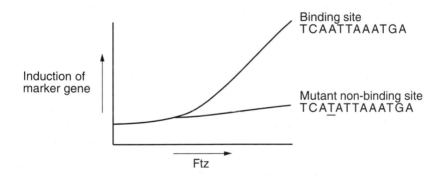

Figure 6.4 Effect of Ubx on various marker genes with or without binding sites (shaded boxes) for the Ubx protein. Note that Ubx can stimulate its own promoter which contains a Ubx-binding site and this effect is abolished by deleting the Ubx-binding site. Similarly the alcohol dehydrogenase (Adh) gene, which is normally unaffected by Ubx, is rendered responsive to Ubx stimulation by addition of Ubx-binding sites. In contrast the *Antennapedia* promoter, which also contains Ubx-binding sites, is repressed by Ubx. Hence binding of Ubx can activate or repress different promoters.

Figure 6.5 Effect of expression of the Ftz protein on the expression of a gene containing its binding site, or a mutated binding site containing a single base-pair change which abolishes binding of Ftz.

promoter activity, however, and this inhibition is dependent upon binding sites for the eve protein within the Ubx promoter. Such findings parallel the ability of a vector expressing eve to repress the Ubx promoter following co-transfection into cultured cells and the genetic evidence which originally led to the definition of eve as a repressor of Ubx (Figure 6.6). Like the thyroid hormone receptor (Section 4.4.4) and the yeast α2 protein (Section 5.4.3), the eve protein acts as a direct transcriptional repressor inhibiting the activity of the basal transcriptional complex (Johnson and Krasnow, 1992) (for further discussion of directly acting transcriptional repressors, see Section 9.3.2).

Hence, in the case of eve, a direct link exists between the genetic data, the results of co-transfections into cells and the *in vitro* data. The fact that the results of experiments in whole organisms or intact cells can be directly reproduced in cell extracts indicates that the homeodomain proteins function directly as transcription factors to control development. Moreover, by showing that these factors can act in a relatively simple *in vitro* system, they provide a means of analysing this effect.

A variety of evidence is therefore available which allows the conclusion that the homeobox proteins are transcription factors which exert their

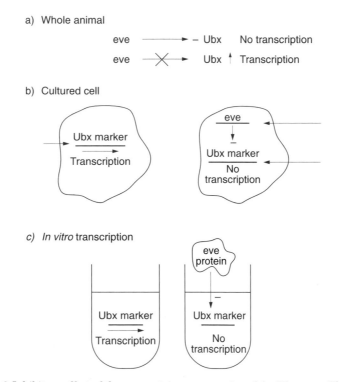

Figure 6.6 Inhibitory effect of the eve protein on expression of the Ubx gene. This inhibitory effect can be observed in the whole animal where mutation of the eve gene enhances Ubx expression (a); in cultured cells, where introduction of a plasmid expressing the eve gene represses a co-transfected Ubx promoter driving a marker gene (b); and in a test tube *in vitro* transcription system, where addition of purified eve protein represses transcription of a marker gene driven by the Ubx promoter (c).

effects on development by activating or repressing the activity of target genes. What remains to be elucidated is the manner in which such relatively simple effects on gene expression can regulate the enormously complicated process of development. The progress which has been made in this area is discussed in the next section.

6.2.3 Homeobox transcription factors and the regulation of development

At first sight it is remarkably difficult to understand how the relatively simple process of transcriptional regulation by homeobox proteins could in turn control development. Although this process is of course not yet fully understood, a number of mechanisms have been defined which indicate ways in which the complex regulatory networks needed to regulate development might be built up.

One such finding has already been discussed in the previous section. Thus, in a number of cases, it has already been shown that a single factor can repress some target genes whilst activating others, thereby increasing the range of effects mediated by one factor. Thus, the Ubx protein, for example, can activate transcription from its own promoter whilst repressing that of the Antp gene. At least in neuronal cells, this effect appears to operate by the Ubx protein binding to the Antp promoter and preventing the binding to the same site of the Antp factor itself which would normally activate the promoter. Hence this represents an example of gene repression by interfering with the binding of another activating factor, rather than by direct repression as in the case of the eve protein discussed above (for a discussion of the mechanisms of negative regulation, see Section 9.3).

Whatever its mechanism the activation and repression of different promoters by the Ubx protein has important consequences in terms of the control of development. Thus the ability of Ubx to induce its own transcription provides a mechanism for the long-term maintenance of Ubx gene expression during development since once expression has been switched on and some Ubx protein made, it will induce further transcription of the gene via a simple positive feedback loop even if the factors which originally stimulated its expression are no longer present (Figure 6.7). This long-term maintenance of Ubx expression is essential since, if the Ubx gene is mutated within the larval imaginal disc cells which eventually produce the adult fly, the cells which would normally produce the haltere (balancer) will produce a wing instead. Thus, although these cells are known to be committed already to form the adult haltere at the larval stage, the continued expression of the Ubx gene is essential to maintain this commitment and allow eventual overt

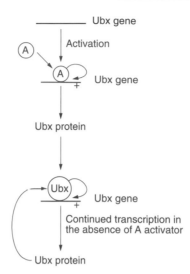

Figure 6.7 The stimulatory effect of the Ubx protein on the transcription of its own gene ensures that, once Ubx gene transcription is initially switched on by an activator protein (A), transcription will continue even if the activator protein is removed.

differentiation (for a review of imaginal discs and their role in *Drosophila* development, see Hadorn (1968)).

Similarly, the inhibition of the Antp gene by Ubx indicates that the homeobox factors do not act simply by activating the transcription of genes for structural proteins or enzymes required in particular cell types. Although this must be one of their functions, they can also regulate the transcription of each other, creating the potential for regulatory networks. Thus, we have already seen how the eve protein can repress Ubx gene transcription whilst Ubx in turn represses the Antp promoter. Since Antp has been shown to stimulate both its own promoter and that of Ubx, this creates the possibility of complex interactions in which the synthesis of one particular factor at a particular time will create changes in the levels of numerous other factors and ultimately result in the activation or repression of numerous target genes.

As well as regulating the expression of other transcription factor genes in this manner, the homeobox proteins evidently also regulate the expression of genes whose protein products are required in a particular cell type such as the cell-surface adhesion molecules (for reviews, see Edelman and Jones, 1993; Gruba *et al.*, 1997). Hence by both regulating each others expression and that of non-homeobox target genes both positively and negatively, the homeobox transcription factors can create regulatory networks of the type which are necessary for the control of development.

The effects which we have described so far involve the activity of single homeobox factors affecting the expression of other genes. In addition, however, it is also possible for the same target gene to be regulated by multiple homeobox factors, with the effects observed with one factor being different depending on whether or not another factor is also present.

In one such mechanism the DNA-binding specificity of one factor is altered in the presence of another factor. Thus several homeobox proteins such as Ubx and Antp bind to the same DNA sequences when tested in isolation *in vitro* (Hoey and Levine, 1988) yet, parodoxically, the effects of mutations which inactivate the genes encoding each of these proteins are different, indicating that they cannot substitute for one another. Similarly, *in vivo*, Ubx can bind to a site in the promoter of the decapentaplegic (dpp) gene and activate its expression, whereas Antp cannot do so.

This paradox is explained by the presence in the dpp promoter of a binding site for another homeobox protein extradenticle (exd), which lies adjacent to the site to which Ubx binds. The exd protein interacts with the Ubx protein and modifies its DNA-binding specificity so it can bind to the dpp gene promoter and activate its expression (Figure 6.8) (for a review, see Mann and Chan, 1996). This interaction with exd is dependent on the homeobox and C-terminal region of Ubx which have several amino-acid differences from the corresponding region of Antp. Because of these differences, Antp does not interact with exd and hence cannot bind to the site in the dpp gene promoter.

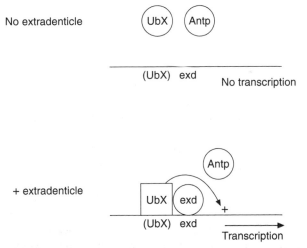

Figure 6.8. The exd protein interacts with the Ubx protein to allow it to bind with high affinity to its potential binding site in the dpp promoter (indicated as (UBX)) and activate its expression. In contrast, the Antp protein cannot interact with exd and so does not bind to the dpp promoter.

The interaction between extradenticle and other homeobox proteins clearly parallels the process in which the yeast α2 homeodomain protein binds to specific binding sites following interaction with the a1 homeodomain protein (see Section 5.4.3). This suggests that the modification of the DNA-binding specificity of a homeobox factor by interaction with other homeobox factors represents an evolutionarily conserved strategy for targeting these factors to different genes in different situations. In agreement with this, similar interactions between mammalian homologues of extradenticle and other mammalian homeo-box-containing proteins have been described (for a review, see Mann and Chan, 1996). Interestingly, a similar interaction promoting DNA binding has recently been described between the homeobox protein fushi-tarazu and a member of the nuclear receptor transcription factor family (see Section 4.3), Ftz-F1 indicating that such interactions of homeobox proteins can also occur with members of other transcription factor families (Guichet *et al.*, 1997; Yu *et al.*, 1997).

As well as interactions which alter DNA binding, other interactions operate at a functional level, with combinations of factors either synergizing with each other or interfering with one another so that functional effects are observed with combinations of factors which are not observable with either factor alone.

For example, it has been shown that the homeobox proteins engrailed (eng), Ftz, paired (prd) and zerknult (zen) can all bind to the sequence TCAATTAAAT (Hoey and Levine, 1988). When plasmids expressing each of these genes are co-transfected with a target promoter carrying multiple copies of this binding site, the Ftz, prd and zen proteins can activate transcription of the target promoter (Jaynes and O'Farrell, 1988; Han *et al.*, 1989). In contrast, the eng protein has no effect on the transcription of such a promoter. It does, however, interfere with the ability of the activating proteins to induce transcription presumably by blocking the binding of the activating factor. Thus, for example, whilst Ftz can stimulate the target promoter when co-transfected with it, it cannot do so in the presence of eng (Jaynes and O'Farrell, 1988). Hence the expression of Ftz alone in a cell would activate particular genes, whereas its expression in a cell also expressing engrailed would not have any effect (Figure 6.9).

Similar types of interaction can also take place between different positively acting factors binding to the same site. Thus, although as noted above, the Ftz, prd and zen products can all activate transcription of a target promoter when transfected alone, this effect is relatively small producing only approximately two-fold activation. In contrast, much larger effects can be obtained by activating the target promoter with two of these proteins in combination, producing 10–20-fold activation, or by all three activators together producing 400-fold activation of the target (Han *et al.*, 1989) (Figure 6.10).

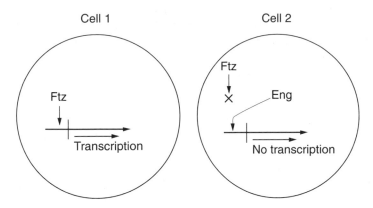

Figure 6.9 Blockage of gene induction by Ftz in cells expressing the engrailed (Eng) protein which binds to the same sequence as Ftz but does not activate transcription.

Transfected DNA				Effect on target promoter
Ftz	prd	zen	Eng	
−	−	−	−	−
+	−	−	−	+
+	+	−	−	+++
+	−	+	−	+++
+	+	+	−	+++++
+	−	−	+	−

Figure 6.10 Activation of a target promoter containing multiple copies of the identical binding site for the Ftz, prd and zen proteins when it is co-transfected with plasmids expressing each of these proteins. Note the synergistic activation of the promoter when more than one of the activator proteins is co-transfected and the inhibitory effect of the Eng protein, which binds to the same site as the other proteins but inhibits gene expression.

This synergistic stimulation of the target promoter by factors which bind the same sequence can be explained on the basis of the observation that the target promoter contains multiple copies of the binding site for these factors. It has been suggested, therefore (Han *et al.*, 1989), that gene activation proceeds via a multi-switch mechanism in which much stronger activation is produced by binding of different factors at each binding site than by the binding of multiple copies of the same factor (Figure 6.11). Interestingly, it has also been suggested that the interaction of extradenticle with other homeobox proteins discussed above may act to stimulate the transcriptional activation ability of the homeobox factor as well as its ability to DNA (Pinsonneault *et al.*, 1997).

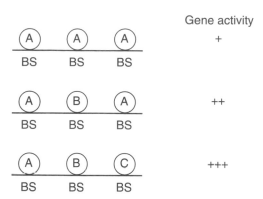

Figure 6.11 Model for the synergistic activation of gene expression by the various homeobox proteins which have the same DNA-binding site as illustrated in Figure 6.10. Gene activation is suggested to be enhanced when different factors (A, B, C) are bound to the multiple copies of their binding site (BS) present in the target promoter compared to the level of activation when the same factor (A) is bound to each of the binding sites.

In this model, extradenticle would act as a co-activator enhancing the transcriptional activation ability of its homeobox-containing partner.

Hence the complex effects of single homeobox factors on the expression of other genes can be rendered still more complex by means of synergistic or inhibitory effects of combinations of factors creating effects which would not be obtained with a single factor alone. Indeed, such interactions of different factors can be used to generate models which predict complex spatial distributions of responder gene activity in response to relatively simple expression patterns of homeobox protein distribution. One such model (Jaynes and O'Farrell, 1988) (Figure 6.12) is based on the interaction of activator and repressor molecules which bind to the same binding site in the manner of the Ftz and eng products. By assuming that target genes vary in the affinity of their binding sites for an activator and two repressor molecules whose areas of expression are overlapping but not identical, it is possible to generate different patterns of responder gene activity in each cell type depending on which particular factors are present (Figure 6.12).

Hence the activation and repression of target genes by different homeobox factors both alone and in combination can generate complex overlapping patterns of target gene expression of the sort which must occur in development. In the model described above and in our discussion so far, however, it has been assumed that a homeobox factor is either present in a particular cell or is entirely absent. In fact, however, a further level of complexity exists since many homeobox factors are not expressed in a simple on–off manner but rather show a concentration

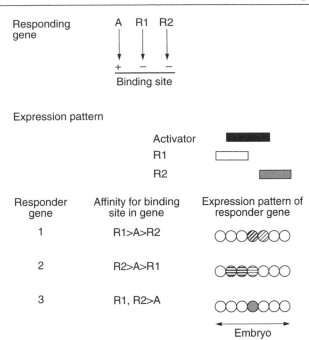

Figure 6.12 Model system producing variation in the spatial expression of three different responder genes along the length of an embryo as indicated by the open or shaded circles. All the responder genes are activated by the activator A and inhibited by the two repressors, R1 and R2, each of which is expressed in different but overlapping regions of the embryo, indicated by the solid, open or shaded boxes. The three genes differ, however, in their relative affinity for the activators and repressors as indicated. This results in variation between the genes in whether they are activated or repressed in the presence of a particular combination of activator and repressor molecules in each part of the embryo as indicated by the open circles (no expression) or the shaded circles (expression). Thus, for example, in the second circle, both A and R1 are present, so genes 1 and 3 are not expressed (since R1 binds more strongly than A to these genes) while gene 2 is expressed (since A binds more strongly than R1 to this gene).

gradient ranging from high levels in one part of the embryo via intermediate levels to low levels in another part. For example, the bicoid (bcd) protein, whose absence leads to the development of a fly without head and thoracic structures, is found at high levels in the anterior part of the embryo and declines progressively posteriorly, being absent in the posterior one third of the embryo (Figure 6.13).

Most interestingly, genes which are activated in response to bicoid contain binding sites in their promoters which have either high affinity or low affinity for the bicoid protein. If these sites are linked to a marker gene, it can be demonstrated that genes with low-affinity binding sites are only activated at high concentrations of bicoid and are therefore expressed only at the extreme anterior end of the embryo. In contrast,

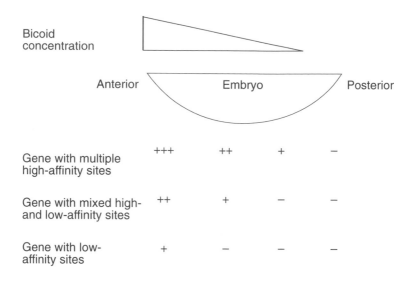

Figure 6.13 The gradient in Bicoid concentration from the anterior to the posterior point of the embryo results in bicoid-dependent genes with only low-affinity binding sites for the protein being active only at the extreme anterior part of the embryo, whereas genes with high-affinity binding sites are active more posteriorly. Note that, in addition to the different posterior boundaries in the expression of genes with high- and low-affinity binding sites, genes with high-affinity binding sites will be expressed at a higher level than genes with low-affinity binding sites at any point in the embryo.

genes which have higher affinity binding sites are active at much lower protein concentrations and will be active both at the anterior end and more posteriorly. Moreover, the greater the number of higher affinity sites the greater the level of gene expression which will occur at any particular point in the gradient (Driever *et al.*, 1989) (Figure 6.13).

The gradient in bicoid expression can be translated therefore into the differential expression of various bicoid-dependent genes along the anterior part of the embryo. Each cell in the anterior region will be able to 'sense' its position within the embryo and respond by activating specific genes. One of the genes activated by bicoid is the homeobox-containing segmentation gene *hunchback*. In turn, this protein regulates the expression of the gap genes *Kruppel* and *giant* (Struhl *et al.*, 1992). All four of these proteins then act on the eve gene, with bicoid and hunchback activating its expression whilst kruppel and giant repress it. The concentration gradients of these four factors thus result in the spatial localization of eve gene expression in a defined region of the embryo where it exerts its inhibitory effects on gene expression (Small *et al.*, 1991) (Figure 6.14). Hence the gradient in bicoid gene expression

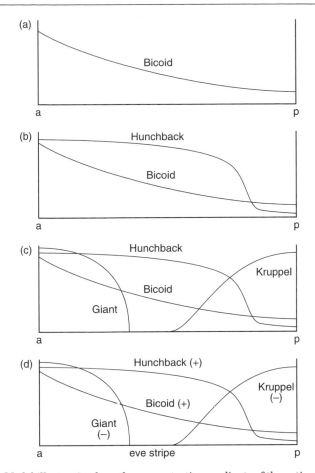

Figure 6.14. Model illustrating how the concentration gradients of the activators bicoid and hunchback, and the repressors kruppel and giant produce a stripe of eve gene expression. Note that the bicoid gradient (a) affects hunchback expression (b), which in turn affects giant and kruppel expression (c). Eve gene activation (+) by hunchback and bicoid, and its repression (−) by giant and kruppel then produces a specific stripe or region of the embryo in which eve is expressed (d).

results in changes in the expression of other genes encoding regulatory proteins leading to the activation of regulatory networks of the type discussed above.

The bicoid factor therefore has all the properties of a morphogen whose concentration gradient determines position in the anterior part of the embryo. This idea is strongly supported by the results of genetic experiments in which the bicoid gradient was artificially manipulated, cells containing artificially increased levels of bicoid assuming a phenotype characteristic of more anterior cells and vice versa.

The anterior to posterior gradient in bicoid levels is required to produce the opposite posterior to anterior gradient in the level of another protein, caudal. However, the caudal mRNA is equally distributed throughout the embryo, indicating that the bicoid gradient does not regulate transcription of the caudal gene. Rather, the bicoid protein binds to the caudal mRNA and represses its translation into protein so that caudal protein is not produced when bicoid levels are high (for reviews see Carr, 1996; Chan and Struhl, 1997). As well as providing further evidence for the key role of the bicoid factor, this finding also shows that homeodomain proteins can bind to RNA as well as to DNA and that they may therefore act at the post-transcriptional level as well as at transcription.

The studies discussed in this section have indicated therefore how by various means, such as affecting each others' expression and that of other target genes both positively and negatively, interacting with each other both synergistically and antagonistically and by affecting gene expression in a concentration-dependent manner, the homeobox transcription factors can be used to build up the complex regulatory networks which are necessary for the regulation of *Drosophila* development.

6.2.4 Homeobox transcription factors in other organisms

The critical role played by the homeobox genes in the regulation of *Drosophila* development suggests that they may also play a similar role in other organisms. Indeed, we have already seen that the a and α mating-type gene products which determine the mating type in yeast contain homeoboxes (Section 5.4). Similarly, in the nematode, *C. elegans*, homeoboxes have been identified in several genes whose mutation affects development such as the *mec-3* gene, which controls the terminal differentiation of specific sensory cells (Way and Chalife, 1988).

As in *Drosophila*, studies in the nematode have been facilitated by the availability of well-characterized mutations affecting development, allowing the corresponding genes to be isolated and the homeobox identified. In higher organisms, where such genetic evidence was unavailable, numerous investigators have used Southern blot hybridization with labelled probes derived from *Drosophila* homeoboxes in an attempt to identify homeobox-containing genes in these species. Thus, for example, Holland and Hogan (1986) used a probe from the Antennapedia homeobox to identify homeobox genes in a wide range of species including not only other invertebrates, such as the molluscs, but also chordates, such as the sea urchin and vertebrates, including the mouse (Figure 6.15). Subsequent studies have resulted in the identification of a large number of different homeobox-containing genes from a wide variety

Mouse

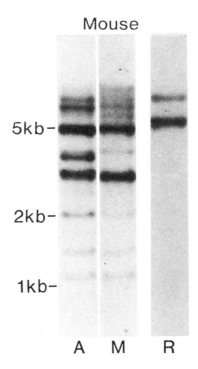

5kb—

2kb—

1kb—

A   M   R

Figure 6.15 Southern blot of mouse DNA hybridized with a probe from the *Drosophila Antennapedia* gene (A), a mouse *Antennapedia*-like gene (M) and mouse ribosomal DNA (R). Note the presence of DNA fragments which hybridize to both *Antennapedia*-like DNAs but not to ribosomal DNA, and which represent *Antennapedia*-like sequences in the mouse genome.

of organisms including both mouse and human and many of these genes have been isolated and their DNA sequence obtained (for reviews, see Kenyon, 1994; Krumlauf, 1994).

It is clear from these studies that homeobox-containing genes are not confined to invertebrates such as *Drosophila* but are found also in vertebrates including mammals such as mouse and human. Interestingly, this evolutionary conservation is not confined to the homeobox portion of these genes. Thus homologues of individual homeobox genes of *Drosophila*, such as engrailed and deformed, have been identified in mouse and human, the fly and mammalian proteins showing extensive sequence homology, which extends beyond the homeobox to include other regions of the proteins.

Moreover, the similarity between the *Drosophila* and mammalian systems extends also to the manner in which the homeobox-containing genes are organized in the genome. Thus in both *Drosophila* and mammals these genes are organized into clusters containing several homeobox-containing genes with homologous genes in the different organisms occupying equivalent positions in the clusters. For example, in a detailed comparison of the genes in the *Drosophila* Bithorax and Antennapedia complexes with those of one mouse homeobox gene complex Hox b (Hox 2), Graham *et al.* (1989) showed that the first gene in the mouse complex, Hox b9 (2.5) was most homologous to the first gene in the *Drosophila* Bithorax complex, Abd-B and so on across the complex (Figure 6.16). Hence both the homeobox genes and their arrangement are highly conserved in evolution, the common ancestor of mammals and insects having presumably possessed a similar cluster of homeobox-containing genes.

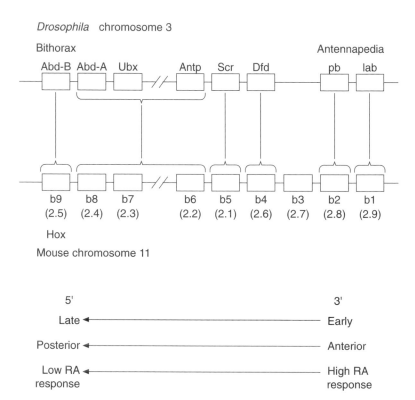

Figure 6.16 Comparison of the *Bithorax–Antennapedia* complex on *Drosophila* chromosome 3 with the Hox b complex on mouse chromosome 11. Individual genes are indicated by open boxes. Note that each gene in the *Drosophila* complex is most homologous to the equivalent gene in the mouse complex as indicated by the vertical lines. The *Drosophila* Abd-A, Ubx and Antp genes are too closely related to each other to be individually related to a particular mouse gene but are most closely related to the Hox b6, b7 and b8 genes which occupy the equivalent positions in the Hox b cluster as indicated by the brackets. The two alternative nomenclatures for mouse Hox genes are indicated. Note that, in moving from the left to the right of the mouse complex, the genes are expressed progressively earlier in development, have a more anterior boundary of expression and a greater responsiveness to retinoic acid (RA).

Most interestingly, in both *Drosophila* and mammals, the position of a gene within a cluster is related to its expression pattern during embryogenesis. Thus, in the mouse Hoxb cluster discussed above, all the genes are expressed in the developing central nervous system of the embryo. However, in moving from the 5′ to the 3′ end of the cluster (i.e. from Hox b9 (2.5) to Hox b1 (2.8) in Figure 6.16), each successive gene is expressed earlier in development and also displays a more anterior boundary of expression within the central nervous system (Figures 6.17 and 6.18). Similar expression patterns have also been observed in *Drosophila* where each successive gene in the Bithorax and Antennapedia clusters is expressed more anteriorly and affects progressively more anterior segments when it is mutated.

In the case of the mouse genes, a possible molecular mechanism for this differential expression pattern is provided by the finding that genes in the 3′ half of the Hoxb cluster are activated in cultured cells by treatment with low levels of retinoic acid whereas genes in the 5′ half of the cluster require much higher levels of retinoic acid for their activation (for reviews, see Conlon, 1995; Tabin, 1995). Considerable evidence exists that retinoic acid can act as a morphogen in vertebrate development and it has been suggested that a gradient of retinoic acid concentration may exist across the developing embryo (for reviews, see Conlon, 1995; Tabin, 1995). Hence the observed difference in expression of the Hoxb genes could be controlled by a retinoic acid gradient in a manner similar to the differential activation of bicoid-dependent genes in the bicoid concentration gradient observed in *Drosophila* (Figure 6.19). In turn, the Hox b genes, like their *Drosophila* counterparts, would switch on other genes required in cells at particular positions in the embryo accounting for the morphogenetic effects of retinoic acid.

This process of gene activation by retinoic acid is of course related to the induction of gene expression in response to treatment with steroid hormones discussed in Section 4.3. Thus retinoic acid functions by binding to specific receptors which are members of the steroid–thyroid hormone receptor super-family and which in turn bind to specific sequences within retinoic acid-responsive genes activating their expression (for reviews, see Mangelsdorf *et al.*, 1995; Mangelsdorf and Evans, 1995). Hence the activation of regulatory genes and the initiation of a regulatory cascade can be achieved by a modification of the processes mediating inducible gene expression discussed in Chapter 4. In agreement with this idea, the treatment of mouse embryos with retinoic acid results, for example, in changes in the expression pattern of the Hox b1 gene, which contains a retinoic acid response element in its 3′ regulatory region. Moreover, the inactivation of this element so that it no longer binds the retinoic acid receptors, abolishes expression of Hox b1 in the neuroectoderm of the early embryo providing direct evidence that the

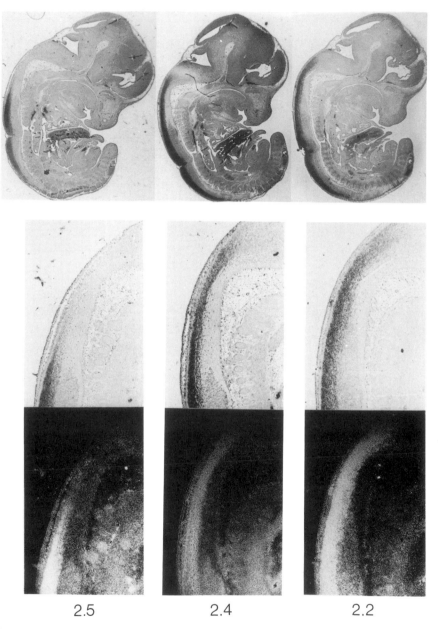

2.5   2.4   2.2

Figure 6.17 Comparison of the expression pattern of the Hox b9 (2.5), b8 (2.4) and b6 (2.2) genes in the 12.5-day mouse embryo. The top panel shows *in situ* hybridization with the appropriate gene probe to a section of the entire embryo, whilst the middle row shows a high-power view of the region in which the anterior limit of gene expression occurs. In these panels, which show the sections in bright field, hybridization of the probe and therefore gene expression is indicated by the dark areas. In the lower panel, which shows the same area in dark field, hybridization is indicated by the bright areas. Note the progressively more anterior boundary of expression of Hox b6 (2.2) compared to Hox b8 (2.4) and to Hox b9 (2.5), and compare with their positions in the Hox b (Hox 2) complex in Figure 6.16.

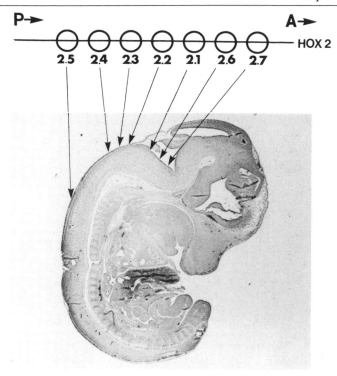

Figure 6.18 Summary of the anterior boundary of expression of the genes in the Hox b (2) complex indicated on a section of a 12.5-day mouse embryo and compared to the position of the gene in the Hox b (2) cluster. Note the progressively more anterior boundary of expression from the 5′ to the 3′ end of the Hox b (2) cluster.

retinoic acid response element is necessary to produce the expression pattern of this gene observed in the developing embryo (Marshall *et al.*, 1994).

Interestingly, the regulation of Hox gene expression by such DNA response elements located adjacent to the individual genes appears to interact with other regulatory processes which operate over the whole gene cluster. Thus, in experiments where individual Hox genes (with their adjacent control elements) were moved to a different position within the gene cluster, their pattern of expression was altered so that they behaved similarly to genes at that position in the cluster, for example, in terms of the time at which they were switched on during development (van der Hoeven *et al.*, 1996). This suggests that the alteration in chromatin structure from a closed inactive state to a more open one compatible with transcription, may begin at one end of the cluster during embryonic development and move progressively to the other end. This would result in the genes at one end of the cluster

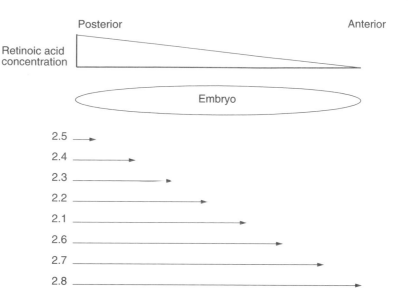

Figure 6.19 Model for the progressively more anterior expression of the genes in the Hox b (2) cluster in which expression is controlled by a posterior to anterior gradient in retinoic acid concentration and the increasing sensitivity to induction by retinoic acid, which occurs from the 5′ to the 3′ end of the cluster. Thus, because genes at the 3′ end of the cluster are inducible by very low levels of retinoic acid they will be expressed in anterior points of the embryo where the retinoic acid level will be too low to induce the genes at the 5′ end of the cluster, which require a much higher level of retinoic acid to be activated.

becoming accessible to activating transcription factors earlier than the genes at the other end of the cluster (Figure 6.20) (for a review, see Mann, 1997).

It is clear, therefore, that homologues of the *Drosophila* homeobox-containing genes do exist in mammals and other vertebrates and that they are expressed in specific cell types in the early embryo. This suggests that they play a similar role to their *Drosophila* counterparts in regulating development by affecting the transcription of other genes. In confirmation of this, a number of cases have been reported in which the inactivation of an individual mouse Hox gene produces developmental abnormalities, many of which show the transformations of specific body regions which are characteristic of such mutations in *Drosophila* (see Section 6.1) (for a review, see Krumlauf, 1994).

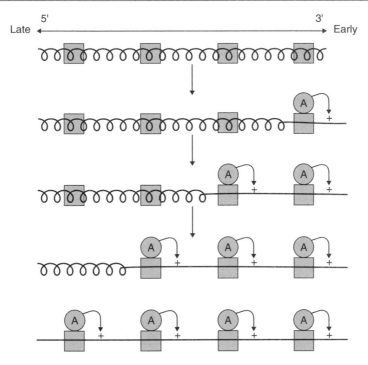

Figure 6.20 Model for the progressively later expression of genes further towards the 5′ end of a Hox gene cluster. During development the chromatin structure changes from a closed (wavy line) to a more open (straight line) structure progressively from the 3′ end of the cluster. This allows activating transcription factors (A) to bind to their binding sites (boxes) and switch on transcription starting with the genes at the 3′ end of the cluster.

6.3 OTHER HOMEOBOX-CONTAINING TRANSCRIPTION FACTORS

6.3.1 POU proteins

As discussed above, the homeobox-containing genes were first identified in *Drosophila* and only subsequently in other organisms. The reverse is true, however, for another set of transcription factors which possess a homeobox as part of a much larger motif and which were first identified in mammalian cells. Thus, as discussed in Section 5.2.2, the octamer-binding transcription factors, Oct-1 and Oct-2, play an important role in regulating the expression of specific genes such as those encoding histone H2B, the SnRNA molecules and the immunoglobulins. Similarly, the transcription factor, Pit-1, which binds to a sequence two bases different from the octamer sequence, plays a critical role in pituitary-specific gene expression (Section 1.2.3).

When the genes encoding these factors were cloned, they were found to share a 150–160-amino-acid sequence which was also found in the protein encoded by the nematode gene, *unc*-86, whose mutation affects sensory neuron development. This common POU (Pit-Oct-Unc) domain contains both a homeobox sequence and a second conserved domain, the POU-specific domain (Figure 6.21) (for reviews, see Verrijzer and Van der Vliet, 1993; Ryan and Rosenfeld, 1997).

Interestingly, whilst the homeoboxes of the different POU proteins are closely related to one another (53 out of 60 homeobox residues are the same in Oct-1 and Oct-2, and 34 out of 60 in Oct-1 and Pit-1), they show less similarity to the homeoboxes of other mammalian genes lacking the POU-specific domain, sharing at best only 21 out of 60 homeobox residues. Hence they represent a distinct class of homeobox proteins containing both a POU-specific domain and a diverged homeodomain.

As with the *Drosophila* homeobox proteins, however, the isolated homeodomains of the Pit-1 and Oct-1 proteins are capable of mediating sequence-specific DNA binding in the absence of the POU-specific domain. The affinity and specificity of binding by such an isolated homeodomain is much lower, however, than that exhibited by the intact POU domain, indicating that the POU-specific domain plays a critical role in producing high-affinity binding to specific DNA sequences. Hence the POU homeodomain and the POU-specific domain form two parts of a DNA-binding element which are held together by a flexible linker sequence. The structure of the POU domain and the manner in which it binds to DNA is discussed further in Section 8.2.

In addition to its role in DNA binding, the POU domain also plays a critical role in several other features of the POU proteins which are not found in the simple homeobox-containing proteins. For example, the ability of both Oct-1 and Oct-2 (like NF1 – see Section 3.3.2) to stimulate DNA replication as well as transcription is also a property of the isolated POU domains of these factors. Similarly, the ability of Oct-1 and not Oct-2 to interact with the herpes simplex virus trans-activator protein VP16 (also known as Vmw65) is controlled by a single difference in the homeodomain region of the POU domains in the two proteins. Thus the replacement of a single amino-acid residue at position 22 in the homeodomain of Oct-2 with the equivalent amino acid of Oct-1 allows Oct-2 to interact with VP16, which is normally a property only of Oct-1 (Lai *et al.*, 1992) (Figure 6.22) (for further discussion of the Oct-1/VP16 interaction, see Sections 8.6 and 9.2.1).

Interestingly, the key role of position 22 in the homeodomain is not confined to the interaction of Oct-1/Oct-2 with VP16. Thus, the closely related mammalian POU factors Brn-3a and Brn-3b differ in that Brn-3a activates the promoter of several genes expressed in neuronal cells, whereas Brn-3b represses them. Alteration of the isoleucine residue found

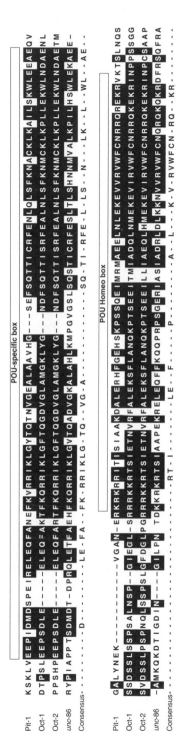

Figure 6.21 Amino-acid sequences of the POU proteins. The homeodomain and the POU-specific domain are indicated. Solid boxes indicate regions of identity between the different POU proteins. The final line shows a consensus sequence obtained from the four proteins.

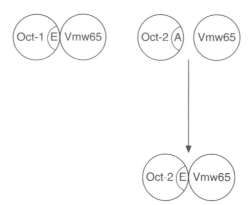

Figure 6.22. Alteration of an alanine residue (A) in the homeodomain of Oct-2 to the glutamic acid residue (E), found at the equivalent position in the homeodomain of Oct-1, allows Oct-2 to interact with the herpes simplex virus transactivator Vmw65, which is normally a property of Oct-1 only.

at position 22 in Brn-3b to the valine found in Brn-3a converts Brn-3b from a repressor into an activator, whereas the reciprocal mutation in Brn-3a converts it into a repressor (Dawson *et al.*, 1996). This effect suggests that the activating/repressing effects of Brn-3a/Brn-3b are mediated by their binding of cellular co-activator or co-repressor molecules whose binding to Brn-3a/Brn-3b is affected by the nature of the amino acid at position 22. More generally, this finding provides the first example of a single amino-acid change which can reverse the functional activity of a transcription factor, from activator to repressor and vice versa.

As in the case of the homeobox-containing proteins, the POU proteins appear to play a critical role in the regulation of developmental gene expression and in the development of specific cell types. Thus the *unc*-86 mutation in the nematode results, for example, in the lack of touch receptor neurons or male-specific cephalic companion neurons, indicating that this POU protein is required for the development of these specific neuronal cell types. Similarly, inactivation of the gene encoding Pit-1 leads to a failure of pituitary gland development resulting in dwarfism in both mice and humans (for a review, see Andersen and Rosenfeld, 1994). Interestingly, however, one type of dwarfism in mice (the Ames dwarf) is produced not by a mutation in Pit-1 but by a mutation in a gene encoding a homeobox-containing factor which was named Prophet of Pit-1 (Sornson *et al.*, 1996). This factor appears to control the activation of the Pit-1 gene in pituitary cells so that Pit-1 is not expressed when this factor is inactivated. This example illustrates how hierarchies of regulatory transcription factors are required in order to control the highly complex process of development.

Following the initial identification of the original four POU factors, a number of other members of this family have been described both in mammals and other organisms such as *Drosophila, Xenopus* and zebra fish. Like the original factors, these novel POU proteins also play a critical role in the regulation of developmental gene expression. Thus, for example, the *Drosophila* POU protein drifter (CFla) has been shown to be of vital importance in the development of the nervous system (Anderson *et al.*, 1995), whilst mutations in the gene encoding the Brn-4 factor appear to be the cause of the most common form of deafness in humans (de Kok *et al.*, 1995). Moreover, all the novel POU domain-containing genes isolated by He *et al.* (1989) from the rat, on the basis of their containing a POU domain, are expressed in the embryonic and adult brain suggesting a similar role for these proteins in the regulation of neuronal-specific gene expression. Such a close connection of POU proteins and the central nervous system is also supported by studies using the original POU domain genes, which revealed expression in the embryonic brain even in the case of Oct-2 which had previously been thought to be expressed only in B cells (He *et al.*, 1989).

It is clear therefore that, like the homeobox proteins, POU proteins occur in a wide variety of organisms and play an important role in the regulation of gene expression in development. Moreover, these proteins may be of particular importance in the development of the central nervous system.

6.3.2 Pax proteins

As well as being found as part of the POU domain which gives the POU factors their name, a homeodomain is also found in some members of another family of transcription factors, the Pax factors (for reviews, see Mansouri *et al.*, 1996; Dahl *et al.*, 1997). These factors are defined on the basis that they contain a common DNA-binding domain, known as the paired domain because it was originally identified in the *Drosophila* paired gene. In addition, however, some Pax proteins also contain a full-size or truncated homeodomain, whilst some, but not all, members of the family contain an eight-amino-acid element known as the octapeptide which is of unknown function. All combinations of the paired domain with or without a homeodomain and/or the octapeptide are found in the various mammalian Pax factors (Figure 6.23).

Obviously in the Pax factors which lack the homeodomain, the paired domain is necessary and sufficient for DNA binding. Hence this case is distinct from that of the POU factors, where the POU-specific and POU homeodomains are both necessary for high-affinity DNA binding. None the less, in factors such as Pax3, which have both a paired domain

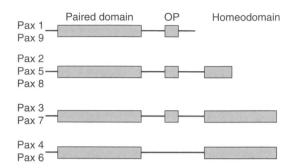

Figure 6.23 Structure of the mammalian Pax factors which contain an N-terminal paired domain linked in some cases to an octapeptide (OP) of unknown function and/or a full-length or truncated homeodomain.

and a full-length homeodomain, both domains participate in DNA binding. This produces very high-affinity binding to a DNA-binding site which contains the recognition sequence for both the DNA-binding domains and the affinity of binding to such sites is greatly reduced when either the paired domain or the homeodomain is deleted. Interestingly, the paired domain itself is distantly related to the homeodomain in terms of its structure and mechanism of DNA binding. This is discussed further in Section 8.2.

As with the POU proteins, Pax factors play a critical role in gene regulation during development, particularly in the developing nervous system. Thus, for example, Pax6 has recently been shown to be of critical importance in specifying which cells will develop into different types of motor neurons during development (Ericson *et al.*, 1997) and also appears to play a critical role in eye development in a wide range of organisms (Harris, 1997). In agreement with the critical role of these genes in development, knock-out mice in which specific Pax genes have been inactivated show defects in the development of the nervous system, whilst the naturally occurring mutant mouse strain splotch, which exhibits spina bifida, exencephaly, and neural crest and limb muscle defects, is due to a mutation in the Pax3 gene. Interestingly, mutations in Pax3 in humans result in Waardenburg syndrome, which is characterized by deafness and eye defects, whilst mutations in Pax6 also result in severe eye defects such as aniridia (for a review, see Latchman, 1996).

Hence the Pax proteins play a particularly critical role in the development of the nervous system. In addition, however, they also play a role in other tissues with mice lacking functional Pax6 showing abnormalities in the development of the pancreas as well as of the nervous system (Sander *et al.*, 1997) whilst, as discussed in Section 5.3.2,

Pax3 is involved in activating the expression of the muscle determining factor, MyoD.

6.4 CONCLUSIONS

The process of development is immensely complex. The modulation of gene expression during this process will clearly involve many of the mechanisms which have been discussed in previous chapters, some genes must become and remain active in all cell types (Chapter 3), others must be induced in response to specific signals (Chapter 4), whilst others must become and remain active in a particular type of differentiated cell (Chapter 5). In turn, such processes must be co-ordinated both temporally and spatially so that each organ forms at the right time and in the right place.

The insights provided by the genetic identification of the *Drosophila* transcription factors which control this process has allowed some understanding of how this might be achieved. Thus gradients of transcription factors such as bicoid which are laid down in the egg can be translated into differential gene activity and hence produce specific cell types at particular points in the embryo. Similarly, retinoic acid can activate the expression of specific genes such as the mammalian homeobox-containing genes in a similar manner to the induction of gene expression in response to steroid hormones discussed in Chapter 4. Hence a gradient in a chemical such as retinoic acid can be used to determine position and initiate a regulatory cascade via its ability to activate regulatory genes differentially.

It is clear therefore that transcription factors play a critical role in the regulation of development and that gradients in these factors or other substances may allow positional information to be determined. Further studies on these factors, their interaction with each other and with their target genes will be necessary, however, before the process of development is fully understood.

REFERENCES

Andersen, B. and Rosenfeld, M.G. (1994). Pit-1 determines cell types during development of the interior pituitary gland. *Journal of Biological Chemistry* **269**: 335–338.

Anderson, M.G., Perkins, G.L., Chittick, P., Shrigley, R.J. and Johnson, W.A. (1995). Drifter, a *Drosophila* POU-domain transcription factor, is required for correct differentiation and migration of tracheal cells and midline glia. *Genes and Development* **9**, 123–127.

Carr, K. (1996). RNA bound to silence. *Nature* **379**, 676.

Chan, S.K. and Struhl, G. (1997). Sequence specific RNA binding by bicoid. *Nature* **388**: 634.

Conlon, R.A. (1995). Retinoic acid and pattern formation in vertebrates. *Trends in Genetics* **11**, 314–319.

Dahl, E., Koseki, H. and Balling, R. (1997) Pax genes and organiogenesis. *BioEssays* **19**, 755–765.

Dawson, S.J., Morris, P.J. and Latchman, D.S. (1996). A single amino acid change converts a repressor into an activator. *Journal of Biological Chemistry* **271**: 11631–11633.

de Kok, Y.J.M., Van der Maarel, S.M., Bitner-Glindzicz, M., Huber, I., Monaco, A.P., Malcolm, S., Pembrey, M.E., Ropers, H.H. and Cremers, F.P.M. (1995). Association between X-linked mixed deafness and mutations in the POU domain gene POU3F4. *Science* **267**, 685–688.

Driever, W., Thoma, G. and Nusslein-Volhard, C. (1989). Determination of spatial domains of zygotic gene expression in the *Drosophila* embryo by the affinity of binding sites for the bicoid morphogen. *Nature* **340**, 363–367.

Edelman, G.M. and Jones, F.S. (1993). Outside and downstream of the homeobox. *Journal of Biological Chemistry* **268**, 20683–20686.

Ericson, J., Rashbass, P., Schedl, A., Brenner-Morton, S., Kawakemi, A., van Heyningen, V., Jessel, T.M. and Briscoe, J. (1997). Pax6 controls progenitor cell identity and neuronal fate in response to graded Shh signalling. *Cell* **90**, 169–180.

Gehring, W.J., Affolter, M. and Burglin, T. (1994). Homeodomain proteins. *Annual Review of Biochemistry* **63**, 487–526.

Graham, A., Papalopulu, N. and Krumlauf, R. (1989). The murine and *Drosophila* homeobox gene complexes have common features of organization and expression. *Cell* **57**, 367–378.

Gruba, Y., Aragnol, D. and Pradel, J. (1997). *Drosophila* Hox complex targets and the function of homeotic genes. *BioEssays* **19**, 379–388.

Guichet, A., Copeland, J.W.R., Erdelyi, M., Hlousek, D., Zavaroszky, P., Ho, J., Brown, S., Percival-Smith, A., Krause, H.M. and Ephrussi, A. (1997). The nuclear receptor homologue Ftz-F1 and the homeodomain protein Ftz are mutually dependent co-factors. *Nature* **385**, 348–552.

Hadorn, E. (1968). Transdetermination in cells. *Scientific American* **219** (Nov) 110–120.

Han, K., Levine, M.S. and Manley, J.L. (1989). Synergistic activation and repression of transcription by *Drosophila* homeobox proteins. *Cell* **56**, 573–583.

Harris, W.A. (1997). Pax–6: where to be conserved is not conservative. *Proceedings of the National Academy of Sciences USA* **94**, 2098–2100.

Hayashi, S. and Scott, M.P. (1990). What determines the specificity of action of *Drosophila* homeodomain proteins. *Cell* **63**, 883–894.

He, X., Treacy, M.N., Simmons, D.M., Ingraham, H.A., Swanson, L.S. and Rosenfeld, M.G. (1989). Expression of a large family of POU-domain regulatory genes in mammalian brain development. *Nature* **340**, 35–42.

Hoey, T. and Levine, M. (1988). Divergent homeobox proteins recognize similar DNA sequences in *Drosophila*. *Nature* **332**, 858–861.

Holland, P.W.H. and Hogan, B.L.M. (1986). Phylogenetic distribution of Antennapedia-like homeoboxes. *Nature* **321**, 251–253.

Ingham, P.W. (1988). The molecular genetics of embryonic pattern formation in *Drosophila*. *Nature* **335**, 25–34.

Jaynes, J.B. and O'Farrell, P.H. (1988). Activation and repression of transcription by homeodomain-containing proteins that bind a common site. *Nature* **336**, 744–749.

Johnson, F.B. and Krasnow, M.A. (1992). Differential regulation of transcription preinitiation complex assembly by activator and repressor homeodomain protein. *Genes and Development* **6**, 2177–2189.

Kenyon, C. (1994). If birds can fly, why can't we? Homeotic genes and evolution. *Cell* **78**, 175–180.

Krumlauf, R. (1994), Hox genes in vertebrate development. *Cell* **78**, 191–201.

Lai, J.S., Cleary, M.A. and Herr, W. (1992). A single amino acid exchange transfers VP16-induced positive control from the Oct-1 to the Oct-2 homeo domain. *Genes and Development* **6**, 2058–2065.

Latchman, D.S. (1996). Transcription factor mutations and human disease. *New England Journal of Medicine* **334**, 28–33.

Lawrence, P.A. and Morata, G. (1994). Homeobox genes: their function in *Drosophila* segmentation and pattern formation. *Cell* **78**, 181–189.

Mangelsdorf, D.J. and Evans, R.M. (1995). The RXR heterodimers and orphan receptors. *Cell* **83**, 841–850.

Mangelsdorf, D.J., Thummel, C., Beato, M., Herrlich, F., Schutz, G., Umesono, K., Blumberg, B., Kastner, P., Mark, M., Chambon, P. and Evans, R.M. (1995). The Nuclear receptor super family: the second decade. *Cell* **83**, 835–839.

Mann, R.S. (1997). Why are Hox genes clustered. *BioEssays* **19**, 661–664.

Mann, R.S. and Chan, S.K. (1996). Extra specificity from extradenticle: the partnership between Hox and PBX/EXD homeodomain proteins. *Trends in Genetics* **12**, 258–262.

Mansouri, A., Hallone, T.M. and Gruss, P. (1996). Pax genes and their roles in cell differentiation and development. *Current Opinion in Cell Biology* **8**, 851–857.

Marshall, H., Studer, M., Popperl, H., Aparicio, S., Kurolwa, A., Brenner, S. and Krumlauf, R. (1994). A conserved retinoic acid response element required for early expression of the homeobox gene Hoxb-1. *Nature* **370**, 567–571.

Pinsonneault, J., Florence, B., Vaessin, H. and McGinnis, W. (1997). A model for extradenticle function as a switch that changes HOX proteins from repressors to activators. *EMBO Journal* **16**, 2033–2042.

Ryan, A.K. and Rosenfeld, M.G. (1997). POU domain family values: flexibility, partnerships and developmental codes. *Genes and Development* **11**, 1207–1225.

Sander, M., Neubuser, A., Kalamara, J., Es, II.C., Martin, G.R. and German, M.S. (1997). Genetic analysis reveals that PAX6 is required for normal transcripton of pancreatic hormone genes and islet development. *Genes and Development* **11**, 1662–1673.

Small, S., Krant, R., Hoey, T., Warrior, R. and Levine, M. (1991). Transcriptional regulation of a pair-rule stripe in *Drosophila*. *Genes and Development* **5**, 827–839.

Sornson, M.W., Wu, W., Dasen, J.S., Flynn, S.E., Norman, D.J., O'Connell, S.M., Gukovsky, I., Carriere, C., Ryan, A.K., Miller, A.P., Zuo, L., Gleiberman, A.S., Anderson, B., Beamer, W.G. and Rosenfeld, M.G. (1996). Pituitary lineage determination by the prophet of Pit-1 homeodomain factor defective in Ames dwarfism. *Nature* **384**, 327–333.

Struhl, G., Johnston, P. and Lawrence, P.A. (1992). Control of *Drosophila* body pattern by the hunchback morphogen gradient. *Cell* 69, 237–249.

Tabin, C. (1995). The initiation of the limb bud: growth factors, Hox genes and retinoids. *Cell* **80**, 671–674.

Van der Hoeven, F., Zakany, J. and Duboule, D. (1996). Gene transpositions in the HoxD complex reveal a hierachy of regulatory controls. *Cell* **85**, 1025–1035.

Verrijzer, C.R. and Van der Vliet, P.C. (1993). POU domain transcription factors. *Biochimica et Biophysica Acta* **1173**, 1–21.

Way, J.C. and Chalfie, M. (1988). *mec-3*, a homeobox-containing gene that specifies differentiation of the touch receptor neurons in *C. elegans*. *Cell* **54**, 5–16.

Yu, Y., Li, W., Su, K., Yussa, M., Han, W., Perriman, N. and Pick, L. (1997). The nuclear hormone receptor Ftz-F1 is a co-factor for the *Drosophila* homeodomain protein Ftz. *Nature* **385**, 552–555.

CHAPTER SEVEN

Transcription factors and human disease

7.1 INTRODUCTION

In previous chapters we have discussed the involvement of transcription factors in normal cellular regulatory processes, for example, constitutive, inducible, cell type-specific or developmentally regulated transcription. It is not surprising that aspects of this complex process can go wrong and that the resulting defects in transcription factors can result in disease (for reviews, see Engelkamp and van Heyningen 1996; Latchman, 1996).

For example, mutations in all the classes of transcription factors discussed in Chapter 6 which control gene expression during development have been shown to result in human developmental disorders. Thus, mutations in the gene encoding the POU family transcription factor Pit-1 (see Section 6.3.1) result in a failure of pituitary gland development leading to congenital dwarfism, whilst mutations in the genes encoding the Pax family transcription factors Pax3 and Pax6 (see Section 6.3.2) result in eye defects. Similarly, mutations in genes encoding homeobox proteins (Section 6.2) result in a variety of congenital abnormalities (for a review see Boncinelli, 1997).

Interestingly, the mutations in Pit-1 and Pax6 discussed above are both dominant, with one single copy of the mutant gene being sufficient to produce the disease even in the presence of a functional copy. However, this dominance arises for different reasons (for a review, see Latchman, 1996). In the case of Pit-1, the mutant Pit-1 can bind to its DNA-binding site but cannot activate gene expression. It therefore not only fails to stimulate transcription of its target genes but can also act as a dominant negative factor inhibiting gene activation by preventing the wild-type protein from binding to DNA (Figure 7.1a). This mechanism is similar to one mode of action of transcriptional repressors, which act by preventing an activator from binding to its DNA target site (see Section 9.3.1). In contrast, the dominant nature of the Pax6 mutation does not reflect any dominant negative action of the mutant protein since such mutations often involve complete deletion of the gene. Rather it reflects a phenomenon known as halpoid insufficency in which the amount of protein produced by a single functional copy of the gene is not enough to allow it to activate its target genes effectively (Figure 7.1b).

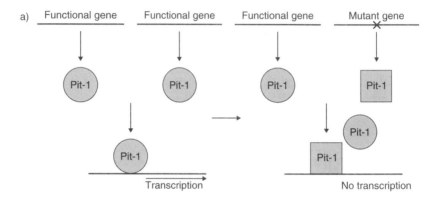

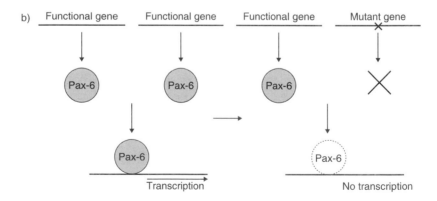

Figure 7.1 Different mechanisms by which mutations in genes encoding transcription factors can be dominant producing disease in the presence of a functional copy of the gene. (a) In the case of Pit-1, the mutation produces a dominant negative form of the factor (square), which binds to the appropriate binding site, and not only fails to activate transcription but also prevents binding and activation by the functional protein (circle) (b). In the case of Pax-6, one functional gene cannot produce enough functional protein (circle) to activate its target genes (dotted circle).

As well as resulting from mutations in the genes encoding DNA-binding factors, developmental disorders can also result from mutations in genes encoding other types of transcription factors such as co-activators or factors which alter chromatin structure. Thus, for example, mutation in the gene encoding the SNF2 factor, which is part of the SWI/SNF chromatin remodelling complex (see Section 1.4.2), results in a lack of α-globin gene expression and a variety of other symptoms such as mental retardation, indicating that this factor is necessary for opening the chromatin structure of the α-globin genes and a number of other genes, so preventing their transcription when it is absent (Gibbons *et al.*, 1995) (Figure 7.2).

Similarly, mutation in the gene encoding the CBP factor which acts as a co-activator for a variety of other transcription factors (see Sections 4.3.2 and 4.4.3) results in the severe developmental disorder known as Rubinstein–Taybi syndrome,which is characterized by mental retardation and physical abnormalities (for a review, see D'Arcangelo and Curran, 1995), indicating that CBP is an important co-activator for developmentally regulated as well as inducible gene expression. Interestingly, no individuals with mutations inactivating both copies of the CBP gene have ever been identified and it is likely that a lack of functional CBP is incompatible with life. Individuals with Rubinstein–Taybi syndrome have a single functional CBP gene and a single mutant gene, indicating that the mutation is dominant. As with Pax6, however, this dominance reflects a haploid insufficency in which a single copy of the CBP gene cannot produce enough functional protein. This is not surprising since as discussed in Section 4.4.4, the amount of CBP in the cell is limited and different transcription factors compete for it.

Hence, developmental disorders can arise from mutations in genes encoding DNA-binding activator proteins such as Pit-1 and Pax-6, co-activators such as CBP, or components of chromatin-modulating complexes such as SNF2 (Figure 7.3). As well as such developmental defects, mutations in the genes encoding the nuclear receptor transcription factor family (see Section 4.4) can produce a failure to respond to the hormone which normally binds to the receptor and regulates

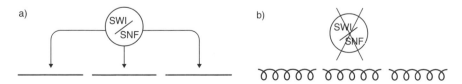

Figure 7.2(a) Functional SWI/SNF alters the chromatin structure of its target genes to a more open configuration (solid line). (b) If SWI/SNF is inactive, these genes remain in a closed chromatin structure (wavy line) and are thus not transcribed.

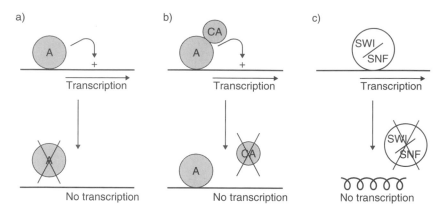

Figure 7.3 Mutations can occur in genes encoding (a) DNA-binding activators (A); (b) co-activators (CA) or (c) factors which alter chromatin structure (SWI/SNF).

transcription. Such mutations have been reported, for example, in the receptors for glucocorticoid, thyroid hormone and vitamin D (for a review, see Latchman, 1996).

Despite the existence of transcription factor mutations producing developmental defects or non-responsiveness to hormone, a special place in the human diseases which can involve alterations in transcription factors is occupied, however, by cancer. Thus, because this disease results from growth in an inappropriate place or at an inappropriate time, it can be caused not only by deficiencies in particular genes, but also by the enhanced expression or activation of specific cellular genes involved in growth-regulatory processes which are normally only expressed at low levels or very transiently. Interestingly, many cancer-causing genes of this type, known as oncogenes (for general reviews, see Bourne and Varmus, 1992; Broach and Levine, 1997; Hunter, 1997), were originally identified within cancer-causing retroviruses, which had picked them up from the cellular genome. Within the virus, the oncogene has become activated either by over-expression or by mutation, and is therefore responsible for the ability of the virus to transform cells to a cancerous phenotype. In contrast, the homologous gene within the cellular genome is clearly not always cancer-causing since all cells are not cancerous. It can be activated, however, into a cancer-causing form either by over-expression or by mutation, and hence these genes can play an important role in the generation of human cancer (Figure 7.4). The form of the oncogene isolated from the retrovirus and from the normal cellular genome are distinguished by the prefixes v and c, respectively, as in v-*onc* and c-*onc*.

Despite this potential to cause cancer, the c-*onc* genes are highly conserved in evolution, being found not only in the species from which

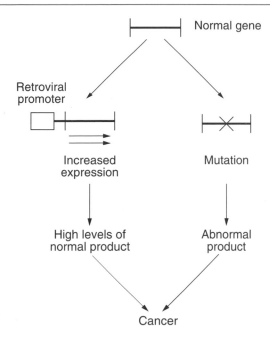

Figure 7.4 A cellular proto-oncogene can be converted into a cancer-causing oncogene by increased expression or by mutation.

the original virus was isolated but in a wide range of other eukaryotes. This indicates that the products of these oncogenes play a critical role in the regulation of normal cellular growth processes, their malregulation or mutation resulting therefore in abnormal growth and cancer. In agreement with this idea, oncogenes identified in this way include genes encoding many different types of protein involved in growth control such as growth factors, growth factor receptors and G proteins. They also include, however, several genes encoding cellular transcription factors which normally regulate specific sets of target genes. Similarly, a number of other genes encoding transcription factors have been identified at the break points of the chromosomal translocations characteristic of human leukemias with their activation being involved in the resulting cancer. Section 7.2 of this chapter therefore discusses several cases of this type and the insights they have provided into the processes regulating gene expression in normal cells and their malregulation in cancer (for reviews, see Rabbits, 1994; Latchman, 1996).

Following the discovery of cellular oncogenes, it subsequently became clear that another class of genes existed whose protein products appeared to restrain cellular growth. The deletion or mutational inactivation of these so called anti-oncogenes therefore results in the abnormal unregulated growth characteristic of cancer cells (for reviews, see Knudson, 1993; Weinberg, 1993). As some of these anti-oncogenes also encode transcription factors, they are discussed in Section 7.3 of this chapter.

7.2 CELLULAR ONCOGENES AND CANCER

7.2.1 Fos, Jun and AP1

As noted in Table 4.1, the AP1-binding site is a DNA sequence that renders genes that contain it inducible by treatment with phorbol esters such as TPA. The activity binding to this site is referred to as AP1 (activator protein 1). It is clear, however, that preparations of AP1 purified by affinity chromatography on an AP1-binding site contain several different proteins (for reviews, see Kerppola and Curran, 1995; Karin *et al.*, 1997).

A possible clue as to the identity of one of these AP1-binding proteins was provided by the finding that the yeast protein GCN4, which induces transcription of several yeast genes involved in amino-acid biosynthesis, does so by binding to a site very similar to the AP1 site (Figure 7.5). In turn, the DNA-binding region of GCN4 shows strong homology at the amino-acid level to v-*jun*, the oncogene of avian sarcoma virus ASV17 (Figure 7.6). This suggested therefore that the protein encoded by the cellular homologue of this gene, c-*jun*, which was known to be a nuclearly located DNA-binding protein, might be one of the proteins which bind to the AP1 site.

In agreement with this, antibodies against the Jun protein react with purified AP1-binding proteins, whilst Jun protein expressed in bacteria can bind to AP1-binding sites. Hence the Jun protein is capable of binding to the AP1-binding site and constitutes one component of purified AP1 preparations which also contain other Jun-related proteins such as Jun B (Figure 7.7). Moreover, co-transfection of a vector expressing the Jun protein with a target promoter resulted in increased transcription if the target gene contained AP1-binding sites but not if it

DNA-binding site

GCN4	5' T G A C/G T C A T 3'
AP1	5' T G A G T C A G 3'

Figure 7.5 Relationship of the DNA-binding sites for the yeast transcription factor GCN4 and the mammalian transcription factor AP1.

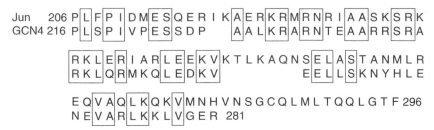

Figure 7.6 Comparison of the carboxyl-terminal amino-acid sequences of the chicken Jun protein and the yeast transcription factor GCN4. Boxes indicate identical residues.

lacked them, indicating that Jun was capable of stimulating transcription via the AP1 site (Figure 7.8). Hence the Jun oncogene product is a sequence-specific transcription factor capable of stimulating transcription of genes containing its binding site.

In addition to Jun and Jun-related proteins, purified AP1 preparations also contain the product of another oncogene, c-*fos*, as well as several Fos-related proteins known as the Fras (Fos-related antigens; see Figure 7.7). Unlike Jun, however, Fos cannot bind to the AP1 site alone but can do so only in the presence of another protein p39 which is identical to Jun (Rauscher *et al.*, 1988). Hence, in addition to its ability to bind to AP1 sites alone, Jun can also mediate binding to this site by the Fos protein. Such

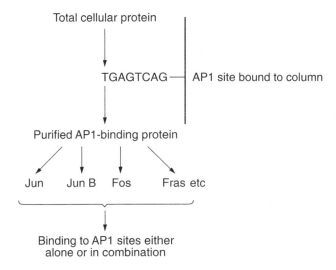

Figure 7.7 Passage of total cellular proteins through a column containing an AP1 site results in the purification of several cellular proteins including Jun, Jun B, Fos and Fos-related antigens (Fras), which are capable of binding to the AP1 site either alone or in combination.

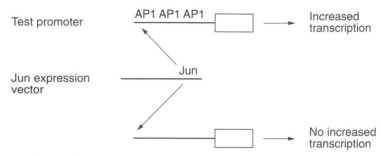

Figure 7.8 Artificial expression of the Jun protein in an expression vector results in the activation of a target promoter containing several AP1-binding sites but has no effect on a similar promoter lacking these sites, indicating that Jun can specifically activate gene expression via AP1-binding sites.

DNA binding by Fos and Jun is dependent on the formation of a dimeric molecule. Although Jun can form a DNA-binding homodimer, Fos cannot do so. Hence DNA binding by Fos is dependent upon the formation of a heterodimer between Fos and Jun, which binds to the AP1 site with approximately 30-fold greater affinity than the Jun homodimer (Figure 7.9).

It is clear therefore that both Fos and Jun which were originally isolated in oncogenic retroviruses are also cellular transcription factors which play an important role in inducing specific cellular genes following phorbol ester treatment (for a review, see Ransone and Verma, 1990). Increased levels of Fos and Jun occur in cells following treatment with phorbol esters (Lamph *et al.*, 1988) indicating that these substances act, at least in part, by increasing the levels of Fos and Jun, which in turn bind to the AP1 sites in phorbol ester-responsive genes and activate their expression.

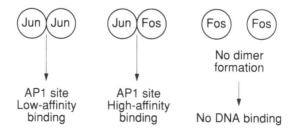

Figure 7.9 Heterodimer formation between Fos and Jun results in a complex capable of binding to an AP1 site with approximately 30-fold greater affinity than a Jun homodimer whilst a Fos homodimer cannot bind to the AP1 site.

Similar increases in the levels of Fos and Jun, as well as Jun-B and the Fos-related protein Fra-1, are also observed when quiescent cells are stimulated to grow by treatment with growth factors or serum (Lamph *et al.*, 1988; Rauscher *et al.*, 1988), indicating that these substances act, at least in part, by increasing the levels of Fos and Jun, which in turn will switch on genes whose products are necessary for growth itself (Figure 7.10). In agreement with this idea, cells derived from mice in which the *c-jun* gene has been inactivated grow very slowly in culture and the mice themselves die early in embryonic development (Johnson *et al.*, 1993). Hence Fos and Jun play a critical role in normal cells, as transcription factors inducing phorbol ester or growth-dependent genes.

Normally, levels of Fos and Jun increase only transiently following growth-factor treatment, resulting in a period of brief controlled growth. Clearly, continuous elevation of these proteins, such as would occur when cells become infected with a retrovirus expessing one of them, would result in cells which exhibited continuous uncontrolled growth and were not subject to normal growth regulatory signals. Since such uncontrolled growth is one of the characteristics of cancer cells, it is relatively easy to link the role of Fos and Jun in inducing genes required for growth with their ability to cause cancer. Normally, however, the transformation of a

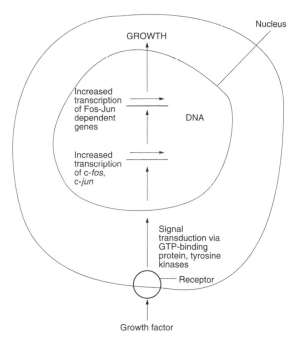

Figure 7.10 Growth factor stimulation of cells results in increased transcription of the c-*fos* and c-*jun* genes which in turn stimulates transcription of genes which are activated by the Fos–Jun complex.

cell to a transformed cancerous phenotype requires more than simply its conversion to a continuously growing immortal cell (for a review, see Land *et al.*, 1983). Since repeated treatments with phorbol esters can promote tumour formation in immortalized cells, the prolonged induction of phorbol ester-responsive genes by elevated levels of Fos and Jun may therefore result in the conversion of already continuously growing cells into the tumourigenic phenotype characteristic of cancer cells (Figure 7.11).

Hence the ability of Fos and Jun to cause cancer represents an aspect of their ability to induce transcription of specific cellular genes. In agreement with this idea, mutations in Fos which abolish its ability to dimerize with Jun and hence prevent it from binding to AP1 sites also abolish its ability to transform cells to a cancerous phenotype. It should be noted, however, that in addition to their over-expression within a retrovirus, there is also some evidence that mutational changes render the viral proteins more potent transcriptional activators than the

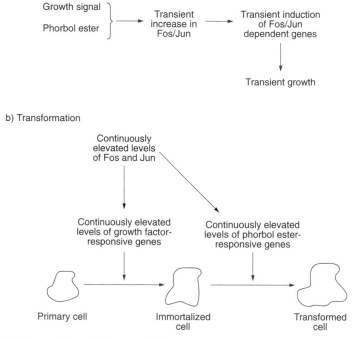

Figure 7.11 Effects of Fos and Jun on cellular growth. In normal cells (a) a brief exposure to a growth signal or phorbol ester will lead to a brief period of growth via the transient induction of Fos and Jun and hence of Fos/Jun-dependent genes. In contrast, the continuous elevation of Fos and Jun produced, for example, by infection with a retrovirus expressing Fos or Jun results in continuous unlimited growth and cellular transformation (b).

equivalent cellular proteins. Thus the v-Jun protein appears to activate transcription more efficiently than c-Jun, owing to a deletion in a region which is involved in targeting the c-Jun protein for degradation (Treier *et al.*, 1994) and which also mediates its interaction with a negatively acting cellular factor (Baichwal and Tjian, 1990).

Interestingly, in addition to its central role in the growth response, the Fos, Jun, AP1 system also appears to represent a target for other oncogenes. Thus, for example, the *ets* oncogene, which, like *fos* and *jun*, encodes a cellular transcription factor, acts via a DNA-binding site known as PEA3 which is located adjacent to the AP1 site in a number of TPA-responsive genes such as collagenase and stromelysin, and the Ets protein co-operates with Fos and Jun to produce high-level activation of these promoters (Wasylyk *et al.*, 1990). In addition to interacting positively with other factors, the Fos/Jun complex can also inhibit the action of other transcription factors. Thus, as described in Section 4.4.4, the Fos/Jun complex requires the CBP co-activator in order to activate transcription. It therefore competes with the activated glucocorticoid receptor for CBP, hence preventing the receptor from activating transcription. Similarly, both Fos and Jun can inhibit the activation of muscle-specific promoters by the MyoD transcription factor (see Section 5.3), thereby preventing cells from differentiating into non-dividing muscle cells and allowing cellular proliferation to continue (Li *et al.*, 1992).

Hence the Fos and Jun oncogene products play a critical role in the regulation of specific cellular genes in normal cells, interacting with the products of other transcription factors to produce the controlled activity of their target genes necessary for normal controlled growth.

7.2.2 v-*erbA* and the thyroid hormone receptor

The v-*erbA* oncogene is one of two oncogenes carried by avian erythroblastosis virus (AEV). The cellular equivalent of this oncogene, c-*erbA*, has been shown to encode the cellular receptor for thyroid hormone (Sap *et al.*, 1986; Weinberger *et al.*, 1986), which is a member of the steroid–thyroid hormone receptor super-family discussed in Section 4.4. Following the binding of thyroid hormone, the receptor–hormone complex binds to its appropriate recognition site in the DNA of thyroid hormone-responsive genes and activates their transcription (Figure 7.12).

Hence the protein encoded by the c-*erbA* gene represents a bona fide cellular transcription factor involved in the activation of thyroid hormone-responsive genes. Unlike the case of the *fos* and *jun* gene products which regulate genes involved in growth, it is not immediately

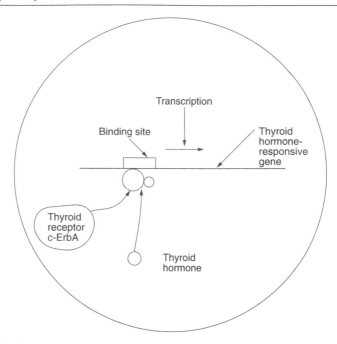

Figure 7.12 The c-*erb*A gene encodes the thyroid hormone receptor and activates transcription in response to thyroid hormone.

obvious how the form of thyroid hormone receptor encoded by the viral v-*erb*A gene can transform cells to a cancerous phenotype.

The solution to this problem is provided by a comparison of the cellular ErbA protein, which is a functional thyroid hormone receptor and the viral ErbA protein encoded by AEV. Thus, in addition to being fused to the retroviral gag protein at its N-terminus, the viral ErbA protein contains several mutations in the regions of the receptor responsible for binding to DNA and for binding thyroid hormone, as well as a small deletion in the hormone-binding domain (Figure 7.13).

Interestingly, although these changes do not abolish the ability of the viral ErbA protein to bind to DNA, they do prevent it from binding thyroid hormone and thereby becoming converted to a form which can activate transcription (Sap *et al.*, 1986, 1989). However, these changes do not affect the inhibitory domain which, as discussed in Section 4.4.4, allows the thyroid hormone receptor to repress transcription. Hence, the viral v-ErbA protein can inhibit the induction of thyroid hormone-responsive genes when cells are treated with thyroid hormone by binding to the thyroid hormone response elements in their promoters and dominantly repressing their transcription, as well as preventing binding of the activating complex of thyroid hormone and the cellular ErbA protein (Figure 7.14). In agreement with this critical role for repression in producing transformation by v-ErbA, a mutation in v-ErbA, which abolishes its ability to repress transcription by preventing it binding its

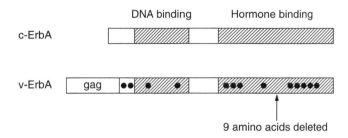

Figure 7.13 Relationship of the cellular ErbA protein and the viral protein. The black dots indicate single amino-acid differences between the two proteins while the arrow indicates the region where nine amino acids are deleted in the viral protein.

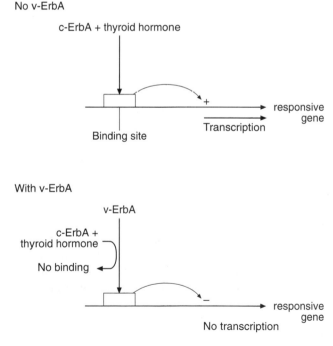

Figure 7.14 Inhibitory effect of the viral ErbA protein on gene activation by the cellular protein in response to thyroid hormone. The viral protein both inhibits binding by the activated c-ErbA protein and also dominantly represses transcription by means of its inhibitory domain. Note the similarity to the action of the α-2 form of the c-ErbA protein, illustrated in Figure 4.30.

co-repressor (see Section 4.4.4), also abolishes its ability to transform cells (Perlmann and Vennstrom, 1995).

Hence the viral ErbA protein acts as a dominant repressor of thyroid hormone-responsive genes, being both incapable of activating transcription itself and able to prevent activation by intact receptor. This mechanism of action is clearly similar to the repression of thyroid hormone-responsive genes by the naturally occurring alternatively spliced form of the thyroid hormone receptor which, as discussed in Section 4.4.4, lacks the hormone-binding domain and therefore cannot bind hormone. Thus the same mechanism of gene repression by a non-hormone-binding receptor is used naturally in the cell and by an oncogenic virus.

One of the targets for repression by the viral ErbA protein is the erythrocyte anion transporter gene (Zenke *et al.*, 1988), which is one of the genes normally induced when avian erythroblasts differentiate into erythrocytes. This differentiation process has been known for some time to be inhibited by the ErbA protein and it is now clear that it achieves this effect by blocking the induction of the genes needed for differentiation. In turn, such inhibition will allow continued proliferation of these cells, rendering them susceptible to transformation into a tumour cell type by the product of the other AEV oncogene, v-*erbB*, which encodes a truncated form of the epidermal growth factor receptor (Downward *et al.*, 1984) and therefore renders cell growth independent of external growth factors (Figure 7.15).

The two cases of Fos/Jun and ErbA therefore represent contrasting examples of the involvement of transcription factors in oncogenesis, in terms of both the mechanism of transformation and the manner in which the cellular form of the oncogene becomes an active transforming gene. Thus, in the case of Fos and Jun, transformation is achieved by the continuous activation of genes necessary for growth in normal cell types. Moreover, it occurs, at least in part, via the natural activity of the cellular oncogene in inducing these genes being enhanced by their over-expression such that it occurs at an inappropriate time or place (Figure 7.16a). In contrast, in the ErbA case, transformation is achieved by inhibiting the expression of genes whose products are required for the differentiation of a particular cell type, therefore allowing growth to continue. Moreover, this occurs via the activity of a mutated form of the transcription factor which, rather than carrying out its normal function more efficiently, actually interferes with the normal role of the thyroid hormone receptor in inducing thyroid hormone-responsive genes required for differentiation (Figure 7.16b).

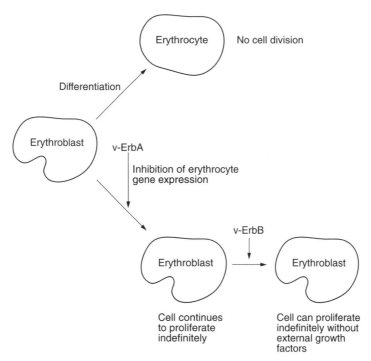

Figure 7.15 Inhibition of erythrocyte-specific gene expression by the v-ErbA protein prevents erythrocyte differentiation and allows transformation by the v-ErbB protein.

7.2.3 The *myc* oncogene

Interestingly, for a considerable period, the techniques of molecular biology failed in the case of the c-*myc* oncogene, which was one of the earliest cellular oncogenes to be identified, with its expression being dramatically increased in a wide variety of transformed cells (for reviews see Spencer and Groudine, 1991; Henriksson and Luscher, 1996). Thus the Myc protein has a number of properties suggesting that it is a transcription factor, notably nuclear localization, the possession of several motifs characteristic of transcription factors, such as the helix–loop–helix and leucine zipper elements (see Section 8.4), and the ability to activate target promoters in co-transfection assays. Despite exhaustive efforts, however, no DNA sequence to which the Myc protein binds could be defined, rendering its mechanism of action uncertain.

The solution to this problem was provided by the work of Blackwood and Eisenman (1991), who identified a novel protein, Max, which can form heterodimers with the Myc protein via the helix–loop–helix motif present in both proteins. Myc/Max heterodimers can bind to DNA and

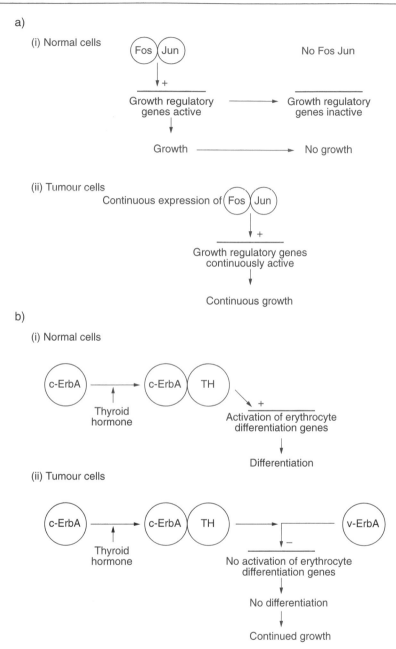

Figure 7.16 Transformation mechanisms of Fos/Jun (a) and ErbA (b). Note that Fos/Jun-induced transformation occurs because the proteins induce the continual activation of growth regulatory genes, which are normally expressed only transiently, whilst v-ErbA-induced transformation occurs because the protein interferes with the action of its cellular homologue and hence inhibits the induction of genes involved in erythrocyte differentiation.

regulate transcription, whereas Myc/Myc homodimers cannot do so (for reviews, see Amati and Land, 1994; Bernards, 1995) (Figure 7.17). This effect evidently parallels the requirement of the Fos protein for dimerization with Jun in order to bind with high affinity to AP1 sites (see Section 7.2.1).

The Max protein therefore plays a critical role in allowing the DNA binding of Myc. Moreover, the ability to interact with Max, bind to DNA and modulate gene expression is critical for the ability of the Myc protein to transform since mutations in Myc which abolish its ability to heterodimerize with Max also abolish its transforming ability. Hence, as was previously speculated, the Myc protein is a transcription factor whose over-expression causes transformation presumably via the activation of genes whose protein products are required for cellular growth (for reviews, see Zornig and Evan, 1996; Grandori and Eisenman, 1997).

Interestingly, the Max protein does not appear to represent a passive partner which merely serves to deliver Myc to the DNA of target genes. Rather, it plays a key role in regulating the activity of target genes containing the appropriate binding site. Thus, it has been shown that whereas Myc/Max heterodimers can activate transcription, Max/Max homodimers can bind to the same site and weakly repress transcription. Moreover, Max can also heterodimerize with another member of the helix–loop–helix family, known as Mad, to form a strong repressor of transcription (for a review, see Bernards, 1995) (Figure 7.18).

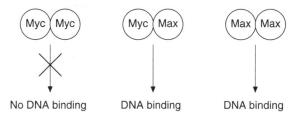

Figure 7.17 Both Myc/Max heterodimers and Max/Max homodimers can bind to DNA, whereas Myc/Myc homodimers cannot.

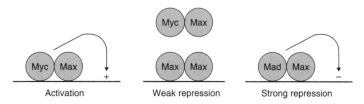

Figure 7.18 Functional effects of Max/Max homodimers and of Myc/Max or Mad/Max heterodimers. Note that Max/Max homodimers repress transcription only weakly by passively blocking activator binding, whereas Mad/Max heterodimers actively repress transcription and therefore have a much stronger effect.

The Max/Max homodimer appears to act as a weak repressor simply by preventing the Myc/Max activator from binding to its appropriate binding sites and thereby preventing it from activating transcription. In contrast, the Mad/Max heterodimer appears to act as an active repressor, which is capable of reducing transcription below that which would be observed in the absence of any activator binding (for discussion of the mechanisms of transcriptional repression, see Section 9.3). Thus, it has recently been shown that the Mad protein can bind the same complex of N-CoR, mSIN-3 and mRPD3, which mediates active repression by nuclear receptors such as the thyroid hormone receptor in the absence of hormone (see Section 4.4.4), (for a review, see Wolffc, 1997). As this complex includes the mRPD3 protein, which has histone deacetylase activity, it is possible that the Mad/Max heterodimer may repress transcription, at least in part, by recruiting a complex which acetylates histones, thereby organizing a more tightly packed chromatin structure (Figure 7.19).

In the case of the nuclear receptors, the switch from the repressed state of target genes to their activation is mediated by the addition of hormone. In contrast, however, in the case of the Myc family it is mediated by signals which produce a rise in Myc expression and a corresponding fall in the expression of Mad. Thus Myc is expressed at very low levels in growing cells and its expression is induced when cells begin to grow, whereas Max is expressed at similar high levels in both resting and proliferating cells, and Mad is expressed at high levels only in resting cells and not in proliferating cells. Hence, in resting cells, Mad and not Myc will be expressed and the expression of Myc-dependent genes will be repressed by Mad/Max homodimers. In contrast, expression will be activated by Myc/Max heterodimers as the cells receive signals to proliferate, resulting in increased Myc expression and decreased Mad expression (Figure 7.20). Clearly the over-expression of the Myc gene which is observed in many cancer cells would result in a similar production of activating Myc/Max heterodimers, leading to gene activation. Hence, as in the case of the Fos/Jun system, transformation by the

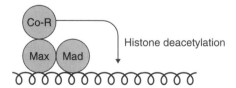

Figure 7.19 The Max/Mad heterodimer can recruit a co-repressor complex (Co-R) with histone deacetylase activity which can produce a more tightly packed chromatin structure (compare with Figure 4.29).

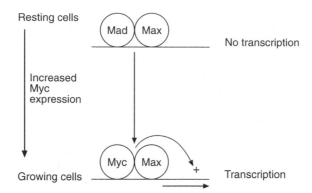

Resting cells

Mad Max — No transcription

Increased
Myc
expression

Growing cells

Myc Max

+ Transcription

Figure 7.20. In resting cells, Myc-dependent genes will be repressed by a Mad/Max heterodimer. As cells begin to grow, the expression of Myc increases, resulting in the formation of Myc/Max heterodimers which activate transcription.

Myc oncogene appears to depend primarily on its over-expression, resulting in the activation of genes required for cellular growth.

Interestingly, it has recently been shown that Myc can also interact with another transcription factor, Miz-1 (Myc-interacting zinc finger protein-1), which is a zinc finger protein (see chapter 8, section 8.3.1 for discussion of this type of protein). Unlike the situation with Max, however, in the absence of Myc, Miz-1 acts as an activator of genes promoting growth arrest. In the presence of Myc, however, this activity of Miz-1 is inhibited, resulting in the repression of these genes (Peukert *et al.*, 1997). Hence the rise in Myc levels in transformed cells stimulates the activity of growth-promoting genes via Myc/Max-mediated gene activation and represses growth inhibitory genes via a repression of Miz-1 activity.

7.2.4 Other oncogenic transcription factors

In view of the likely need for multiple different trancription factors to regulate genes involved in cellular growth processes, it is not surprising that several other genes encoding transcription factors have also been identified as oncogenes as well as playing a key role in gene expression in specific cell types. Thus, for example, the *myb* oncogene, and the *maf* oncogene, both of which were originally isolated from avian retroviruses, play key roles in gene regulation in monocytes and erythroid cells, respectively (for reviews, see Graf, 1992; Blank and Andrews, 1997; Motohashi *et al.*, 1997). Similarly, the *rel* oncogene of the avian retrovirus

Rev-T is a member of the NFκB family of transcription factors discussed in Section 5.2.2 (for a review, see Baeurale and Baltimore, 1996), whilst the *bcl-3* oncogene is a member of the IκB family, which interacts with the NFκB proteins (Bours *et al.*, 1993). Interestingly, the Bcl-3 factor illustrates another facet of the mechanisms by which transcription factor genes become oncogenic. Thus this factor was not identified as a retroviral oncogene but on the basis that it was located at the break point of chromosomal rearrangements which resulted in its translocation to a position adjacent to the immunoglobulin gene in some B-cell chronic leukemias. A number of other transcription factors have also been shown to be capable of causing cancer when translocated in this way. This can occur because their expression is increased owing to their being translocated to a highly expressed locus, such as the immunoglobulin gene loci in B cells or the T-cell receptor gene loci in T cells (Figure 7.21a). Alternatively it can occur because the translocation results in the production of a novel form of the transcription factor owing to its truncation or its linkage to another gene (encoding either another transcription factor or another class of protein) following the translocation (Figure 7.21b).

Factors translocated in these ways include both factors which were originally identified in oncogenic retroviruses and others which had not previously been shown to have oncogenic potential (for reviews, see Rabbits, 1994; Latchman, 1996; Look, 1997). Thus, for example, expression of the c-*myc* oncogene (Section 7.2.3) is dramatically increased by its translocation into the immunoglobulin heavy-chain locus which occurs in the human B-cell malignancy known as Burkitt's lymphoma (for a review, see Spencer and Groudine, 1991), whilst the Ets transcription factor gene discussed above (Section 7.2.1) is fused to the gene for the platelet-derived growth factor receptor to create a novel oncogenic fusion protein in patients with chronic myelomonocytic leukemia (for a review, see Sawyers and Denny, 1994). Similarly, expression of the homeobox gene Hox11 (see Section 6.2.4) is activated in cases of acute childhood T-cell leukemia, whilst the CBP co-activator (see Section 4.3.2) is fused to the MLL gene in acute myeloid leukemia (Sobulo *et al.*, 1997). Interestingly, the PBX factors, which are the mammalian homologues of the *Drosophila* extradenticle factor discussed in Section 6.2.3, were originally identified on the basis of the fact that the gene encoding PBX1 was found fused to the E2A gene (which encodes the E12 and E47 proteins discussed in Section 5.2.2) in a human leukaemia (for a review see Mann and Chan, 1996).

These findings provide further evidence that transcription factor genes are not only rendered oncogenic by transfer into a retrovirus but are also involved in the causation of human cancers playing a key role, for example, in the oncogenic effects of the chromosome translocations which are characteristic of specific cancers.

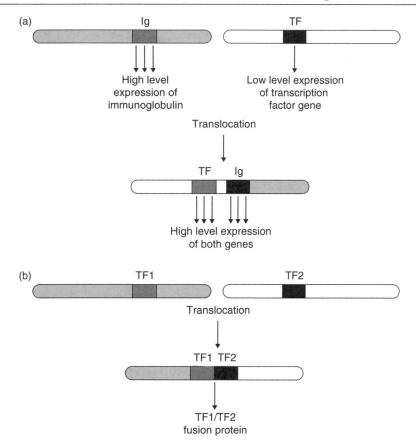

Figure 7.21 Chromosomal translocations can result in cancer when (a) the gene encoding a transcription factor is translocated next to a highly transcribed locus such as the immunoglobulin gene (Ig) and is therefore expressed at a high level, or (b) the translocation results in the fusion of the genes encoding two different transcription factors, resulting in a fusion protein with oncogenic properties.

7.3 ANTI-ONCOGENES AND CANCER

7.3.1 Nature of anti-oncogenes

As noted in Section 7.1, a number of genes exist whose normal function is to encode proteins that function in an opposite manner to those of oncogenes, acting to restrain cellular growth. The deletion or mutational inactivation of these anti-oncogenes (also known as tumour supressor genes) therefore results in cancer (for reviews, see Knudson, 1993; Fearon, 1997; Hunter, 1997) (Figure 7.22). This effect evidently parallels the production of cancer by the over-expression or mutational activation of cellular proto-oncogenes (compare Figures 7.4 and 7.22).

A number of anti-oncogenes of this type have been defined and several encode transcription factors. The two best characterized of these act by different mechanisms. Thus, p53 acts by binding to the DNA of its target genes and regulating their expression, whereas the retinoblastoma gene product (Rb-1) acts primarily via protein–protein interactions with other DNA-binding transcription factors. The p53 and Rb-1 proteins are therefore discussed in Sections 7.3.2 and 7.3.3 as examples of these two mechanisms of action. Other anti-oncogenes encoding transcription factors are discussed in Section 7.3.4.

7.3.2 p53

The gene encoding the 53-kDa protein known as p53 is mutated in a very wide variety of human tumours, especially carcinomas (for reviews, see Ko and Prives, 1996; Levine, 1997). In normal cells, expression of this protein is induced by agents which cause DNA damage, and its over-expression results in growth arrest of cells containing such damage or their death by the process of programmed cell death (apoptosis). Hence

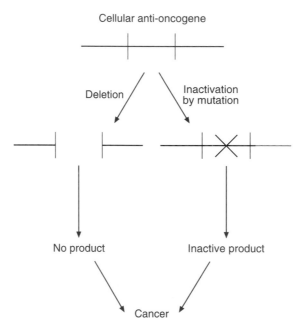

Figure 7.22 Cancer can result from the deletion of specific anti-oncogenes or their inactivation by mutation.

p53 has been called the 'guardian of the genome' (Lane, 1992), which allows cells to proliferate only if they have intact undamaged DNA. This would prevent the development of tumours containing cells with mutations in their DNA and the inactivation of the p53 gene by mutation would therefore result in an enhanced rate of tumour formation. In agreement with this idea, mice in which the p53 gene has been inactivated do not show any abnormalities in normal development but do exhibit a very high rate of tumour formation leading to early death (for a review see Berns, 1994).

The molecular analysis of the p53 gene product showed that it contains a DNA-binding domain and a region capable of activating transcription. The majority of the mutations in p53 which occur in human tumours are located in the DNA-binding domain (Friend, 1994; Anderson and Tegtmeyer, 1995). These mutations result in a failure of the mutant p53 protein to bind to DNA, indicating that this ability is crucial for the ability of the normal p53 protein to regulate cellular growth and suppress cancer.

The p53 protein therefore functions, at least in part, by activating the expression of genes whose protein products act to inhibit cellular growth (Figure 7.23a). The absence of functional p53 either due to gene deletion (Figure 7.23b) or to its inactivation by mutation (figure 7.23c) results in a failure to express these genes leading to uncontrolled growth.

In addition, functional p53 can also be prevented from activating gene transcription by interaction with the MDM2 oncoprotein (Figure 7.23d). Thus MDM2 masks the activation domain of p53, preventing it from activating transcription (Figure 7.24a). Moreover, MDM2, when bound to p53, also actively inhibits transcription by interacting with the basal transcriptional complex to reduce its activity (Thut *et al.*, 1997) (Figure 7.24b). This effect is supplemented by a further mechanism in which the interaction of MDM2 with p53 also results in the rapid degradation of p53 (Haupt *et al.*, 1997; for a review, see Lane and Hall, 1997) (Figure 7.24c). Thus several different mechanisms are involved in the inhibitory effect of MDM2 on p53 (Figure 7.24).

The inhibitory effect of MDM2 on p53 brought about by these multiple mechanisms is of particular importance in many human soft-tissue sarcomas where the p53 gene is intact and encodes wild-type p53, but the protein is functionally inactivated owing to the high levels of MDM2 resulting from amplification of the *mdm2* gene encoding it. Indeed, the major function of MDM2 even in normal cells may be to inhibit the action of p53 by interacting with it. Thus mice in which the *mdm2* gene is inactivated are non-viable but can be rendered viable by the additional inactivation of the p53 gene (de Oca Luna *et al.*, 1995).

A similar interaction of p53 with oncogenic proteins also occurs with the transforming proteins of several DNA viruses. Indeed, p53 was

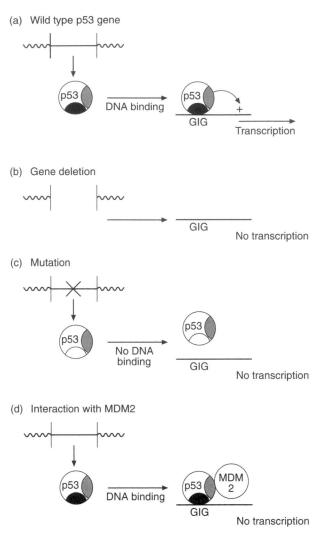

Figure 7.23 The ability of wild-type p53 to activate genes encoding growth inhibiting proteins (GIG) (a) can be abolished by deletion of the p53 gene (b), mutations in the DNA-binding domain (solid), which prevent it binding to DNA (c), or by the interaction of functional p53 with the MDM2 protein, which prevents it from activating transcription (d).

originally discovered as a protein which interacted with the large T oncoprotein of the DNA tumour virus SV40. The functional inactivation of p53 produced by this interaction appears to play a critical role in the ability of these DNA viruses to transform cells paralleling the similar action of MDM2. These interactions suggest that functional antagonism between oncogene and anti-oncogene products is likely to be critical for

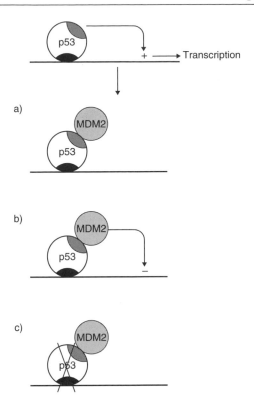

Figure 7.24 Multiple mechanisms by which MDM2 inhibits p53 involving (a) masking of its activation domain (shaded); (b) direct inhibition of transcription by MDM2 itself and (c) targeting p53 for degradation by proteolytic enzymes.

the control of cellular growth with changes in this balance which activate oncogenes or inactivate anti-oncogenes resulting in cancer.

These considerations evidently focus attention on the genes which are activated by p53. One such gene is that encoding a 21-kDa protein (p21) which acts as an inhibitor of cyclin-dependent kinases (for a review, of p53-dependent genes, see Ko and Prives (1996). As the cyclin-dependent kinases are enzymes that stimulate cells to enter cell division, the finding that p53 stimulates the expression of an inhibitor of these enzymes is entirely consistent with its role in restraining growth, since the inhibition of the cyclin-dependent kinases will prevent cells replicating their DNA and undergoing cell division (Figure 7.25).

Interestingly, p53 also stimulates expression of the *mdm2* gene, whose protein product interferes with the activity of p53 as described above. This effect is likely to be part of a negative feedback loop in which p53, having fulfilled its function, activates *mdm2* expression, resulting in p53

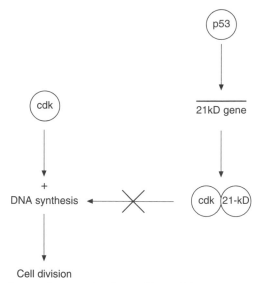

Figure 7.25 p53 activates the gene for the 21 kDa inhibitor of cyclin-dependent kinases (cdk). This inhibitor then prevents the cyclin-dependent kinases from stimulating DNA synthesis and consequent cell division.

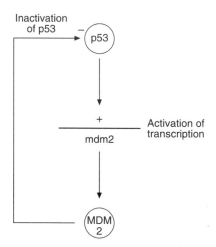

Figure 7.26 p53 activates the gene for the MDM2 protein, which acts in a negative feedback loop to inactivate p53 by the mechanisms illustrated in Figure 7.24.

inactivation (Figure 7.26). This would allow, for example, cells which had repaired the damage to their DNA to inactivate p53 and resume cell division. Similarly, p53 also stimulates the expression of the *bax* gene, whose protein product stimulates programmed cell death or apoptosis, allowing p53 to promote the death of cells whose damaged DNA is irreparable. Recent studies have also identified several genes involved in the generation of toxic reactive oxygen species whose expression is induced by p53 during this process, indicating that p53 may also promote apoptosis by inducing the production of these species (Wyllie, 1997). As with AP1 (Section 7.2.1) transcriptional activation by p53 requires the CBP co-activator or the closely related p300 protein (Avantaggiati *et al.*, 1997). Hence as with AP1 and the steroid receptors (see Section 4.4.4), AP1 and p53 can compete for CBP/p300, resulting in antagonism between the oncogenic activity of AP1 and the anti-oncogenic activity of p53.

Hence the p53 gene product plays a key role in regulating cellular growth by binding to DNA and activating the expression of specific genes (for a review, see Almog and Rotler, 1997). Its inactivation by mutation or by interaction with oncogene products is likely to play a critical role in most human cancers. Interestingly, a novel p53-related protein, which is also a transcription factor and is known as p73, has recently been described (Clurman and Groudine, 1997; Oren, 1997), whilst a factor known as p33, which appears to be required for activation of gene expression by p53, has also been reported (Garkavtsev *et al.*, 1998; Oren, 1998). Both of these factors are likely to play a role in controlling cellular growth in specific cell types and it has been suggested that inactivation of p73 and/or p33 is likely to be involved in the development of human neuroblastomas which contain normal p53 protein. These findings further emphasize the importance of p53, the proteins related to it and the factors with which it interacts in the regulation of cellular proliferation and the development of cancer.

7.3.3 The Retinoblastoma protein

The Retinoblastoma gene (Rb-1) was the first anti-oncogene to be defined and is so named because its inactivation in humans results in the formation of eye tumours known as retinoblastomas (for reviews, see Bartek *et al.*, 1996; Herwig and Strauss, 1997). Like p53, the Rb-1 gene product is a transcription factor which exerts its anti-oncogenic effect by modulating the expression of specific target genes. In contrast to p53, however, it exerts this effect via protein–protein interactions with other transcription factors rather than by direct DNA binding.

One of the major targets for Rb-1 is the transcription factor E2F, which plays a critical role in stimulating the expression of genes encoding growth promoting proteins such as Myc (Section 7.2.3), DNA polymerase α and thymidine kinase. The association of Rb-1 and E2F does not inhibit the DNA binding of E2F but prevents it from stimulating the transcription of these growth-promoting genes (Figure 7.27a). Moreover,

a) Resting cells

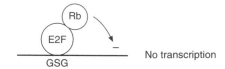

No transcription

b) Dividing cells

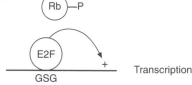

Transcription

c) Deletion or inactivation of Rb

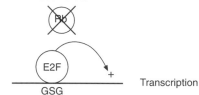

Transcription

d) Interaction with tumour virus oncogene

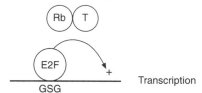

Transcription

Figure 7.27 In resting cells, the Rb-1 protein binds to E2F and prevents it activating the transcription of genes encoding growth-stimulating proteins (GSG), as well as directly inhibiting transcription of these genes by inhibiting the activity of other positively acting factors (a). In normal dividing cells, the Rb-1 protein is phosphorylated at the G1/S transition in the cell cycle, which prevents it from interacting with E2F and hence allows E2F to activate transcription (b). This release of E2F can also occur in tumour cells where the Rb-1 gene is deleted or inactivated by mutation (c), or following the interaction of Rb-1 with tumour virus oncogenes (T) (d).

Rb-1 does not simply act as a passive repressor by preventing E2F from stimulating transcription. Rather, when bound to E2F, it also interacts with other positively acting transcription factors such as c-Myc (see Section 7.2.3) bound at adjacent DNA-binding sites and inhibits their ability to stimulate transcription (Weintraub *et al.*, 1995). Hence Rb-1 exerts its anti-oncogenic effect by inhibiting the transcription of growth-promoting genes rather than, as with p53, promoting the transcription of growth-inhibiting genes.

In normal dividing cells, this interaction of Rb-1 and E2F is inhibited as cells move from G1 to S phase in the cell cycle. This effect is dependent on the phosphorylation of Rb-1, which prevents it interacting with E2F (Figure 7.27b). Hence the controlled growth of normal cells can be regulated by the regulated phosphorylation of Rb-1, which in turn regulates its ability to interact with E2F and modulate its activity.

Interestingly, the phosphorylation of Rb-1 in the cell cycle is carried out by the cyclin-dependent kinases (for a review, see Sherr, 1994). This provides a link between p53 and the regulation of Rb-1 activity since, as noted above (Section 7.3.2), p53 activates the gene encoding the p21 protein which inhibits cyclin-dependent kinases and would thus prevent the phosphorylation of Rb-1 and cell cycle progression (Figure 7.28). To add to the complexity still further, it appears that the activity of both p53 itself and E2F is also altered following phosphorylation by cyclin-dependent kinases, indicating that a complex network of interacting transcription factors, kinases and their inhibitors regulates cellular growth (for a review, see Dynlacht, 1997).

Clearly, abolishing the activity of Rb-1, either by deletion of its gene or by mutation, will result in the unregulated activity of E2F, leading to the uncontrolled growth that is characteristic of cancer cells (Figure 7.27c). Interestingly, the inactivation of Rb-1 can also be achieved by the transforming proteins of DNA tumour viruses, such as SV40 or

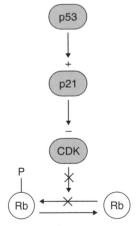

Figure 7.28 By activating the p21 gene, whose protein product inhibits cyclin-dependent kinases (CDK), p53 produces a fall in CDK activity, which results in more Rb being in the growth inhibitory unphosphorylated form.

adenovirus. These proteins bind to the Rb-1 protein, resulting in the dissociation of the Rb-1/E2F complex releasing free E2F, which can activate gene expression (Figure 7.27d).

Although E2F is a major target of Rb-1, there are also other factors with which Rb-1 interacts. Thus Rb-1 has recently been shown to inactivate the UBF factor which plays a critical role in transcription of the ribosomal RNA genes by RNA polymerase I (see Chapter three, Section 3.2.2). Due to the need for these ribosomal RNAs for the effective functioning of the ribosomes, the inactivation of UBF by Rb-1 will lead to a decrease in the levels of total protein synthesis which would in turn lead to the arrest of cell growth. In agreement with this idea, the inactivation of UBF by Rb-1 appears to play a critical role in the growth arrest and associated differentiation of U937 monocytic cells (for a review, see Dynlacht, 1995).

Obviously, the 5S ribosomal RNA and the transfer RNAs, which are produced by RNA polymerase III, are also necessary for ribosomal function and protein synthesis. Indeed, it has recently been shown that Rb can also inhibit RNA polymerase III transcription by interacting directly with the polymerase III transcription factor TFIIIB (see Section 3.2.3) and inhibiting its activity (Larminie *et al.*, 1997). Hence Rb-1 can directly inhibit transcription of genes involved in cellular growth both by inhibiting the transcription of E2F-dependent genes by RNA polymerase II and the transcription of all the genes transcribed by RNA polymerase I and III. It therefore has a remarkable ability to modulate transcription by all three RNA polymerases (for a review, see White, 1997) (Figure 7.29), and is likely to play a critical role not only in preventing cancer but also in normal cells by promoting the growth arrest which is necessary for terminal differentiation (Figure 7.27). In agreement with this idea, mice in which the Rb-1 gene has been inactivated die before birth and show gross defects in cellular differentiation (Lee *et al.*, 1992). This indicates that Rb-1 plays a key role in normal development as well as

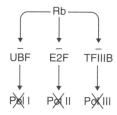

Figure 7.29 Rb can inhibit all transcription by RNA polymerases I and III by inhibiting the activity of UBF and TFIIIB, as well as inhibiting the activity of E2F and hence inhibiting the ability of RNA polymerase II to transcribe genes whose protein products stimulate growth.

acting as an anti-oncogene and contrasts with the viability of mice in which the p53 gene has been inactivated (see Section 7.3.2).

Hence the Rb protein plays a key role in regulating cellular growth and differentiation by interacting with transcription factors involved in transcription by RNA polymerases I, II and III. Its inactivation, either by mutation or by specific oncogenes, therefore, results in uncontrolled proliferation and cancer. When taken together with the similar role of p53 in growth regulation and as a target for oncogenes, this suggests that anti-oncogenes are likely to play a key role in regulating cellular growth, which is likely to be controlled by the balance between the antagonistic effects of oncogene and anti-oncogene products.

7.3.4 Other anti-oncogenic transcription factors

Normally, anti-oncogenes are identified on the basis of their inactivation in specific human cancers and their functional role is subsequently characterized. For some time, only three anti-oncogene products were known to be transcription factors, namely the p53 and Rb-1 proteins discussed in previous sections and the Wilm's tumour gene product (for a review, see Hastie, 1994).

More recently, however, other oncogene products have also been implicated in transcriptional control. Thus, whilst the BRCA-1 and BRCA-2 anti-oncogenes, which are mutated in many cases of familial breast cancer, appear to function primarily in controlling the repair of damaged DNA, there is also evidence that they may influence transcription. For example, both BRCA-1 and BRCA-2 contain regions which can act as activation domains and stimulate transcription (for a review, see Marx, 1997) (for discussion of such domains, Section 9.2). Moreover, BRCA-1 appears to be a component of the RNA polymerase holoenzyme which also contains RNA polymerase II and basal transcription factors (see Section 3.2.4) again suggesting that this factor is involved in transcriptional control (Scully *et al.*, 1997).

In contrast to these features suggesting that BRCA-1 can influence transcription rates within the nucleus, the adenomatous polyposis coli (APC) anti-oncogene, which is mutated in most human colon tumours (for a review, see Moon and Miller, 1997), appears to influence transcription indirectly. Thus APC acts within the cytoplasm by interacting with a protein known as β-catenin, which is involved both in cell adhesion and also acts as a transcription factor (for a review, see Peifer, 1997). This interaction between APC and β-catenin results in the rapid degradation of β-catenin within the cytoplasm, preventing it from entering the nucleus and influencing transcription (Figure 7.30a).

In normal cells, specific secreted proteins known as WNT proteins (or

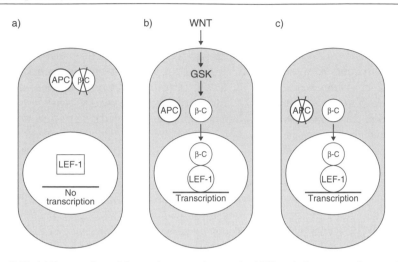

Figure 7.30 (a) Interaction of the anti-oncogenic protein APC and the oncogenic protein β-catenin (B−C) resulting in degradation of β-catenin. (b) Following activation of glycogen synthase kinase (GSK) by WNT proteins, β-catenin is stablized. It then moves to the nucleus and interacts with the LEF-1 transcription factor, promoting its ability to stimulate transcription. (c) In cancer, the APC factor is inactivated resulting in the constitutive activation of β-catenin.

wingless proteins after the first member of the family which was discovered in *Drosophila*) activate a kinase enzyme glycogen synthase kinase. This kinase phosphorylates and thereby stabilizes β-catenin, preventing it from being degraded (for reviews see Hunter, 1997; Nusse, 1997; Dale, 1998). The β-catenin then moves to the nucleus and interacts with the LEF-1 transcription factor discussed in Section 1.3 and stimulates its ability to activate transcription (Figure 7.30b).

In a normal situation, therefore, this ability of β-catenin to interact with LEF-1 and stimulate its activity is tightly regulated by the presence or absence of WNT proteins. Any change which causes this pathway to become constitutively active results in cancer. For example, if the APC gene is mutated so that APC cannot inactivate β-catenin, cancer will result from the constitutive activation of β-catenin (Figure 7.30c). Hence APC acts as an anti-oncogene whose inactivation by mutation causes cancer.

As well as illustrating how an anti-oncogene can act indirectly to influence transcription, this example also illustrates how oncogene products interact with one another. Clearly, mutations in the β-catenin gene that enhance β-catenin stability or mutations in the WNT genes that result in their over-expression will also cause cancer, and hence the genes encoding β-catenin or the WNT proteins are oncogenes whose products act in the same pathway as the APC anti-oncogene product.

Like the majority of transcription factors, the anti-oncogenic proteins discussed so far act by directly or indirectly altering the rate at which transcription is initiated by RNA polymerase. This is not the case, however, for the von Hippel–Lindau anti-oncogene protein which is mutated in multiple forms of cancer. Thus, the normal role of this factor is to modulate the rate at which transcriptional elongation occurs to produce a full-length transcript (for a review, see Krumm and Groudine, 1995).

As with transcriptional initiation, specific proteins are required for transcriptional elongation (for reviews of transcriptional elongation, see Reines *et al.*, 1996; Uptain *et al.*, 1997). One of these factors, elongin, consists of three subunits, elongin A, B and C, with elongin A being the catalytic subunit which is necessary for transcriptional elongation. The von Hippel–Lindau protein binds to elongin B and C and prevents binding of elongin A. This results in an inactive complex which does not stimulate elongation (Figure 7.31).

Interestingly, the mutant forms of the von Hippel–Lindau protein found in human tumours do not interact with elongin B/C, indicating that the anti-oncogenic action of the protein is mediated by its effect on transcriptional elongation. This is likely to be because several oncogenes, such as c-*fos* and c-*myc*, are regulated at the stage of transcriptional elongation with many of the RNA transcripts which are initiated not being elongated to produce a full-length functional mRNA (for further discussion of transcriptional elongation and its regulation, see Section

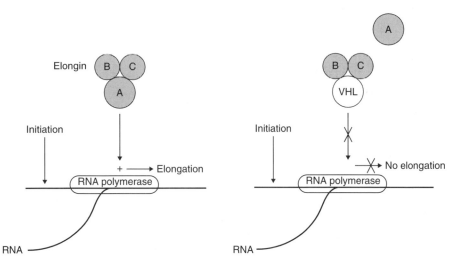

Figure 7.31 The von Hippel–Lindau gene product (VHL) interacts with the B and C subunits of elongin and prevents the catalytic A subunit from binding, thereby preventing stimulation of transcriptional elongation by the intact elongin complex.

9.2.5). It is possible therefore that in the absence of the von Hippel–Lindau protein, too much full-length mRNA is produced, resulting in over-production of the corresponding oncogenic proteins.

Interestingly, the gene encoding another transcriptional elongation factor ELL is found at the break point of chromosomal translocations in several leukaemias (see Section 7.2.4), indicating that it can be oncogenic under certain circumstances (for a review, see Reines *et al.*, 1996). The existence of both oncogenic and anti-oncogenic transcription factors which modulate transcriptional elongation indicates that this is an important target for processes which regulate normal cellular growth and hence for malregulation in cancer (for a review, see Li and Green, 1996).

More generally, the examples given in this section add considerable variety to the three 'classical' anti-oncogenes encoding transcription factors (p53, Rb-1 and the Wilm's tumour gene) and indicate the key role of such gene products in different forms of transcriptional regulation in normal cells and in cancer.

7.4 CONCLUSIONS

The ability to affect cellular transcriptional regulatory processes is crucial to the ability of many different viruses to transform cells. Thus, for example, the large T oncogenes of the small DNA tumour viruses SV40 and polyoma and the Ela protein of adenovirus can all affect cellular gene expression, and this ability is essential for the transforming ability of these viruses (for a review, see Moran, 1993).

In this chapter we have seen that several RNA viruses also have this ability, containing transcription factors which can act as oncogenes either by promoting the expression of genes required for growth or by inhibiting the expression of genes required for the production of non-proliferating differentiated cells.

Although the oncogenes of both DNA and some RNA tumour viruses can therefore affect transcription, their origins are completely different. Thus, whilst the oncogenes of the DNA viruses do not have equivalents in cellular DNA and appear to have evolved within the viral genome, the oncogenes of retroviruses have, as we have seen, been picked up from the cellular genome. The fact that, despite their diverse origins, both types of oncogenes can affect transcription, indicates therefore that the modulation of transcription represents an effective mechanism for the transformation of cells.

In addition, however, the origin of retroviral oncogenes from the cellular genome allows several other features of transcription to be studied. Thus, for example, the conversion of a normal cellular transcription factor into a cancer-causing viral oncogene allows insights to be obtained into the processes whereby oncogenes become activated.

In general, such oncogenes, whether they encode growth factors, growth factor receptors or other types of protein can be activated within a virus either by over-expression driven by a strong retroviral promoter or by mutation. The transcription factors we have discussed in this chapter illustrate both these processes. Thus the Fos, Jun and Myc oncogenes, for example, become cancer-causing both by continuous expression of proteins which are normally made only transiently, leading to constitutive stimulation of genes required for growth as well as in some cases by mutations in the viral forms of the protein which render them more potent transcriptional activators. Similarly, the *erbA* oncogene is activated by deletion of a part of the protein coding region leading to a protein with different or enhanced properties. Although such effects of mutation or over-expression have initially been defined in tumorigenic retroviruses, it is likely such changes can also occur within the cellular genome, over-expression of the c-*myc* oncogene, for example, being characteristic of many different human tumours (for a review, see Spencer and Groudine, 1991), whilst several other transcription factor genes are activated by the translocations characteristic of particular human leukemias (see section 7.2.4).

In addition, since cellular oncogenes clearly also play an important role in the regulation of normal cellular growth and differentiation, their identification via tumorigenic retroviruses has, paradoxically, greatly aided the study of normal cellular growth regulatory processes. Thus, for example, the prior isolation of the c-*fos* and c-*jun* genes greatly aided the characterization of the AP1-binding activity and of its role in stimulating genes involved in cellular growth.

A similar boost to our understanding of growth regulation in normal cells has also emerged from studies of the anti-oncogene proteins. Thus studies on the Rb-1 gene, which was orginally identified on the basis of its inactivation in retinoblastomas, have led to an understanding of its key role in regulating the balance between cellular growth and differentiation. Similarly, work on p53, which was originally identified as a protein interacting with the product of the SV40 large T oncogene, has led to the identification of its key role as the so-called 'guardian of the genome'.

The interaction of p53 and SV40 large T indicates another aspect of anti-oncogenes, namely their antagonistic interaction with oncogene products. Thus both p53 and Rb-1 have been shown to bind cellular and viral oncogene proteins, with the activity of the anti-oncogene product being inhibited by this interaction. Such interactions are not confined to the oncogenes and anti-oncogenes which encode factors directly regulating transcription. Thus, as discussed in Section 7.3.4, the APC anti-oncogene protein does not act directly as a transcription factor. Rather, it interacts with the β-catenin oncogene product to promote its degradation. Hence cancer can result from mutations in the β-catenin oncogene, which enhance the stability of its protein product and therefore its ability to stimulate transcription, or from mutations in the APC protein, which inactivate it and prevent it interfering with the function of β-catenin. Hence the interaction between oncogene and anti-oncogene products is likely to play a key role in regulating cellular growth. The uncontrolled growth characteristic of cancer cells therefore results from changes in this balance due either to over-expression or mutational activation of oncogenes or to deletion or mutational inactivation of anti-oncogenes.

It is clear therefore that, as with other oncogenes and anti-oncogenes, the study of the oncogenes and anti-oncogenes which encode transcription factors can provide considerable information on both the processes regulating normal growth and differentiation, and on how these processes are altered in cancer. When taken together with the involvement of transcription factor mutations in disorders of development or hormone responses discussed in Section 7.1, they illustrate the key role played by transcription factors and the manner in which alterations in their activity can result in disease.

REFERENCES

Almog, N. and Rotter, V. (1997). Involvement of p53 in differentiation and development. *Biochemica et Biophysica Acta* **1333**, F1–F27.

Amati, B. and Land, H. (1994). Myc-Max-Mad: a transcription factor network controlling differentiation and death. *Current Opinion in Genetics and Development* **4**, 102–108.

Anderson, M.E. and Tegtmeyer, P. (1995). Giant leap for p53, small step for drug design. *Bioessays* **17**, 3–7.

Avantaggiati, M.L., Ogryzko, V., Gardner, K., Giordano, A., Levine, A.S. and Kelly, K. (1997). Recruitment of p300/CBP in p53 dependent signalling pathway. *Cell* **89**, 1175–1184.

Baeurale, P.A. and Baltimore, D. (1996). NFκB: Ten years after. *Cell* **87**, 13–20.

Baichwal, U.R. and Tjian, R. (1990). Control of c-Jun activity by interaction of a cell-specific inhibitor with regulatory domain delta: differences between v- and c-Jun. *Cell* **63**, 815–825.

Bartek, J., Bartkova, J. and Lukas, J. (1996). The retinoblastoma protein pathway and the restriction point. *Current Opinion in Cell Biology* **8**, 805–814.

Bernards, R. (1995). Flipping the myc switch. *Current Biology* **5**, 859–861.

Berns, A. (1994). Is p53 the only real tumour suppressor gene? *Current Biology* **4**, 137–139.

Blackwood, E.M. and Eisenman, R.N. (1991). Max: a helix–loop–helix zipper protein that forms a sequence–specific–DNA binding complex with myc. *Science* **251**, 1211–1217.

Blank, V. and Andrews, N.C. (1997). The Maf transcription factors: regulators of differentiation. *Trends in Biochemical Sciences* **22**, 437–441.

Boncinelli, E. (1997). Homeobox genes and disease. *Current Opinion in Genetics and Development* **7**, 331–337.

Bourne, H.R. and Varmus, H.E. (eds) (1992). Oncogenes and cell proliferation. *Current Opinion in Genetics and Development* **2**, 1–57.

Bours, V., Franzoso, G., Azarenko, V., Parks Karno, T., Brown, K. and Siebenlist, U. (1993). The oncoprotein Bcl-3 directly transactivates through κB motifs via association with DNA-binding p50B homo-dimers. *Cell* **72**, 729–739.

Broach, J.R. and Levine, A.J. (1997). Oncogenes and cell proliferation. *Current Opinion in Genetics and Development* **7**, 1–6.

Clurman, B. and Groudine, M. (1997). Killer in search of a motive? *Nature* **389**, 122–123.

Dale, T.C. (1998). Signal transduction by the Wnt family of ligands. *Biochemical Journal* **329**, 209–223.

D'Arcangelo, G. and Curran, T. (1995). Smart transcription factors. *Nature* **389**, 149–152.

de OcaLuna, R.M., Wagner, D.S. and Lozano, G. (1995). Rescue of early embryonic lethality in mdm2-deficent mice by deletion of p53. *Nature* **378**, 203–206.

Downward, J., Yarden, Y., Mayes, E., Scrace, G., Totty, N., Stockwell, P., Ullrich, A., Schlessinger, J. and Watefield, M.D. (1984). Close similarity of epidermal growth factor receptor and v-*erb-B* oncogene protein sequences. *Nature* **307**, 521–527.

Dynlacht, B.D. (1995). PolI gets repressed. *Nature* **374**, 114.

Dynlacht, B.D. (1997). Regulation of transcription by proteins that control the cell cycle. *Nature* **389**, 149–152.

Engelkamp, D. and van Heyningen, V. (1996). Transcription factors in disease. *Current Opinion in Genetics and Development* **6**, 334–342.

Fearon, E.R. (1997). Human cancer syndromes: clues to the origin and nature of cancers. *Science* **278**, 1043–1050.

Friend, S. (1994) p53: a glimpse at the puppet behind the shadow play. *Science* **265**, 334–335.

Garkavtsev, I., Grigorian, I.A., Ossovskaya, V.S., Chernov, M.V., Chuma-kov, P.M. and Gudnov, A.V. (1998). The candidate tumour supressor gene p33^{ING1} cooperates with p53 in cell growth control. *Nature*, **391**, 295–298.

Gibbons, R.J., Picketss, D.S., Villard, L. and Higgs, D.R. (1995). Mutations in a putative global regulator cause X-linked mental retardation with a-Thalassemia (ATR-X) syndrome. *Cell* **80**, 837–845.

Graf, T. (1992). Myb, a transcriptional activator linking proliferation and differentiation in haematopoetic cells. *Current Opinion in Genetics and Development* **2**, 249–255.

Grandori, C. and Eisenman, R.N. (1997). Myc target genes. *Trends in Biochemical Sciences* **22**, 177–181.

Hastie, N.D. (1994). The genetics of Wilms' tumour, a case of disrupted development. *Annual Review of Genetics* **28**, 523–558.

Haupt, Y., Maya, R., Kazaz, A. and Oren, M. (1997). Mdm2 promotes rapid degradation of p53. *Nature* **387**, 296–299.

Henriksson, M. and Luscher, B. (1996). Proteins of the Myc network: essential regulators of cell growth and differentiation. *Advances in Cancer Research* **68**, 109–182.

Herwig, S. and Strauss, M. (1997). The retinoblastoma protein, a master regulator of cell cycle differentiation and apoptosis. *European Journal of Biochemistry* **246**, 581–601.

Hunter, T. (1997) Oncoprotein networks. *Cell* **88**, 333–346.

Johnson, R.S., Van Lingen, B., Papaivannou, V.E. and Speigelmann, B.M. (1993). A null mutation at the c-*jun* locus causes embryonic lethality and retarded cell growth in culture. *Genes and Development* **7**, 1309–1317.

Karin, M., Liu, Z.G. and Zandi, E. (1997). AP-1 function and regulation. *Current Opinion in Cell Biology* **9**, 240–246.

Kerppola, T. and Curran, T. (1995). Zen and the art of Fos and Jun. *Nature* **373**, 199–200.

Knudson, A.G. (1993). Anti-oncogenes and human cancer. *Proceedings of the National Academy of Sciences USA* **90**, 10914–10921.

Ko, L.J. and Prives, C. (1996) p53: puzzle and paradigm. *Genes and Development* **10**, 1054–1072.

Krumm, A. and Groudine, M. (1995). Tumor suppression and transcription elongation: the dire consequences of changing partners. *Science* **269**, 1400–1401.

Lamph, W.W., Wamsley, P., Sassone-Corsi, P. and Verma, I.M. (1988). Induction of proto-oncogene Jun/AP-1 by serum and TPA. *Nature* **334**, 629–631.

Land, H., Parada, L.F. and Weinberg, R. (1983). Cellular oncogenes and multistep carcinogenesis. *Science* **222**, 771–778.

Lane, D.P. (1992). p53, guardian of the genome. *Nature* **358**, 15–16.

Lane, D.P. and Hall, P.A. (1997). MDM2 – arbitor of p53's destruction. *Trends in Biochemical Sciences* **22**, 373–374.

Larmine, C.G.C., Cairns, C.A., Mital, R., Martin, K., Kouzarides, T., Jackson, S.P. and White, R.J. (1997). Mechanistic analysis of RNA polymerase III regulation by the retinoblastoma protein. *EMBO Journal* **16**, 2061–2071.

Latchman, D.S. (1996). Transcription factor mutations and human diseases. *New England Journal of Medicine* **334**, 28–33.

Lee, E.Y.H.P., Chang, C.Y., Hu, N., Wang, Y.C., Lai, C.C., Herrup, K., Lee, W.H. and Bradley, A. (1992). Mice Deficient for Rb are non viable and show defects in neurogenesis and haematopoesis. *Nature*, **359**, 289–294.

Levine, A.J., (1997). p53, the cellular gatekeeper for growth and division. *Cell* **88**, 323–331.

Li, L., Chambard, J.C., Karin, M. and Olson, E.M. (1992). Fos and Jun repress transcriptional activation by myogenin and MyoD: the amino terminus of Jun can mediate repression. *Genes and Development* **6**, 676–689.

Li, X.Y. and Green, M.R. (1996). Transcriptional elongation and cancer. *Current Biology* **6**, 943–944.

Look, A.T. (1997). Oncogenic transcriptin factors in the human acute leukaemias. *Science* **278**, 1059–1064.

Mann, R.S. and Chan, S.K. (1996). Extra specificity from extradenticle: the partnership between Hox and PBX/EXD homeodomain proteins. *Trends in Genetics* **12**, 258–262.

Marx, J. (1997). Possible function found for breast cancer genes. *Science* **276**, 531–532.

Moon, R.T. and Miller, J.R. (1997). The APC tumour supressor in development and cancer. *Trends in Genetics* **13**, 256–258.

Moran, E. (1993). DNA tumour virus transforming proteins and the cell cycle. *Current Opinion in Genetics and Development*, **3**, 63–70.

Motohashi, H., Shavit, J.A., Igarashi, K., Yamamoto, K. and Engel, J.D. (1997). The world according to MAF. *Nucleic Acids Research* **25**, 2953–2959.

Nusse, R. (1997). A versatile transcriptional effector of wingless signalling. *Cell* **89**, 321–323.

Oren, M. (1997). Lonely no more, p53 finds its kin in a tumour supressor haven. *Cell* **90**, 829–832.

Oren, M. (1998). Teaming up to restrain cancer. *Nature* **391**, 233–234.

Peifer, M. (1997). β-Caterin as oncogene: the smoking gun. *Science* **275**, 1752–1753.

Perlmann, T. and Vennstrom, B. (1995). Nuclear receptors: the sound of silence. *Nature* **377**. 387–388.

Peukert, K., Staller, P., Schneider, A., Carmichael, G., Hanel, F. and Eilers, M. (1997). An alternative pathway for gene regulation by Myc. *EMBO Journal* **16**, 5672–5686.

Rabbits, T.H. (1994). Chromosomal translocations in human cancer. *Nature* **372**, 143–149.

Ransone, L.J. and Verma, I.M. (1990). Nuclear proto-oncogenes. Fos and Jun. *Annual Review of Cell Biology* **6**, 531–557.

Rauscher, F.J., Cohen, D.R., Curran, T., Bos, T.J., Bogt, P.K., Bohmann, D., Tjian, R. and Franza, B.R. (1988). Fos-associated protein p39 is the product of the c-*jun* oncogene. *Science* **240**, 1010–1016.

Reines, D., Conaway, J.W. and Conaway, R.C. (1996). The RNA polymerase II general elongation factors. *Trends in Biochemical Sciences* **21**, 351–355.

Sap, J., Munoz, A., Damm, K., Goldberg, Y., Ghysdael, J., Leutz, A., Beug, H. and Vennstrom, B. (1986). The c-*erb-A* protein is a high-affinity receptor for thyroid hormone. *Nature* **324**, 635–640.

Sap, J., Munoz, A., Schmitt, A., Stunnenberg, H. and Vennstrom, B. (1989). Repression of transcription at a thyroid hormone response element by the v-*erbA* oncogene product. *Nature* **340**, 242–244.

Sawyers, C.L. and Denny, C.T. (1994). Chronic myelomonocytic leukemia. Tel-a-kinase what ets all about. *Cell* **77**, 171–173.

Scully, R., Anderson, S.F., Chao, D.M., Wei, W., Ye, L., Young, R.A., Livingston, D.M. and Parvin, J.D. (1997). BRCA1 is a component of the RNA polymerase II holoenyzme. *Proceedings of the National Academy of Sciences USA* **94**, 5605–5610.

Sherr, C.J. (1994). G1 phase progression: cycling on cue. *Cell* **79**, 551–555.

Sobulo, O.M., Borrow, J., Tomek, R., Reshmi, S., Harden, A., Schlegelberger, B., Houseman, D., Doggett, N.A., Rowley, J.D. and Zeleznik-Le, N.J. (1997). MLL is fused to CBP a histone acetyltransferase, in therapy-related acute myeloid leukaemia with a t (11; 16) q23; p13.3). *Proceedings of the National Academy of Sciences USA* **94**, 8732–8737.

Spencer, C.A. and Groudine, M. (1991). Control of c-*myc* regulation in normal and neoplastic cells. *Advances in Cancer Research* **56**, 1–48.

Thut, C.J., Goodrich, J.A. and Tjian, R. (1997). Repression of p53 mediated transcription by MDM2: a dual mechanism. *Genes and Development* **11**, 1974–1986.

Treier, M., Staszewski, L.M. and Bohmann, D. (1994). Ubiquitin-dependent c-Jun degradation *in vivo* is mediated by the δ domain. *Cell* **78**, 787–798.

Uptain, S.M., Kane, C.M. and Chamberlin, M.J. (1997). Basic mechanisms of transcriptional elongation and its regulation. *Annual Review of Biochemistry* **66**, 117–172.

Wasylyk, B., Wasylyk, C., Flores, P., Begue, A., Leprince, D. and Stehelin, D. (1990). The c-*ets* proto-oncogenes encode transcription factors that copopperate with c-Fos and c-Jun for transcriptional activation. *Nature* **346**, 191–193.

Weinberg, R.A. (1993). Tumour suppressor genes. *Neuron* **11**, 191–196.

Weinberger, C., Thompson, C.C., Ong, E.S., Lebo, R., Gruol, D.J. and Evans, R.M. (1986). The c-*erb*-A gene encodes a thyroid hormone receptor. *Nature* **324**, 641–646.

Weintraub, S.J., Chow, K.N.B., Luo, R.X., Zhang, S.H., He, S. and Dean, D.C. (1995). Mechanism of active transcriptional repression by the retinoblastoma protein. *Nature* **375**, 812–815.

White, R.J. (1997). Regulation of RNA polymerases I and II by the retinoblastoma protein: a mechanism for growth control. *Trends in Biochemical Sciences* **22**, 77–80.

Wolffe, A. (1997). Sinful repression. *Nature* **387**, 16–17.

Wyllie, A. (1997). Clues in the p53 murder mystery. *Nature* **389**, 237–238.

Zenke, M., Kahn, D., Disela, C., Vennstrom, B., Leutz, A., Keegan, K., Hayman, M.J., Choe, H.R., Yew, N., Engel, J.D. and Beug, H. (1988). v-*erb* A specifically suppresses transcription of the avian erythrocyte anion transporter (Band 3) gene. *Cell* **52**, 107–119.

Zornig, M. and Evan, G.I. (1996). On target with Myc. *Current Biology* **6**, 1553–1556.

CHAPTER EIGHT

DNA binding by transcription factors

8.1 INTRODUCTION

In previous chapters we have considered the role of transcription factors in various processes such as constitutive, inducible and tissue-specific gene expression. In order to fulfil these roles, transcription factors must possess certain features allowing them to modulate gene expression. These features will be discussed in the remaining chapters of this book. Clearly the first feature that many of these factors require is the ability to bind to DNA in a sequence-specific manner and this is discussed in this chapter. Following binding, the factor must interact with other factors or with the RNA polymerase itself in order to influence transcription either positively or negatively and this aspect is discussed in Chapter 9. Finally, in the case of factors modulating inducible, tissue-specific or developmentally regulated gene expression, some means must exist to regulate the synthesis or activity of the factor so that it is active only in a particular situation. This regulation of factor synthesis or activity is discussed in Chapter 10.

Following the cloning of many different eukaryotic transcription factors, the domain-mapping experiments described in Section 2.3.3 have led to the identification of several distinct structural elements in different factors which can mediate DNA binding. These motifs will be discussed in turn using transcription factors which contain them to illustrate their properties (for reviews, see Harrison, 1991; Pabo and Sauer, 1992; Travers, 1993).

8.2 THE HELIX–TURN–HELIX MOTIF IN THE HOMEOBOX

As discussed in Section 6.2.2, the DNA-binding abilities of the homeobox-containing transcription factors are mediated by the homeobox region of the protein. Thus this region, when synthesized without the remainder of the protein either by expression in bacteria or by chemical synthesis, can bind to DNA in the identical sequence-specific manner exhibited by the intact protein.

This ability to define the 60-amino-acid homeodomain as the region binding to DNA has led to intensive study of its structure in the hope of elucidating how the protein binds to DNA in a sequence-specific manner (for reviews, see Kornberg, 1993; Gehring *et al.*, 1994). In particular, the crystal structure of the Antennapedia (Antp) homeodomain bound to DNA has been determined by nuclear magnetic resonance spectroscopy (NMR), whilst similar structural studies of the engrailed (eng) and MATα2 homeodomains bound to DNA have been carried out by X-ray crystallography.

By this means the Antp homeodomain was shown to contain a short N-terminal arm of six residues followed by four α-helical regions (Figure 8.1). The first two helices are virtually anti-parallel to each other, with the other two helices arranged at right angles to the first. Most interestingly, helices II and III are separated by a β-turn, forming a helix–turn–helix motif (Figure 8.2). The eng and MATα2 homeodomains also have a similar structure with an N-terminal arm and a subsequent

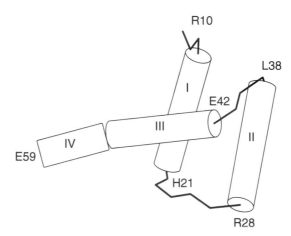

Figure 8.1 Structure of the Antennapedia homeodomain as determined by nuclear magnetic resonance spectroscopy. Note the four α-helical regions (I–IV) represented as cylinders with the amino acids at their ends indicated by numbers and the one-letter amino-acid code.

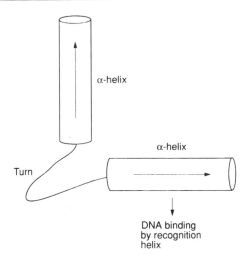

Figure 8.2 The helix–turn–helix motif.

helix–turn–helix motif. In this case, however, the third and fourth helices observed in Antp form a single helical region. Interestingly, the helix–turn–helix stucture typical of the homeodomain is very similar to the DNA-binding motif of several bacteriophage regulatory proteins such as the λ cro protein or the phage 434 repressor, which have also been crystallized and subjected to intensive structural study.

In these bacteriophage proteins X-ray crystallographic studies have shown that the helix–turn–helix motif does indeed contact DNA. One of the two helices lies across the major groove of the DNA, whilst the other lies partly within the major groove, where it can make sequence-specific contacts with the bases of DNA. It is this second helix (known as the recognition helix) that therefore controls the sequence-specific DNA-binding activity of these proteins (Figure 8.3).

The similarity in structure of helices II and III in the eukaryotic homeodomains to the two helices of the bacteriophage proteins led to the suggestion that these two helices in the homeodomain are similarly aligned relative to the DNA, with helix III constituting the recognition helix responsible for sequence-specific DNA binding. Hence the precise amino-acid sequence in the recognition helix in different homeodomain proteins would determine the DNA sequence that they bound (for a review, see Treisman *et al.*, 1992).

In agreement with this idea, exchanging the recognition helix in the Bicoid (Bcd) homeodomain for that of Antp resulted in a protein with the DNA-binding specificity of Antp and not that of Bicoid. Most interestingly, a Bcd protein with the DNA-binding specificity of Antp could also

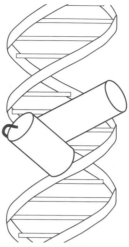

Figure 8.3 Binding of the helix–turn–helix motif to DNA with the recognition helix in the major groove of the DNA.

be obtained by exchanging only the ninth amino acid in the recognition helix, replacing the lysine residue in Bcd with the glutamine residue found in the Antp protein (Figure 8.4), whereas the exchange of other residues which differ between the two proteins has no effect on the DNA-binding specificity. Hence the ninth amino acid within the recognition helix of the homeodomain plays a critical role in determining DNA-binding specificity.

It is likely that the amino group of lysine found at the ninth position in the Bcd protein makes hydrogen bonds with the 06 and N7 positions of a

	Recognition helix										Binding to	
											Bicoid site	Antp site
											TCTAATCCC	TCAATTAAAT
Bicoid	T	A	Q	V	K	I	W	F	K	N	+	−
	A	–	–	–	–	–	–	–	–	–	+	−
	–	–	–	–	A	–	–	–	–	–	+	−
	–	–	–	–	–	–	–	–	A	–	−	−
	A	–	–	–	A	A	–	–	A	–	−	−
	E	R	–	–	–	–	–	–	Q	–	−	+
	E	R	–	–	–	–	–	–	–	–	+	−
	–	–	–	–	–	–	–	–	Q	–	−	+
Antp	E	R	Q	I	K	I	W	F	Q	N	−	+

Figure 8.4 The effect of changing the amino-acid sequence in the recognition helix of the Bicoid protein on its binding to its normal recognition site and that of the Antennapedia (Antp) protein. Note the critical effect of changing the ninth amino acid in the helix, which completely changes the specificity of the Bicoid protein.

guanine residue in the Bcd-specific DNA-binding site, whereas the amide group of glutamine found at the corresponding position in the Antp recognition helix forms hydrogen bonds with the N6 and N7 positions of an adenine residue at the equivalent position within the Antp-specific DNA binding site. Hence the replacement of lysine with glutamine results in the loss of two potential hydrogen bonds to a Bcd site and the gain of two potential hydrogen bonds to an Antp site, explaining the observed change in DNA binding specificity (Figure 8.5).

A similar critical role for the ninth amino acid in determining the precise DNA sequence that is recognized is also seen in other homeobox-containing proteins, replacement of the serine found at this position in the paired protein with the lysine found in Bicoid or the glutamine found

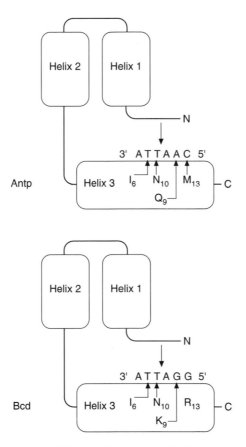

Figure 8.5. Contacts between DNA and the Antp or Bcd homeodomains. Note that the change in the ninth amino acid of the recognition helix (helix 3) alters the base, which is preferentialy bound from a G for Bcd to an A for Antp, as discussed in the text, whilst the N-terminal arm of the homeodomain contacts the ATTA sequence common to the recognition site of both proteins.

in Antp, allowing the paired protein to recognize respectively Bcd- or Antp-specific DNA sequences. Hence the DNA sequence recognized by a homeobox-containing protein appears to be primarily determined by the ninth amino acid in the recognition helix, proteins with different amino acids at this position recognizing different DNA sequences, whereas proteins such as Antp and fushi-tarazu, which have the same amino acid at this position, recognize the same DNA sequence.

This critical role of the ninth amino acid is in contrast to the situation in the bacteriophage proteins in which the helix–turn–helix motif was originally defined. In these proteins, the most N terminal residues (1–3) in the recognition helix play a critical role in determining DNA-binding specificity (for a review, see Pabo and Sauer, 1992). As shown in Figure 8.4, however, these amino acids appear to play little or no role in determining the DNA-binding specificity of eukaryotic helix–turn–helix proteins, suggesting therefore that the recognition helix of these proteins is oriented differently in the major groove of the DNA.

This idea is in agreement with the structural studies of the eukaryotic homeodomains bound to DNA which have identified the actual protein–DNA contacts. These studies have shown that, as in the bacteriophage proteins, the recognition helix directly contacts the bases of DNA in the major groove. However, in the eukaryotic homeobox proteins, this helix is oriented within the major groove somewhat differently such that the critical base-specific contacts are, as predicted, made by the C-terminal end of the helix, which contains residue nine (Figure 8.5).

It is clear therefore that the helix–turn–helix motif in the homeobox mediates both the DNA binding of the protein and also, via the recognition helix, controls the precise DNA sequence that is recognized. Interestingly, however, the short N-terminal arm of the homeodomain also contacts the bases of the DNA, although it makes contact in the minor groove rather than the major groove. Removal of this short N-terminal arm dramatically reduces the DNA-binding affinity of the homeodomain, indicating that this region contributes significantly to the DNA-binding ability of the homeodomain, probably by contacting the ATTA bases common to the DNA-binding sites of several homeodomain proteins (Figure 8.5).

This critical role for the helix–turn–helix motif in the homeodomain is also seen in the POU proteins which, as discussed in Section 6.3.1, contain the homeobox as part of a much larger conserved domain, which also contains a POU-specific region. The crystal structure of the Oct-1 POU domain bound to DNA (Klemm *et al.*, 1994) has shown that the Oct-1 homeodomain binds in a similar manner to the classical homeobox proteins, with the recognition helix lying in the major groove and the N-terminal arm in the minor groove. As noted in Chapter 6 however, whilst the isolated homeodomain can bind to DNA in a sequence-specific manner, both the binding affinity and sequence specificity are greatly increased in the presence of the POU-specific domain, which is separated from the homeodomain by a short, flexible linker sequence. Like the homeodomain, the POU-specific domain forms a helix–turn–helix motif, which allows it to bind to the bases within the DNA adjacent to those contacted by the homeodomain with binding of the two regions occurring on opposite sides of the DNA double helix (Figure 8.6).

As noted in Section 6.3.2, members of the Pax family of transcription factors contain both a paired domain and a homeodomain both of which contribute to high-affinity DNA binding (for a review, see Mansouri *et al.*, 1996). Interestingly, as well as the helix–turn–helix motif in the homeodomain region, the paired domain also binds to DNA via a helix–turn–helix motif. Structural analysis of this motif, however, reveals that it is more similar to that in the bacteriophage proteins than in the eukaryotic homeodomain proteins with the residues at the N-terminus of the recognition helix being critical for DNA binding (Xu *et al.*, 1995). Indeed, one form of Waardenburg's syndrome which results from inactivation of PAX3 (see Section 7.1) is due to mutation in a glycine residue at the N-terminus of the PAX3 recognition helix, resulting in a failure of the factor to bind to DNA.

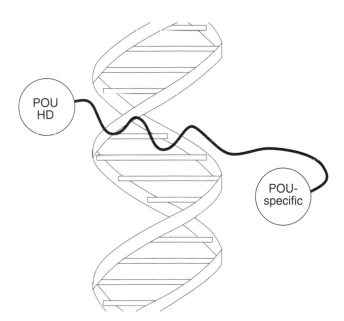

Figure 8.6 Binding of the POU-specific domain and POU homeodomain to opposite sides of the DNA double helix. Note the flexible linker region joining the two DNA-binding motifs.

Hence the helix–turn–helix motif is a widely used DNA-binding domain which exists in at least two different forms that differ in the manner in which the recognition helix contacts the DNA.

8.3 THE ZINC FINGER MOTIF

8.3.1 The two-cysteine two-histidine finger

Transcription factor TFIIIA plays a critical role in regulating the transcription of the 5S ribosomal RNA genes by RNA polymerase III (see Section 3.2.3). When this transcription factor was purified, it was found to have a repeated structure and to be associated with between 7 and 11 atoms of zinc per molecule of purified protein (Miller *et al.*, 1985). When the gene encoding TFIIIA was cloned, it was shown that this repeated structure consisted of the unit, Tyr/Phe-X-Cys-X-Cys-$X_{2,4}$-Cys-X_3-Phe-X_5-Leu-X_2-His-$X_{3,4}$-His-X_5, which is repeated nine times within the TFIIIA molecule. This repeated structure therefore contains two invariant cysteine and two invariant histidine residues, which were predicted to bind a single zinc atom, accounting for the multiple zinc atoms bound by the intact molecule.

This motif is referred to as a zinc finger on the basis of its proposed structure in which a loop of twelve amino acids containing the conserved leucine and phenylalanine residues, as well as several basic amino acids, projects from the surface of the molecule, being anchored at its base by the cysteine and histidine residues which tetrahedrally co-ordinate an

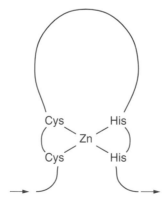

Figure 8.7 Schematic representation of the zinc finger motif. The finger is anchored at its base by the conserved cysteine and histidine residues which tetrahedrally co-ordinate an atom of zinc.

atom of zinc (Figure 8.7). The proposed interaction of zinc with the conserved cysteine and histidine residues in this structure was subsequently confirmed by X-ray adsorption spectroscopy of the purified TFIIIA protein.

Following its identification in the RNA polymerase III transcription factor TFIIIA, similar Cys_2-His_2-containing zinc finger motifs were identified in a number of RNA polymerase II transcription factors, such as Sp1, which contains three contiguous zinc fingers (Kadonaga *et al.*, 1987), and the *Drosophila* Kruppel protein, which contains four finger motifs (see Section 6.1). A list of zinc finger-containing transcription factors is given in Table 8.1 (for reviews, see Evans and Hollenberg, 1988; Struhl, 1989; Klug and Schwabe, 1995).

In all cases studied, the zinc finger motifs have been shown to constitute the DNA binding domain of the protein, with DNA-binding being dependent upon their activity. Thus, in the case of TFIIIA, DNA binding is dependent on the presence of zinc, allowing the finger structures to form, whilst progressive deletion of more and more zinc finger repeats in the molecule results in a parallel loss of DNA-binding activity. Similarly, in the case of Sp1, DNA binding is dependent on the presence of zinc and, most importantly, the sequence-specific binding activity of the intact protein can be reproduced by a protein fragment containing only the zinc finger region (Kadonaga *et al.*, 1987).

Table 8.1 Transcriptional regulatory proteins containing Cys_2–His_2 zinc fingers

Organism	Gene	Number of fingers
Drosophila	Kruppel	4
	Hunchback	6
	Snail	4
	Glass	5
Yeast	ADR 1	2
	SW15	3
Xenopus	TFIIIA	9
	Xfin	37
Rat	NGF-1A	3
Mouse	MK1	7
	MK2	9
	Egr 1	3
	Evi 1	10
Human	Sp1	3
	TDF	13

A similar dependence of DNA binding on the zinc finger motif is also seen in the *Drosophila* Kruppel protein which, as discussed in Chapter 6, is essential for correct thoracic and abdominal development. In this case a single mutation in one of the conserved cysteine residues in the finger, replacing it with a serine which cannot bind zinc, results in the production of a mutant fly indistinguishable from that produced by a complete deletion of the gene (Redemann *et al.*, 1988) indicating the vital importance of the zinc finger (Figure 8.8).

As with the helix–turn–helix motif of the homeobox therefore, the zinc finger motif forms the DNA-binding element of the transcription factors which contain it. Interestingly, however, a single zinc finger taken from the yeast ADRI protein is unable to mediate sequence-specific DNA binding in isolation, whereas a protein fragment containing both the two fingers present in the intact protein can do so. This suggests therefore that DNA binding by the zinc finger is dependent upon interactions with adjacent fingers and explains why zinc finger-containing transcription factors always contain multiple copies of the zinc finger motif (see Table 8.1).

In the zinc finger structure, the zinc co-ordination via cysteine and histidine serves as a scaffold for the intervening region which makes direct contact with the DNA. Detailed structural analysis has shown that these intervening amino acids do not form a simple loop structure as proposed in the original model (for reviews, see Rhodes and Klug, 1993; Klug and Schwabe, 1995). Rather, the finger region forms a motif consisting of two anti-parallel β-sheets with an adjacent α-helix packed against one face of the β-sheet (Figure 8.9; see Plate 5) (Lee *et al.*, 1989). Upon contact with DNA, the α-helix lies in the major groove of the DNA

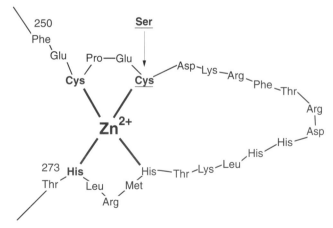

Figure 8.8 Zinc finger in the *Drosophila* Kruppel protein indicating the cysteine to serine change which abolishes the ability to bind zinc and results in a mutant fly indistinguishable from that obtained when the entire gene is deleted.

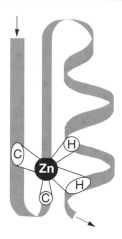

Figure 8.9 Structure of the zinc finger in which two anti-parallel β-sheets (straight lines) are packed against an adjacent α-helix (wavy line).

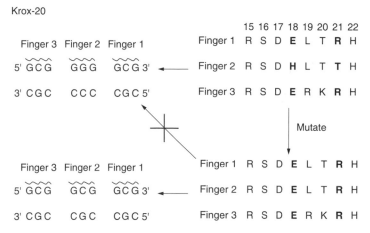

Figure 8.10 DNA-binding specificity and amino-acid sequence of the three cysteine–histidine zinc fingers in the *Drosophila* Krox 20 protein. Note that each finger binds to three specific bases in the recognition sequence and that finger 2, which differs from fingers 1 and 3 in the DNA sequence it recognizes, also differs in the amino acids at positions 18 and 21 in the finger (bold letters). Mutating these amino acids to their equivalents in fingers 1 and 3 changes the DNA-binding specificity of finger 2 to that of fingers 1 and 3, indicating that these amino acids play a critical role in determining the DNA sequence that is recognized.

and makes sequence-specific contacts with the bases of DNA, whilst the β-sheets lie further away from the helical axis of the DNA and contact the DNA backbone.

Most interestingly, this structure indicates that a critical role in sequence-specific DNA binding will be played by amino acids at the amino-terminus of the α-helix, most notably the amino acids immediately preceding the first histidine residue. In agreement with this idea, two amino acids in this region play a critical role in determining the DNA-binding specificity of the *Drosophila* Krox-20 transcription factor (Nardelli *et al.*, 1991). Thus this factor contains three zinc fingers and interacts with the DNA sequence 5'-GCGGGGGCG-3'. If each finger contacts three bases within this sequence, then the central finger must recognize the sequence GGG whereas the two outer fingers will each recognize the sequence GCG (Figure 8.10). When the amino-acid sequence of each of the Krox-20 fingers was compared, it was found that the two outer fingers contain a glutamine residue at position 18 of the finger and an arginine at position 21, whereas the central finger differs in that it has histidine and threonine residues at these positions. As expected, if these two amino-acid differences are critical in determining the DNA sequence that is recognized, altering these two residues in the central finger to their equivalents in the outer two fingers resulted in a factor which failed to bind to the normal Krox-20 binding site but instead bound to the sequence 5'-GCGGCGGCG-3' in which each finger binds the sequence GCG. This experiment therefore indicates the critical role of two amino acids at the amino-terminus of the α-helix in producing the DNA-binding specificity of zinc fingers of this type and also shows that at least in the case of Krox-20, each successive finger interacts with three bases of DNA within the recognition sequence. The importance of these amino acids has also been confirmed in experiments in which the amino acids at different positions in the zinc finger were randomly altered and their interaction with a wide range of DNA sequences assessed (Choo and Klug, 1994; Rebar and Pabo, 1994). Clearly such an important role for the amino acids at the amino-terminus of an α-helix, parallels the similar critical role for the equivalent amino acids in the recognition helix of the bacteriophage DNA recognition proteins and in the paired domain (see Section 8.2).

Hence, like the helix–turn–helix motif, the cysteine–histidine zinc finger plays a critical role in mediating the DNA-binding abilities of transcription factors that contain it, with sequence-specific recognition of DNA being determined in both cases by amino acids within an α-helix.

8.3.2 The multi-cysteine zinc finger

A similar zinc-binding domain to that discussed above has also been identified in the DNA-binding domains of the members of the nuclear receptor family, which includes the receptors for steroid hormones and for other related substances such as thyroid hormone or retinoic acid (for a discussion of these receptors, see Section 4.4, and Mangelsdorf *et al.*, (1995) and Perlmann and Evans (1997)). As with the cysteine–histidine fingers, this motif has been shown by X-ray adsorption spectroscopy to bind zinc in a tetrahedral configuration. However, in this case, co-ordination is achieved by four cysteine residues rather than the two-cysteine–two-histidine structure discussed above. Similar multi-cysteine motifs have also been identified in several other DNA-binding transcription factors, such as the yeast proteins GAL4, PRRI and LAC9, as well as in the adenovirus transcription factor E1A (see Table 8.2) (for reviews, see Evans and Hollenberg, 1988; Klug and Schwabe, 1995), indicating that this type of motif is not confined to the nuclear receptors.

In the case of the nuclear receptors, the DNA-binding domain has the consensus sequence $Cys-X_2-Cys-X_{13}-Cys-X_2-Cys-X_{15,17}-Cys-X_5-Cys-X_9-Cys-X_2-Cys-X_4-Cys$. This motif is therefore capable of forming a pair of fingers each with four cysteines co-ordinating a single zinc atom (Figure 8.11) and, as with the cysteine–histidine finger proteins, DNA binding of the receptors is dependent on the presence of zinc.

Table 8.2 Transcriptional regulatory proteins with multiple cysteine fingers

Finger type	Factor	Species
Cys_4–Cys_5	Steroid, thyroid receptors	Mammals
Cys_4	E1A	Adenovirus
Cys_6	Gal4, PPRI, LAC9	Yeast

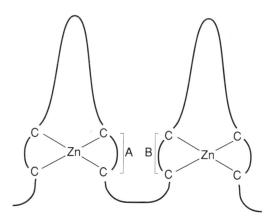

Figure 8.11 Schematic representation of the four-cysteine zinc finger. Regions labelled A and B are of critical importance in determining, respectively the DNA sequence that is bound by the finger and the optimal spacing between the two halves of the palindromic sequence that is recognized.

However, the multi-cysteine finger cannot be converted into a functional cysteine–histidine finger by substituting two of its cysteine residues with histidines, indicating that the two types of finger are functionally distinct (Green and Chambon, 1987). Moreover, unlike the cysteine–histidine zinc finger, which is present in multiple copies within the proteins that contain it, the unit of two multi-cysteine fingers present in the steroid receptors is found only once in each receptor. Interestingly, structural studies of the two multi-cysteine fingers in the glucocorticoid and oestrogen receptors (for reviews, see Schwabe and Rhodes, 1991; Klug and Schwabe, 1995) have indicated that the two fingers form one single structural motif consisting of two α-helices perpendicular to one another with the cysteine–zinc linkage holding the base of a loop at the N-terminus of each helix (Figure 8.12; see Plate 6) (Hard *et al.*, 1990). This is quite distinct from the modular structure of the two-cysteine–two-histidine finger, where each finger constitutes an independent structural element whose configuration is unaffected by the presence or absence of adjacent fingers.

Thus, although these two DNA-binding motifs are similar in their co-ordination of zinc, they differ in the lack of histidines and of the conserved phenylalanine and leucine residues in the multi-cysteine finger as well as structurally. It is clear therefore that they represent distinct functional elements and are unlikely to be evolutionarily related (for reviews see Schwabe and Rhodes, 1991; Rhodes and Klug, 1993; Klug and Schwabe, 1995).

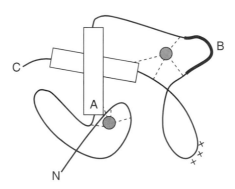

Figure 8.12 Schematic model of a pair of zinc fingers in a single molecule of the oestrogen receptor. Note the helical regions (indicated as cylinders) with the critical residues for determining the DNA sequence that is bound located at the terminus of the recognition helix (indicated as A), the zinc atoms (shaded), conserved basic residues (+++), and the region which interacts with another receptor molecule and determines the optimal spacing between the two halves of the palindromic sequence that is recognized (indicated as B). Note that A and B indicate the same regions as in Figure 8.11.

Whatever the precise relationship between these motifs, it is clear that the multi-cysteine finger mediates the DNA binding of the nuclear receptors. Thus mutations which eliminate or alter critical amino acids in this motif interfere with DNA binding by the receptor (Figure 8.13).

The role of the cysteine fingers in mediating DNA binding by the nuclear receptors can also be demonstrated by taking advantage of the observation (discussed in Section 4.4.1) that the different steroid receptors bind to distinct but related palindromic sequences in the DNA of hormone responsive genes (see Beato *et al.* (1995) and Mangelsdorf and Evans (1995) for reviews and Table 4.2 for a comparison of these binding sites). Thus, if the cysteine-rich region of the oestrogen receptor is replaced by that of the glucocorticoid receptor, the resulting chimaeric receptor has the DNA binding specificity of the glucocorticoid receptor but continues to bind oestrogen, since all the other regions of the molecule are derived from the oestrogen receptor (Green and Chambon, 1987) (Figure 8.14). Hence the DNA-binding specificity of the hybrid receptor is determined by its cysteine-rich region, resulting in the hybrid receptor inducing the expression of glucocorticoid responsive genes (which carry its DNA-binding site) in response to oestrogen (to which it binds).

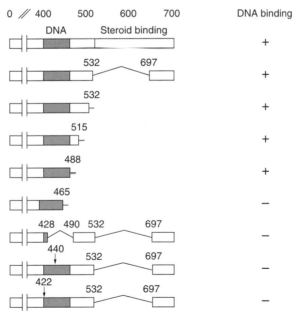

Figure 8.13 Effect of various deletions or mutations on the DNA binding of the glucocorticoid receptor. Note that DNA-binding is only prevented by deletions which include part of the DNA binding domain (shaded) or by mutations within it (arrows), but not by deletions in other regions such as the steroid-binding domain. Numbers indicate amino-acid residues.

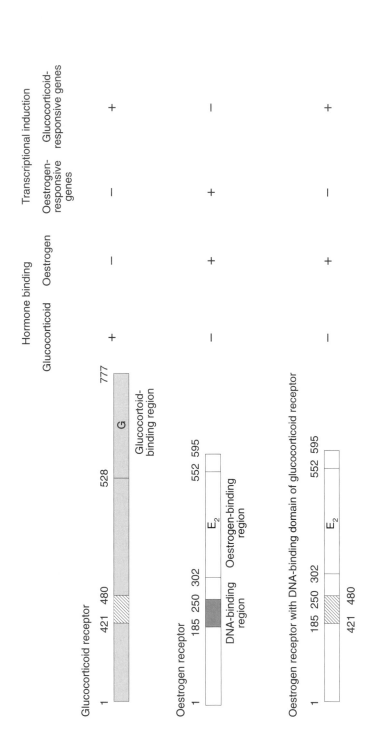

Figure 8.14 Effect of exchanging the DNA-binding domain (shaded) of the oestrogen receptor with that of the glucocorticoid receptor on the binding of hormone and gene induction by the hybrid receptor.

These so-called 'finger-swap' experiments therefore provide further evidence in favour of the critical role for the multi-cysteine fingers in DNA-binding, exchanging the fingers of two receptors exchanging the DNA binding specificity. In addition, however, because of the existence of short distinct DNA-binding regions of this type in receptors which bind to distinct but related DNA sequences, they provide a unique opportunity to dissect the elements in a DNA-binding structure which mediate binding to specific sequences.

Thus, by exchanging one or more amino acids between two different receptors, it is possible to investigate the effects of these changes on DNA-binding specificity and hence elucidate the role of individual amino-acid differences in producing the different patterns of sequence-specific binding. For example, the alteration of the two amino acids between the third and fourth cysteines of the N-terminal finger in the glucocorticoid receptor for their equivalents in the oestrogen receptor changes the DNA-binding specificity of the chimaeric receptor to that of the oestrogen receptor (Figure 8.15). Hence the exchange of two amino acids in a critical region of a protein of 777 amino acids (indicated as A in Figure 8.11) can completely change the DNA-binding specificity of the glucocorticoid receptor resulting in it binding to and activating genes which are normally oestrogen responsive. The specificity of this hybrid receptor, for such oestrogen-responsive genes can be further enhanced by exchanging another amino acid located between the two fingers (Figure 8.15), indicating that this region also plays a role in controlling the specificity of DNA binding.

As discussed in Section 4.4.1, the steroid receptors bind to palindromic recognition sequences within DNA, with the receptor binding to DNA as a homodimer in which each receptor molecule interacts with one half of the palindrome. In addition to differences in the actual sequence recognized, steroid–thyroid hormone receptors can also differ in the optimal spacing between the two separate halves of the palindromic DNA sequence which is recognized (Table 4.2a). Thus the oestrogen receptor and the thyroid hormone receptor both recognize the identical palin-dromic sequence in the DNA but differ in that, in the thyroid receptor binding sites, the two halves of the palindrome are adjacent, whereas in the oestrogen receptor binding sites, they are separated by three extra bases. The further alteration of the chimaeric receptor, illustrated in Figure 8.15 by changing five amino acids in the second finger to their thyroid hormone receptor equivalents, is sufficient to allow the receptor to recognize thyroid hormone receptor binding sites (Umesono and Evans, 1989) (Figure 8.15). These amino acids in the second finger (indicated as B in Figure 8.11) appear to play a critical role therefore in determining the optimal spacing of the palindromic sequence that is recognized.

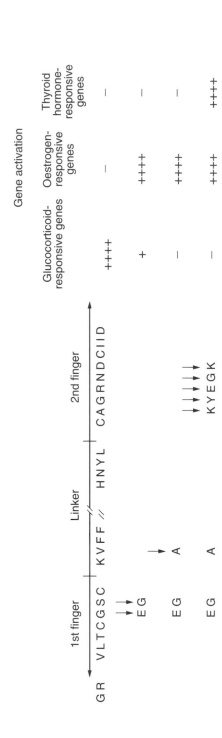

Figure 8.15 Effect of amino-acid substitutions in the zinc finger region of the glucocorticoid receptor on the ability to bind to and activate genes that are normally responsive to different steroid hormones.

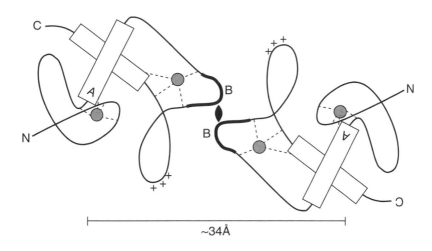

Figure 8.16 Interaction of two oestrogen receptor molecules to form a DNA-binding dimer. Compare with Figure 8.12 and note the interaction of the B regions on each molecule. The resulting dimer has a spacing of 34 Å between the two DNA-binding regions, allowing binding in successive major grooves of the DNA molecule.

As discussed above, structural studies of the two zinc fingers in the oestrogen and glucocorticoid receptors suggest that they form a single structural motif with two perpendicular α-helices (Figure 8.12). In this structure, the critical amino acids for determining the spacing in the palindromic sequence recognized are located on the surface of the molecule, allowing them to interact with equivalent residues on another receptor monomer during dimerization (indicated as B in Figure 8.16; see Plate 7) (Schwabe *et al.*, 1993). Hence differences in the interaction of these regions in the different receptors determine the spacing of the two monomers within the receptor dimer and thus the optimal spacing in the palindromic DNA sequence that is recognized.

Interestingly, within this structure, the critical residues for determining the precise DNA sequence that is recognized are located at the N-terminus of the first α-helix (indicated as A in Figure 8.16), further supporting the critical role of such helices in DNA binding. Moreover, in the proposed structure of the oestrogen receptor dimer, the DNA-binding helices in each monomer will be separated by 34Å allowing each of these recognition helices to make sequence-specific contacts in adjacent major grooves of the DNA molecule.

Differences in the DNA-binding domain also regulate the binding of members of the nuclear receptor family to directly repeated sequences with different spacings between the two halves of the repeat (see Table 4.2b). Thus, when the direct repeats are separated by only one base, they can bind a homodimer of the retinoid X-receptor (RXR) and hence confer a response to 9-*cis*-retinoic acid, which binds to this receptor (Figure 8.17). In contrast, the RXR homodimer cannot bind to the direct repeats when they are separated by between two and five base pairs. Rather, on these elements, RXR forms a heterodimer with other members of the nuclear receptor family (Figure 8.17).

Moreover, the nature of the heterodimers which form on a particular response element controls the response it mediates with the nature of the non-RXR component determining the response. Thus a spacing of two or five base pairs binds a heterodimer of RXR and the retinoic acid receptor (RAR) and therefore mediates responses to all *trans*-retinoic acid which binds to RAR. In contrast, a spacing of four base pairs binds a heterodimer of RXR and thyroid hormone receptor (TR), and therefore can mediate responses to thyroid hormone (for reviews, see Gronemeyer and Moras, 1995; Mangelsdorf and Evans, 1995).

As on the palindromic repeats, it is the DNA-binding domain of the receptors which controls which heterodimers can form on particular spacings of the direct repeat (for reviews, see Gronemeyer and Moras, 1995; Mangelsdorf and Evans, 1995). Interestingly, the crystal structure of the RXR–TR heterodimer bound to a direct repeat with a four-base spacing indicates that the dimerization interface involves amino acids in the first finger of the thyroid hormone receptor and the second finger of RXR rather than only residues in the second finger as occurs for homodimerization of receptors on palindromic repeats (Rastinejad *et al.*, 1995) (Figure 8.18).

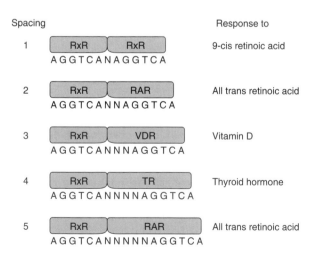

Figure 8.17 Binding of different nuclear receptor heterodimers to directly repeated elements with different spacings between the repeats determines the response mediated by each element.

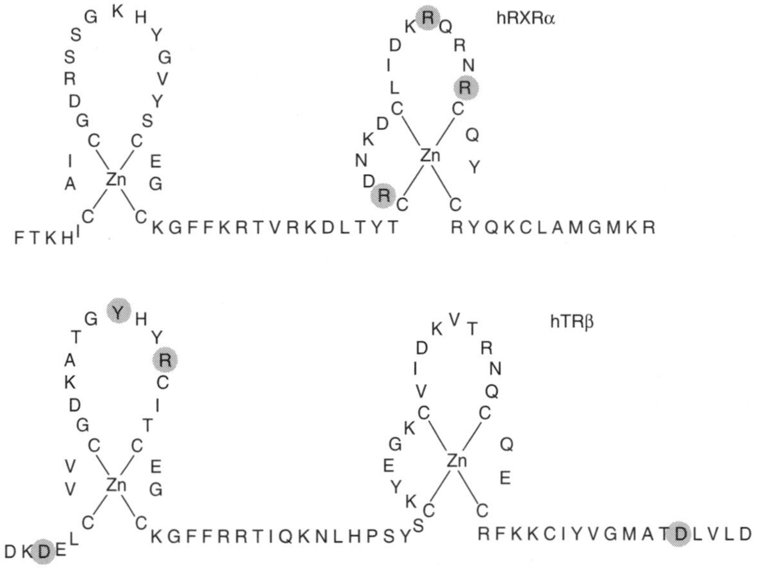

Figure 8.18 Zinc fingers in the retinoid X-receptor α and the thyroid hormone receptor β. The residues in each receptor which are involved in heterodimer formation with the other receptor are indicated.

The definition of the DNA-binding domain of the nuclear receptors as a short sequence containing two multi-cysteine fingers has therefore allowed the elucidation of the features in this motif which mediate the different sequence specificities of the different receptors and their relationship to the structure of the motif. In particular, a helical region of the first finger plays a critical role in determining the precise DNA sequence, which is recognized by binding in the major groove of the DNA. Similarly, other regions in either the first or second fingers control the spacing of adjacent palindromic or directly repeated sequences, which is optimal for the binding of receptor homo- or heterodimers by interacting with another receptor monomer and hence affecting the structure of the receptor dimer that forms.

8.4 THE LEUCINE ZIPPER AND THE BASIC DNA-BINDING DOMAIN

As discussed in the preceding sections of this chapter, the study of motifs common to several different transcription factors has led to the identification of the role of these motifs in DNA-binding. A similar approach has led to the identification of the leucine zipper motif (for reviews, see Lamb and McKnight, 1991; Hurst, 1996; Kerppola and Curran, 1995). Thus this structure has been detected in several different transcription factors such as the CAAT box-binding protein C/EBP (see Section 3.3.3), the yeast factor GCN4 and the oncogene products Myc, Fos and Jun (see Sections 7.2.1 and 7.2.3). It consists of a leucine-rich region in which successive leucine residues occur every seventh amino acid (Figure 8.19).

In all these cases, the leucine-rich region can be drawn as an α-helical structure in which adjacent leucine residues occur every two turns on the same side of the helix. Moreover, these leucine residues appear to play a critical role in the functioning of the protein. Thus, with one exception (a single methionine in the Myc protein), the central leucine residues of the motif are conserved in all the factors that contain it (Figure 8.19). It was therefore proposed (Landshultz *et al.*, 1988) that the long side chains of the leucine residues extending from one polypeptide would interdigitate with those of the analogous helix of a second polypeptide, forming a motif known as the leucine zipper, which would result in the dimerization of the factor (Figure 8.20). This effect could also be achieved by a methionine residue, which, like leucine, has a long side chain with no lateral methyl groups, but not by other hydrophopic amino acids such as valine or isoleucine, which have methyl groups extending laterally from the β-carbon atom.

In agreement with this idea, substitutions of individual leucine residues in C/EBP or other leucine zipper containing proteins such as Myc, Fos and Jun with isoleucine or valine, abolish the ability of the intact protein to form a dimer indicating the critical role of this region in dimerization. A comparison of the effects of various mutations of this type on the ability of the mutant protein to dimerize, suggested that the two leucine-rich regions associate in a parallel manner with both helices

```
C/EBP    L T S D N D R L R K R V E Q L S R E L D T L R G I F R Q L
Jun B    L E D K V K T L K A E N A G L S S A A G L L R E Q V A Q L
Jun      L E E K V K T L K A Q N S E L A S T A N M L R E Q V A Q L
GCN 4    L E D K V E E L L S K N Y H L E H E V A R L K K L V G E R
Fos      L Q A E T D Q L E D E K S A L Q T E I A N L L K E K E K L
Fra 1    L Q A E T D K L E D E K S G L Q R E I I E L Q K Q K E R L
c-Myc    V Q A E E Q K L I S E E D L L R K R R E Q L K H K L E Q L
n-Myc    L Q A E E H Q L L L L E K E K L Q A R Q Q Q L L K K I E H A
l-Myc    L V G A E K K M A T E K R Q L R C R Q Q Q L Q K R I A Y L
```

Figure 8.19 Alignment of the leucine-rich region in several cellular transcription factors. Note the conserved leucine residues (L) that occur every seven amino acids.

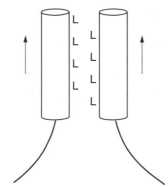

Figure 8.20 Model of the leucine zipper and its role in the dimerization of two molecules of a transcription factor.

oriented in the same direction (as illustrated in Figure 8.20) rather than in an anti-parallel configuration as originally suggested (Landshultz *et al.*, 1989). This idea was confirmed by structural studies of the leucine zipper regions in GCN4 and in the Fos/Jun dimer bound to DNA (Glover and Harrison, 1995). These studies indicated that each zipper motif forms a right-handed α-helix with dimerization occurring via the association of two parallel helices that coil around each other to form a coiled-coil motif similar to that found in fibrous proteins such as the keratins and myosins (Figure 8.21).

In addition to its role in dimerization, the leucine zipper is also essential for DNA binding by the intact molecule. Thus mutations in the zipper which prevent dimerization also prevent DNA binding from occurring (Landshultz *et al.*, 1989). Unlike the zinc finger or helix–turn–helix motifs, however, the zipper is not itself the DNA-binding

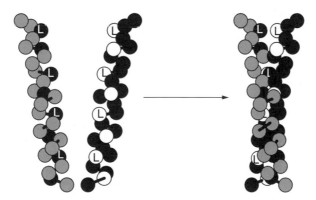

Figure 8.21 Coiled-coil structure of the leucine zipper formed by two helical coils wrapping around each other. L indicates a leucine residue.

domain of the molecule and does not directly contact the DNA. Rather, it facilitates DNA binding by an adjacent region of the molecule which, in C/EBP, Fos and Jun, is rich in basic amino acids and can therefore interact directly with the acidic DNA. The leucine zipper is believed therefore to serve an indirect structural role in DNA binding, facilitating dimerization, which in turn results in the correct positioning of the two basic DNA-binding domains in the dimeric molecule for DNA binding to occur (Figure 8.22).

In agreement with this idea, mutations in the basic domain abolish the ability to bind to DNA without affecting the ability of the protein to dimerize, as expected for mutations which directly affect the DNA-binding domain (Landshultz *et al.*, 1989). Similarly, exchange of the basic region of GCN4 for that of C/EBP, results in a hybrid protein with the DNA-binding specificity of C/EBP whilst exchange of the leucine zipper region has no effect on the DNA binding specificity of the hybrid molecule (Figure 8.23).

Hence the DNA-binding specificity of leucine zipper-containing transcription factors is determined by the sequence of their basic domain with the leucine zipper allowing dimerization to occur and hence facilitating DNA binding by the basic domain. As expected from this idea, the basic DNA binding domain can interact with DNA in a sequence-specific manner in the absence of the leucine zipper, if it is first dimerized via an inter-molecular disulphide bond (Figure 8.24). Interestingly, the basic DNA-binding domain can bind to DNA as a monomer in the case of the Skn-1 factor, which lacks a leucine zipper (Blackwell *et al.*, 1994). In this factor however, the basic domain is part of a composite DNA-binding domain, which also contains a region homologous to the N-terminal arm of the homeobox (see Section 8.2).

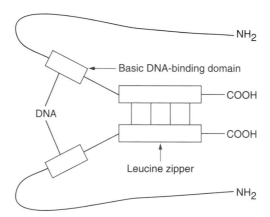

Figure 8.22 Model for the structure of the leucine zipper and the adjacent DNA-binding domain following dimerization of the transcription factor C/EBP.

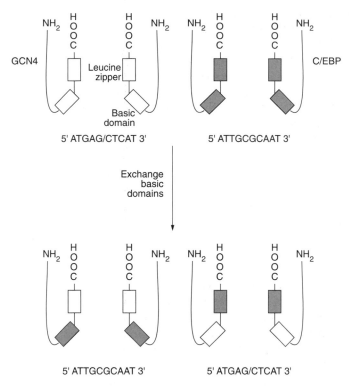

Figure 8.23 Effect of exchanging the basic domains of GCN4 and C/EBP on the DNA-binding specificity. Note that the DNA-binding specificity is determined by the origin of the basic domain and not that of the leucine zipper.

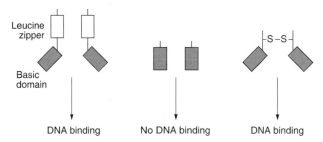

Figure 8.24 DNA binding of molecules containing basic DNA-binding domains can occur following dimerization mediated by leucine zippers or by a disulphide bridge (S–S) but cannot be achieved by unlinked monomeric molecules.

In factors having a simple basic DNA-binding domain, following dimerization via the leucine zipper, the intact transcription factor will form a rotationally symmetric dimer that contacts the DNA via the bifurcating basic regions (Figure 8.22), which form α-helical structures.

These two helices then track along the DNA in opposite directions corresponding to the dyad symmetric structure of the DNA recognition site and form a clamp or scissors grip around the DNA, similar to the grip of a wrestler on his opponent, resulting in very tight binding of the protein to DNA (Glover and Harrison, 1995). Most interestingly, structural studies have suggested that the basic region does not assume a fully α-helical structure until it contacts the DNA when it undergoes a configurational change to a fully α-helical form. Hence the association of the transcription factor with the appropriate DNA sequence results in a conformational change in the factor leading to a tight association with that sequence (for a discussion see Sauer, 1990).

Although originally identified in the leucine zipper-containing proteins, the basic DNA-binding domain has also been identified by a homology comparison in a number of other transcription factors which do not contain a leucine zipper (Prendergast and Ziff, 1989). These factors include the immunoglobulin enhancer binding proteins E12 and E47 (discussed in Section 5.2), the muscle determining gene MyoD (Section 5.3) and the *Drosophila* daughterless protein.

In these cases, the basic DNA-binding domain is juxtaposed to a region that can form a helix–loop–helix motif (for a review, see Littlewood and Evan, 1995). This helix–loop–helix motif is distinct from the helix–turn–helix motif in the homeobox (Section 8.2) in that it can form two amphipathic helices, containing all the charged amino acids on one side of the helix, which are separated by a non-helical loop (Murre *et al.*, 1989a). This helix–loop–helix motif plays a similar role to the leucine zipper, allowing dimerization of the transcription factor molecule and thereby facilitating DNA binding by the basic motif (Murre *et al.*, 1989b) (for a discussion, see Jones, 1990).

In agreement with this, deletion or mutations in the basic domain of the MyoD protein do not abolish dimerization but do prevent DNA-binding, paralleling the effect of similar mutations in C/EBP (Figure 8.25). Similarly, mutations or detetions in the helix–loop–helix region abolish both dimerization and DNA binding, parallelling the effects of similar mutations in leucine zipper-containing proteins. Moreover, the DNA-binding ability of MyoD from which the basic DNA-binding domain has been deleted can be restored by substituting the basic domain of the E12 protein (Davis *et al.*, 1990). However, such substitution does not allow the hybrid protein to activate muscle-specific gene expression suggesting that, in addition to mediating DNA binding, the basic region of MyoD also contains elements involved in the activation of muscle-specific genes (Davis *et al.*, 1990) (Figure 8.25).

Interestingly, it has been shown that the conversion of three amino acids within the E12 basic region to their MyoD equivalents allows the E12 basic region to activate muscle-specific gene expression following

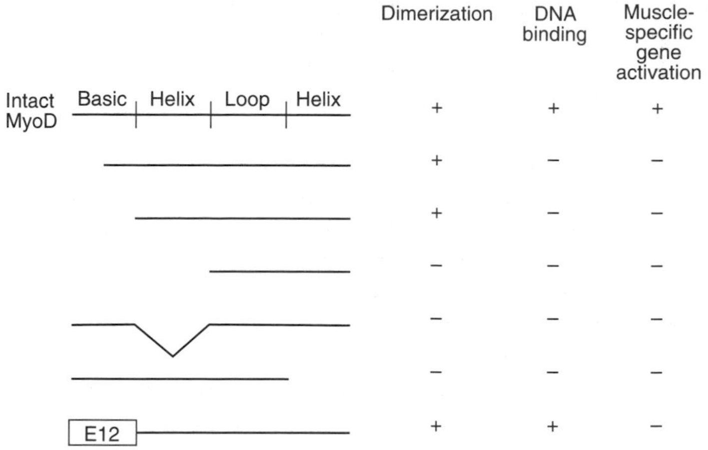

	Dimerization	DNA binding	Muscle-specific gene activation
Intact MyoD — Basic | Helix | Loop | Helix	+	+	+
	+	−	−
	+	−	−
	−	−	−
	−	−	−
	−	−	−
E12 —	+	+	−

Figure 8.25 The effect of deleting the basic domain or the adjacent helix–loop–helix motif on dimerization, DNA binding and activation of muscle-specific gene expression by the MyoD transcription factor. Note that deletion of any part of the helix–loop–helix motif abolishes dimerization and consequent DNA binding and gene activation, whilst deletion of the basic domain directly abolishes DNA binding and consequent gene activation. Substitution of the basic domain of the constitutive factor E12 for that of MyoD restores DNA binding but not the ability to activate muscle-specific gene expression.

		114 115		124		DNA binding	Muscle-specific gene activation	Interaction with MEF2A
MyoD	——	A T	——	K	——	+	+	+
E12	-------	N N	--------	D	-----	+	−	−
E12(M)	-------	A T	--------	K	-----	+	+	+

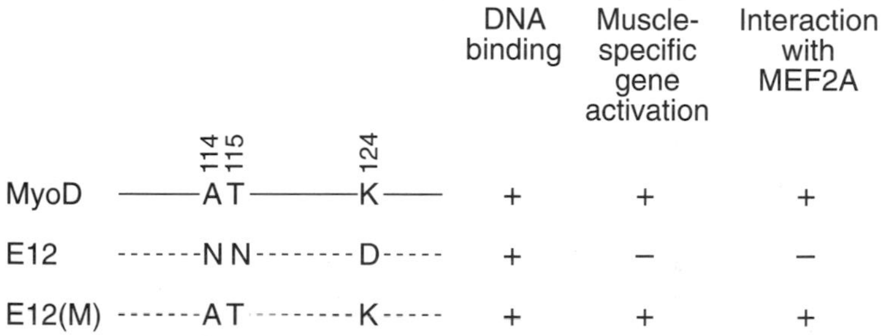

Figure 8.26. Alterations of three amino acids (positions 114, 115 and 124) in the E12 basic domain to their MyoD equivalents confers on the resulting protein (E12(M)) the ability to interact with the muscle-specific transcription factor MEF2A and activate muscle-specific genes following DNA binding, which are normally properties of MyoD alone.

DNA binding (Figure 8.26) (Davis and Weintraub, 1992). The crystal structure of MyoD bound to DNA (Ma *et al.*, 1994) suggests that these amino acids may play a critical role in allowing the MyoD basic region to assume a particular structural configuration in which it can interact with other activating transcription factors. In agreement with this idea, the substitution of these same three amino acids in E12 for their MyoD equivalents allows the mutant E12 protein to bind to another muscle-specific transcription factor MEF2A which is normally a property of

MyoD alone (Figure 8.26) (Kanshal *et al.*, 1994). Hence, like the POU domain (see Section 6.3.1), the basic domain appears to function both as a DNA-binding domain and as a site for protein–protein interactions critical for transcriptional activation.

Both the leucine zipper and the helix–loop–helix motif therefore act by causing dimerization, allowing DNA binding by the adjacent basic motif. Interestingly, the Myc oncoproteins contain both a helix–loop–helix motif and a leucine zipper region adjacent to the basic DNA-binding region (Landshultz *et al.*, 1988; Murre *et al.*, 1989a). Moreover, the leucine zipper can also be found as a dimerization motif in proteins which use DNA-binding motifs other than the basic region. For example, in the *Arabidopsis* Athb-1 and -2 proteins, the leucine zipper facilitates dimerization, with DNA binding being produced by the adjacent homeobox (Sessa *et al.*, 1993). Thus individual DNA binding and dimerization motifs can be combined in different combinations to produce molecules capable of dimerizing and binding to DNA.

The essential role of dimerization (mediated by the leucine zipper or the helix–loop–helix motifs) in allowing DNA binding by basic DNA-binding domain proteins provides an additional aspect to the regulation of these factors (for a discussion, see Jones, 1990; Lamb and McKnight, 1991). Thus, in addition to the formation of homodimers, it is possible to hypothesize that heterodimers will also form between two different leucine zipper or two different helix–loop–helix-containing factors allowing the production of dimeric factors with novel DNA-binding specificities or affinities for different sites. One example of this process is seen in the oncogene products Fos and Jun. Thus, as discussed in Section 7.2.1, the Fos protein cannot bind to AP1 sites in DNA when present alone but can form a heterodimer with the Jun protein, which is capable of binding to such sites with 30-fold greater affinity than a Jun homodimer (Figure 8.27). The formation of Jun homodimers and Jun/Fos heterodimers is dependent upon the leucine zipper regions of the proteins. Moreover, the failure of Fos to form homodimers is similarly dependent on its leucine zipper region. Thus, if the leucine zipper domain of Fos is replaced by that of Jun, the resulting protein can dimerize and the chimaeric protein can bind to DNA through the basic DNA-binding region of Fos which is therefore a fully functional DNA-binding domain. Hence the ability of leucine zipper proteins to bind to DNA is determined both by the nature of the leucine zipper that facilitates homodimerization and/or heterodimerization as well as by the basic DNA-binding motif which allows DNA-binding following dimerization (for a discussion, see Kerppola and Curran, 1995).

In addition to its positive role in allowing DNA binding by factors which cannot do so as homodimers, heterodimerization between two related factors can also have an inhibitory role. Thus, as discussed in

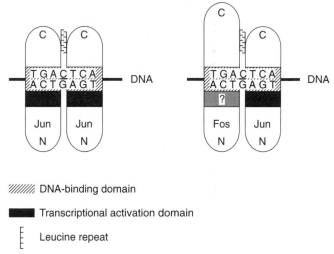

%%%%% DNA-binding domain

■■■■ Transcriptional activation domain

〔 Leucine repeat

Figure 8.27 Model for DNA binding by the Jun homodimer and the Fos/Jun heterodimer.

Section 5.3.2, the DNA-binding ability of functional helix–loop–helix proteins which contain a basic DNA-binding domain can be inhibited by association with the Id protein. This protein contains a helix–loop–helix motif, allowing it to associate with other members of this family, but lacks the basic DNA-binding domain. The heterodimer of Id and a functional protein therefore lacks the dimeric basic regions necessary for DNA binding and the activity of the functional transcription factor is thereby inhibited by Id.

Hence the role of the leucine zipper and helix–loop–helix motifs in dimerization can be put to use in gene regulation in both positive and negative ways, either allowing DNA binding by factors which could not do so in isolation or inhibiting the binding of fully functional factors.

8.5 OTHER DNA-BINDING MOTIFS

Although the majority of DNA-binding domains which have been identified in known transcription factors fall into the three classes we have discussed in the preceding sections, not all do so. Thus, for example, the DNA-binding domains of transcription factors, such as AP2 and the CAAT box-binding factor CTF/NFI (see Section 3.3.3) are distinct from the known motifs and from each other. As more and more factors are cloned, it is likely that other factors with DNA-binding motifs similar to those of these proteins will be identified and that they will become founder members of new families of DNA-binding motifs. Indeed, this process is already under way, for example, the UBF ribosomal RNA transcription factor (see Section 3.2.2) contains a DNA-binding domain which has also been identified in several other factors including high-mobility group (HMG) proteins and which is therefore known as the HMG box (Grosschedel *et al.*, 1994), whilst the DNA-binding domain in the p53 protein discussed in Section 7.3.2 has been shown to be related to that of the NFκB family discussed in Section 5.2.2 (Muller *et al.*, 1995; for a review, see Baltimore and Beg, 1995).

Interestingly, however, as the structure of more and more DNA-binding domains is understood, relationships have emerged between different domains which were originally thought to be entirely distinct. For example, structural analysis of the Ets DNA-binding domain which is found in the Ets-1 proto-oncogene protein (Section 7.2.1) and the mouse PU-1 factor has revealed it to be identical to the winged helix–turn–helix motif originally identified in the *Drosophilia* fork head factor and in the mammalian liver transcription factor HNF-3 (Donaldson *et al.*, 1996).

Moreover, as its name suggests, this domain contacts the DNA via a helix–turn–helix motif which is similar to that found in homeobox proteins discussed in Section 8.2, although the winged helix–turn–helix motif contains an additional β-sheet structure with two loops that appear as wings protruding from the DNA-bound factor, giving this motif its name. Similarly, as discussed in Section 8.2, both the POU-specific domain of the POU factors and the paired box of the PAX proteins bind to DNA via helix–turn–helix motifs, indicating that this is one of the most commonly used motifs mediating the DNA-binding of factors whose DNA binding domains appear distinct at first sight.

8.6 CONCLUSIONS

In this chapter we have discussed a number of different DNA-binding motifs common to several different transcription factors which can mediate DNA binding. These motifs are listed in Table 8.3.

As we have seen, it has proved possible in many cases to define the precise amino acids in a particular motif that mediate binding to a particular DNA sequence. It should be noted, however, that many transcription factors have the ability to bind to several dissimilar sequences using the same DNA-binding domain. For example, as discussed in Chapter 4, the glucocorticoid receptor binds to a specific DNA sequence in genes which are induced by glucocorticoid and to a

Table 8.3 DNA-binding motifs

Motif	Structure	Factors containing domain	Comments
Homeobox	Helix–turn–helix	Numerous *Drosophila* homeotic genes, related genes in other organisms	Structurally related to similar motif in bacteriophage proteins
POU	Helix–turn–helix and adjacent helical region	Mammalian Oct-1, Oct-2, Pit-1, nematode *unc86*	Related to homeodomain
Paired	Helix–turn–helix	Mammalian Pax factors, *Drosophila* paired factor	Often found in factors which also contain a homeobox
Cysteine–histidine zinc finger	Multiple fingers, each co-ordinating a zinc atom	TFIIIA, Kruppel, Sp1, etc.	May form β-sheet and adjacent α-helical structure
Cysteine–cysteine zinc finger	Single pair of fingers each co-ordinating a zinc atom	Steroid–thyroid hormone receptor family	Related motifs in EIA, GAL4, etc.
Basic domain	α-Helical	C/EBP *c-fos*, *c-jun*, *c-myc*, MyoD, etc.	Associated with leucine zipper and/ or helix–loop–helix dimerization motifs
Winged HTH	Helix–turn–helix	Fork head, HNF 3A *c-ets*, *c-erg*, *Drosophila* E74, PU.1	Binds purine-rich sequences

distinct DNA sequence in genes which are repressed by this hormone, the same DNA-binding domain of the protein being used in each case (Sakai *et al.*, 1988). Similarly, the yeast CYCI and CYC7 genes contain entirely distinct sequences, both of which bind the HAPI transcription factor allowing gene activation to occur (Pfeifer *et al.*, 1987).

This phenomenon of a single factor binding to highly divergent sequences has been most extensively analysed in the case of the POU family, octamer binding protein Oct-1. Thus this factor binds to a sequence in the SV40 enhancer which shares less than 30% homology (4 out of 14 bases) or little more than a random match with another Oct-1-binding sequence in the herpes simplex virus (HSV) immediate-early (IE) gene promoters (Figure 8.28). By analysing a series of other Oct-1-binding elements, however, Baumruker *et al.* (1988) were able to show that the two apparently unrelated Oct-1-binding sites could be linked by a smooth progression via a series of other binding sites which were related to one another (Figure 8.28). This suggests therefore that Oct-1 can bind to very dissimilar sequences because there are few, if any, obligatory contacts with specific bases in potential binding sites. Rather, specific binding to a particular sequence can occur via many possible independent interactions with DNA, only some of which will occur with any particular binding site. Hence the binding to apparently unrelated sequences does not reflect two distinct binding specificities but indicates that the protein can make many different contacts with DNA, the sequences which can specifically bind the protein being those with which it can make a certain proportion of these possible contacts.

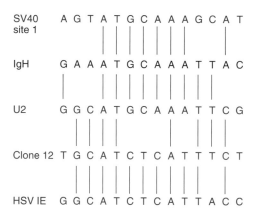

Figure 8.28 The relationship between the various diverse sequences bound by the Oct-1 transcription factor in the simian virus 40 (SV40) enhancer, the immunoglobulin IgH chain gene enhancer (IgH), the U2 snRNA gene, clone 12 (a mutated version of a site in the SV40 enhancer which binds Oct-1) and the herpes simplex virus immediate–early genes (HSV IE).

Interestingly, it has been shown that the secondary structure of Oct-1 bound to these sites differs so that its configuration when bound to the HSV IE sequence is different to that observed when it is bound to the other sequences (Walker *et al.*, 1994). Moreover, this configurational change allows the Oct-1 bound to the HSV promoter to be recognized by the HSV VP16 protein, whereas this does not occur with Oct-1 bound to other sequences. As discussed in Section 9.2.1, VP16 is a much stronger trans-activator than Oct-1 alone and this therefore results in the strong activation of the HSV IE promoters by the Oct-1/VP16 complex, whereas other promoters in which Oct-1 has bound to different sequences are insensitive to such *trans*-activation by VP16. Hence this provides a novel example of gene regulation in which the nature of the sequence bound by a factor controls its recognition by another factor, resulting in strong *trans*-activation only from a sub-set of sequences bound by Oct-1 (Figure 8.29).

In addition to the ability of one factor to bind to different sequences and have different effects, it is also possible for the same sequence to be bound by more than one factor. Thus, as well as the Oct-1 factor, several other proteins binding to the octamer motif exist, including the B-cell-specific factor Oct-2 (see Section 5.2.2) and several others expressed specifically in the brain (for a review, see Ryan and Rosenfeld, 1997). In all these factors, DNA binding is mediated by the POU domain discussed above (Section 8.2). In contrast, however, whilst the transcription factors CTF/NFI and C/EBP both bind to the CAAT box sequence (see Section 3.3.2), they do so via completely different DNA-binding domains, with C/EBP having a basic

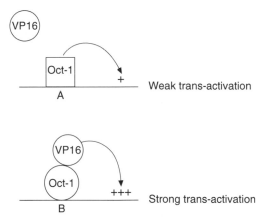

Figure 8.29. The octamer-binding protein Oct-1 binds to most binding sites (A) in a configuration that is not recognized by VP16. This results in only the weak *trans*-activation characteristic of Oct-1 alone. In contrast when it binds to its binding sites in the HSV IE promoters (B), Oct-1 undergoes a conformational change allowing it to be recognized by the strong trans-activator VP16, leading to strong trans-activation.

DNA-binding domain (Section 8.4), whilst CTF/NFI has a DNA-binding domain distinct from that of any other factor (Section 8.5).

It is unlikely therefore that the existence of several distinct DNA-binding domains reflects the need of the factors which contain them to bind to distinct types of DNA sequences. Rather, it seems perfectly possible that one DNA-binding motif could be present in all factors with variations of it in different factors producing the observed binding to different DNA sequences. This is particularly so in view of the fact that in diverse DNA binding motifs such as the helix–turn–helix, the basic DNA-binding domain and the two types of zinc fingers, the amino acids which determine sequence-specific binding to DNA are all located within similar α-helical structures. This idea evidently begs the question of why different DNA-binding motifs exist.

It is possible that this situation has occurred simply by different motifs that could produce DNA binding having arisen in particular factors during evolution and having been retained since they efficiently fulfilled their function. Alternatively, it may be that the existence of different motifs reflects other differences in the factors containing them other than the specific DNA sequence which is recognized. For example, the highly repeated zinc finger motif may be of particular use where, as in the case of transcription factor TFIIIA, the factor must contact a large regulatory region in the DNA. Similarly, a motif such as the basic domain which can only bind to DNA following dimerization will be of particular use where the activity of the factor must be regulated whether positively or negatively via dimerization with another factor.

Whatever the case, it is clear that DNA binding by transcription factors is dependent upon specific domains of defined structure within the molecule. Following such DNA binding, the bound factor must influence the rate of transcription either positively or negatively. The manner in which this occurs and the regions of the factors which achieve this effect are discussed in the next chapter.

REFERENCES

Baltimore, D. and Beg, A.E. (1995). DNA-binding proteins: a butterfly flutters by. *Nature* **373**, 287–288.

Baumruker, T., Sturm, R. and Herr, W. (1988). OBP 100 binds remarkably degenerate octamer motifs through specific interaction with flanking sequences. *Genes and Development* **2**, 1400–1413.

Beato, M., Herrlich, P. and Schutz, G. (1995). Steroid hormone receptors: many actors in search of a plot. *Cell* **83**, 851–857.

Berg, J.M. (1989). DNA binding specificity of steroid receptors. *Cell* **57**, 1065–1068.

Blackwell, T.K., Bowerman, B., Priess, J.R. and Weintraub, H. (1994). Formation of a monomeric DNA binding domain by Skn–1 bZip and homeodomain elements. *Science* **266**, 621–628.

Choo, Y. and Klug, A. (1994). Toward a code for the interaction of zinc fingers with DNA: selection of randomized fingers displayed on phage. *Proceedings of the National Academy of Sciences USA* **91**, 11163–11167.

Cilberto, G., Castagnoli, L. and Cortese, R. (1983). Transcription by RNA polymerase III. *Current Topics in Development Biology* **18**, 59–88.

Davis, R.L., Cheng, P.F., Lassar, A.B. and Weintraub, H. (1990). The MyoD DNA binding domain contains a recognition code for muscle-specific gene activation. *Cell* **60**, 733–746.

Davis, R.L. and Weintraub, H. (1992). Acquisition of myogenic specificity by replacement of three amino acid residues from MyoD into E12. *Science* **256**, 1027–1030.

Donaldson, L.W., Peterson, J.M., Graves, B.J. and McIntosh, P. (1996). Solution structure of the ETS domain from murine Ets-1: a winged helix–turn–helix DNA binding motif. *EMBO Journal* **15**, 125–134.

Evans, R.M. and Hollenberg, S.M. (1988). Zinc fingers: gilt by association. *Cell* **52**, 1–3.

Gehring, W.J., Qian, Y.Q., Billeter, M., Furukubo-Tukunagu, K., Schier, A.F., Resendez-Perez, D., Affolter, M., Otting, G. and Wuthrich, K. (1994). Homeodomain-DNA recognition. *Cell* **78**, 211–223.

Glover, J.N.M. and Harrison, S.C. (1995). Crystal structure of the heterodimeric bZip transcription factor c-Fos-c-Jun bound to DNA. *Nature* **373**, 257–261.

Green, S. and Chambon, P. (1987). Oestradiol induction of a glucocorticoid-response gene by a chimaeric receptor. *Nature* **325**, 75–78.

Gronemeyer, H. and Moras, P. (1995). How to finger DNA. *Nature* **375**, 190–191.

Grosschedel, R., Giese, K. and Pagel, J. (1994) HMG proteins: achitectural elements in the assembly of nucleoprotein structures. *Trends in Genetics* **10**, 94–100.

Hard, T., Kellenbach, E., Boelens, R., Maler, B.A., Dahlam, K., Freedman, L.P., Carlstedt-Duke, J., Yamamoto, K.R., Gustafsson, J.A. and Kaplein, R. (1990) Solution structure of the glucocortoid receptor DNA-binding domain. *Science* **249**, 157–160.

Harrison, S.C. (1991). A structural taxonomy of DNA binding domains. *Nature* **353**, 715–719.

Hurst, H.C. (1996). bZIP proteins. *Protein Profile* **3**, 1–72.

Jones, N. (1990). Transcriptional regulation by dimerization: two sides to an incestuous relationship. *Cell* **61**, 9–11.

Kadonaga, J.T., Carner, K.R., Masiarz, F.R. and Tjian, R. (1987). Isolation of cDNA encoding the transcription factor Sp1 and functional analysis of the DNA binding domain. *Cell* **51**, 1079–1090.

Kanshal, S., Schneider, J.W., Nudal-Ginard, B. and Mahdavi, V. (1994). Activation of the myogenic lineage by MEF2A, a factor that induces and cooperates with MyoD. *Science* **266**, 1236–1240.

Kerppola, T. and Curran, T. (1995). Zen and the art of Fos and Jun. *Nature* **373**, 199–200.

Klemm, J.D., Rould, M.A., Aurora, R., Herr, W. and Pabo, C.O. (1994). Crystal structure of the Oct-1 POU domain bound to an octamer site: DNA recognition with tethered DNA-binding molecules. *Cell* **77**, 21–23.

Klug, A. and Schwabe, J.R. (1995). Zinc fingers. *FASEB Journal* **9**, 597–604.

Kornberg, T.B. (1993). Understanding the homeodomain. *Journal of Biological Chemistry* **268**, 26813–26816.

Lamb, P. and McKnight, S.L. (1991). Diversity and specificity in transcriptional regulation: the benefits of heterotypic dimerization. *Trends in Biochemical Sciences* **16**, 417–422.

Landschulz, W.H., Johnson, P.F. and McKnight, S.L. (1988). The leucine zipper: a hypothetical structure common to a new class of DNA binding proteins. *Science* **240**, 1759–1764.

Landschulz, W.H., Johnson, P.F. and McKnight, S.L. (1989). The DNA binding domain of the rat liver nuclear protein C/EBP is bipartite. *Science* **243**, 1681–1688.

Lee, M.S., Gippert, G.P., Soman, K.V., Case, D.A. and Wright, P.E. (1989). Three-dimensional solution structure of a single zinc finger DNA binding domain. *Science* **245**, 635–637.

Littlewood, T. and Evan, G. (1995). Helix–loop–helix. *Protein profile* **2**, 621–702.

Ma, P.C.M., Rould, M.A., Weintraub, H. and Pabo, C.O. (1994). Crystal structure of MyoD bHLH domain–DNA complex: perspectives on DNA recognition and implications for transcriptional activation. *Cell* **77**, 451–459.

Mangelsdorf, D.J. and Evans, R.M. (1995). The RXR heterodimers and orphan receptors. *Cell* **83**, 841–850.

Mangelsdorf, D.J., Thummel, C., Beato, M., Herrlich, F., Schutz, G., Umesono, K., Blumberg, B., Kustner, P., Mark, M., Chambon, P. and Evans, R.M. (1995). The nuclear receptor superfamily: the second decade. *Cell* **83**, 835–839.

Mansouri, A., Hallonet, M. and Gruss, P. (1996). Pax genes and their roles in cell differentiation and development. *Current Opinion in Cell Biology* **8**, 851–857.

Miller, J., McLachlan, A.D. and Klug, A. (1985). Repetitive zinc-binding domains in the protein transcription factor III A from *Xenopus* oocytes. *EMBO Journal* **4**, 1609–1614.

Muller, C.W., Rey, F.A., Sodeoka, M., Verdine, G.L. and Harrison, S.C. (1995). Structure of the NF-κB p50 homodimer bound to DNA. *Nature* **373**, 311–317.

Murre, C., McCaw, P.S. and Baltimore, D. (1989a). A new DNA binding and dimerization motif in immunoglobulin enhancer binding, daughterless, MyoD and myc proteins. *Cell* **56**, 777–783.

Murre, C., McCaw, P.S., Vaessin, H., Caudy, M., Jan, L.Y., Jan, Y.N., Cabera, C.V., Buskin, J.N., Hauschka, S.D., Lassar, A.B., Weintraub, H. and Baltimore, D. (1989b). Interactions between heterologous helix–loop–helix proteins generate complexes that bind specifically to a common DNA sequence. *Cell* **58**, 537–544.

Nardelli, J., Gibson, T.J., Vesque, C. and Charnay, P. (1991). Base sequence discrimination by zinc-finger DNA binding domains. *Nature* **349**, 175–178.

Pabo, C. and Sauer, R.T. (1992). Transcription factors: structural families and principles of DNA recognition. *Annual Review of Biochemistry* **61**, 1053–1095.

Perlmann, T. and Evans, R.M. (1997). Nuclear receptors in Sicily: fun in the famiglia. *Cell* **90**, 391–397.

Pfeifer, K., Prezant, T. and Guarente, L. (1987). Yeast HAPI activator binds to two upstream sites of different sequence. *Cell* **49**, 19–27.

Prendergast, G.C. and Ziff, E.B. (1989). DNA-binding motif. *Nature* **341**, 392.

Rastinejad, F., Perlmann, T., Evans, R.M. and Sigler, P.B. (1995). Structural determinants of nuclear receptor assembly on DNA direct repeats. *Nature* **375**, 203–211.

Rebar, E.J. and Pabo, C.O. (1994). Zinc finger phage: affinity selection of fingers with new DNA-binding specificities. *Science* **263**, 671–673.

Redemann, N., Gaul, U. and Jackle, H. (1988). Disruption of a putative Cys-zinc interaction eliminates the biological activity of the Kruppel finger protein. *Nature* **332**, 90–92.

Rhodes, D. and Klug, A. (1993). Zinc finger structure. *Scientific American* **268**(4), 32–39.

Ryan, A.K. and Rosenfeld, M.G. (1997). POU domain family values: flexibility partnership and developmental codes. *Genes and Development* **11**, 1207–1225.

Sakai, D.D., Helms, S., Carlstedt-Duke, J., Gustafsson, J.A., Rottman, F.M. and Yamamoto, K.R. (1988). Hormone-mediated repression: a negative glucocorticoid response element from the bovine prolactin gene. *Genes and Development* **2**, 1144–1154.

Sauer, R.T. (1990). Scissors and helical forks. *Nature* **347**, 514–515.

Schwabe, J.W.R., Chapman, L., Finch, T. and Rhodes, D. (1993). The crystal structure of the estrogen receptor DNA binding domain bound to DNA – how receptors discriminate between their response elements. *Cell* **75**, 567–578.

Schwabe, J.W.R. and Rhodes, D. (1991). Beyond zinc fingers: steroid hormone receptors have a novel structural motif for DNA recongition. *Trends in Biochemical Sciences* **16**, 291–296.

Sessa, G., Morelli, G. and Ruberti, I. (1993). The Athb-1 and -2 HD-ZIP domains homodimerize forming complexes of different DNA binding specificities. *EMBO Journal* **12**, 3507–3517.

Struhl, K. (1989). Helix–turn–helix, zinc finger, and leucine zipper motifs for eukaryotic transcriptional regulatory proteins. *Trends in Biochemical Sciences* **14**, 137–140.

Travers, A. (1993). *DNA–Protein Interactions*. London: Chapman and Hall.

Treisman, J., Harris, E., Wilson, D. and Desplan, C. (1992). The homeodomain: a new face for the helix–turn–helix. *Bio essays* **14**, 145–150.

Walker, S., Hayes, S. and O'Hare, P. (1994). Site-specific conformational alteration of the Oct-1 POU domain-DNA complex as the basis for differential recognition by Vmw65 (VP16). *Cell* **79**, 841–852.

Xu, W., Rould, M.A., Jun, S., Desplan, C. and Pabo, C.O. (1995). Crystal structure of a paired domain–DNA complex at 2.5Å resolution reveals structural basis for Pax developmental mutations. *Cell* **80**, 639–650.

CHAPTER NINE

Activation and repression of gene expression by transcription factors

9.1 INTRODUCTION

Although binding to DNA is often a necessary prerequisite for a factor to affect transcription, it is not in itself sufficient. Thus, following binding to DNA, the bound factor must interact with other transcription factors or with the RNA polymerase itself in order to affect the rate of transcription. Although such an interaction very often results in the activation of transcription, this does not always occur and a number of cases have now been defined in which the binding of a factor can result in the repression of transcription. Activation and repression of transcription will therefore be discussed in turn.

9.2 ACTIVATION OF TRANSCRIPTION

9.2.1 Activation domains

Extensive studies on a variety of transcription factors have shown that they have a modular structure in which distinct regions of the protein mediate particular functions such as DNA binding (see Chapter 8) or interaction with specific effector molecules such as steroid hormones (see Section 4.4.1). It is likely therefore that a specific region of each individual transcription factor will be involved in its ability to activate transcription following DNA binding. As described in Section 2.3.3, such activation domains have been identified by so-called 'domain-swap' experiments in which various regions of one factor are linked to the DNA-binding domain of another factor and the ability to activate transcription assessed.

In general, these experiments have confirmed the modular nature of transcription factors with distinct domains mediating DNA binding and transcriptional activation. Thus, in the case of the yeast factor, GCN4, two distinct regions, each of 60 amino acids, have been identified which mediate, respectively, DNA binding and transcriptional activation (Figure 9.1a) (Hope and Struhl, 1986). Similarly, domain-swap experiments have identified two regions of the glucocorticoid receptor, one at the N-terminus of the molecule and the other near the C-terminus, which can independently mediate gene activation when linked to the DNA-binding domain of another transcription factor (Hollenberg and Evans, 1988) and both of these are distinct from the DNA-binding domain of the molecule. Interestingly, the C-terminal activation domain is located close to the hormone-binding domain of the receptor (Figure 9.1b), and can mediate the activation of transcription only following hormone addition. It therefore plays an important role in the steroid-dependent activation of transcription following hormone addition (see Section 10.3.3).

Studies on a variety of transcription factors have therefore strongly indicated their modular nature with distinct regions of the molecule

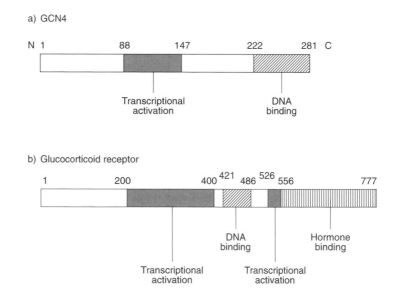

Figure 9.1 Domain structure of the yeast GCN4 transcription factor (a) and the mammalian glucocorticoid receptor (b). Note the distinct domains which are active in DNA binding or transcriptional activation.

mediating DNA binding and transcriptional activation. An extreme example of this modularity is provided by the interaction of the cellular transcription factor Oct-1 (see Sections 8.2 and 8.6) and the herpes simplex virus *trans*-activating protein VP16 (for a review, see Goding and O'Hare, 1989). Thus, although VP16 contains a very strong activating region, which can strongly induce transcription when artificially fused to the DNA-binding domain of the yeast GAL4 transcription factor, it contains no DNA-binding domain and cannot therefore bind to DNA itself. Transcriptional activation by VP16 following viral infection is therefore dependent upon its ability to form a protein–protein complex with the cellular Oct-1 protein. This complex then binds to the octamer-related TAATGARAT (R = purine) motif in the viral immediate-early genes via the DNA-binding domain of Oct-1 and transcription is activated by the activation domain of VP16. Hence, in this case, the DNA-binding and transcriptional activation domains are actually located on different proteins in the DNA-binding complex (Figure 9.2). A similar example, in which the constitutively expressed Oct-1 recruits a non-DNA-binding B-cell-specific, activating molecule, OCA-B, to the immunoglobulin promoter resulting in its B cell-specific activation, was discussed in Section 5.2.2 indicating that this effect is not confined to viral *trans*-activating molecules.

9.2.2 Nature of activation domains

Following the identification of activation domains in different transcription factors, it rapidly became clear that they fell into several distinct families with common features which will be discussed in turn (for a typical example of each of the major classes of activation domain, see

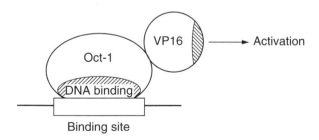

Figure 9.2 Activation of gene transcription by interaction of the cellular factor Oct-1, which contains a DNA-binding domain, and the herpes simplex virus VP16 protein, which contains an activation domain but cannot bind to DNA.

GAL4 acidic domain

D S A A A H H **D** N S T I P L **D** F M P R **D**

A L H G F **D** W S **E E D D** M S **D** G L P F L

K T **D** P N N N G F

Sp1 glutamine-rich domain B

Q G **Q** T P **Q** R V S G L **Q** G S D A L N I **Q**

Q N **Q** T S G G S L **Q** A G **Q Q** K E G E **Q** N

Q Q T **Q Q Q Q** I L I **Q** P **Q** L V **Q** G G **Q** A

L **Q** A L **Q** A A P L S G **Q** T F T T **Q** A I S

Q E T L **Q** N L **Q** L **Q** A V P N S G P I I I

R T P T V G P N G **Q** V S W **Q** T L **Q** L **Q** N

L **Q** V **Q** N P **Q** A **Q** T I T L A P M **Q** G V S

L G **Q**

CTF/NFI Proline-rich domain

P P H L N **P** Q D **P** L K D L V S L A C D **P**

A S Q Q **P** G R L N G S G Q L K M **P** S H C

L S A Q M L A **P P P P** G L **P** R L A L **P P**

A T K **P** A T T S E G G A T S **P** S Y S **P P**

D T S **P**

Figure 9.3 Structure of typical members of each of the three classes of activation domains. Acidic, glutamine or proline residues are highlighted in the appropriate case.

Figure 9.3) (for a review, see Mitchell and Tjian, 1989; Triezenberg, 1995).

Acidic domains

Comparison of several different activation domains, including those of the yeast factors GCN4 and GAL4 as well as the activation domain at the N-terminus of the glucocorticoid receptor and that of VP16, which were discussed above (section 9.2.1), indicated that, although they do not show any strong amino-acid sequence homology to each other, they all have a large proportion of acidic amino acids, producing a strong net negative charge (Figure 9.3). Thus the 82-amino-acid activating region of the glucocorticoid receptor contains 17 acidic residues (Hollenberg and Evans, 1988), whilst the same number of negatively charged amino acids is found within the 60-amino-acid activating region of GCN4 (Hope and Struhl, 1986). These findings indicated therefore that these activation regions consist of so-called 'acid blobs' or 'negative noodles' with a high proportion of negatively charged amino acids which are involved in the activation of transcription (for a review see Hahn, 1993a).

In agreement with this idea, mutations in the activation domain of GAL4 which increase its net negative charge, increase its ability to activate transcription. Similarly, if recombination is used to create a GAL4 protein with several more negative charges, the effect on gene activation is additive, a mutant with four more negative charges than the parental wild type, activating transcription nine-fold more efficiently than the wild type. Thus the acidic nature of these domains is likely to be important in their function. It has been suggested that, in the case of VP16, the negative charge of its acidic domain allows it to establish long-range electrostatic interactions with the $TAF_{II}31$ component of TBP (see Section 9.2.4) with which it interacts to stimulate transcription (Uesugi *et al.*, 1997) (Figure 9.4a).

Although the acidic nature of the activation domain is clearly important for its function, it is not the only feature required since it is possible to decrease the activity of the GAL4 activation domain without reducing the number of negatively charged residues. Indeed, recent evidence indicates that conserved hydrophobic residues in the acidic activation domains play a key role in their ability to stimulate transcription. Thus, when the VP16 activation domain interacts with the $TAF_{II}31$ component of TBP, it undergoes a conformational change from a random coil to an α-helix,

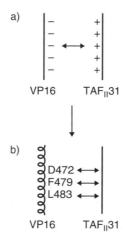

Figure 9.4 (a) The negatively charged acidic residues in the VP16 activation domain allow its initial long-distance interaction with $TAF_{II}31$. (b) Interaction with $TAF_{II}31$ induces a conformational charge in the domain to an α-helical structure in which the hydrophobic residues asparagine (D) at position 472, phenylalanine (F) at position 479 and leucine (L) at position 483 are brought close to one another and bind to $TAF_{II}31$.

which brings together three hydrophobic residues within the acidic domain which then interact directly with TAF$_{II}$31 (Uesugi *et al.*, 1997) (Figure 9.4b). Hence the acidic domain would interact with TAF$_{II}$31 via a two-step process in which the initial long-range attraction produced by the acidic residues allows a subsequent structural change facilitating a close interaction of the hydrophobic residues within the acidic domain with TAF$_{II}$31. Hence both the acidic and hydrophobic residues are of importance for the activity of this domain.

Although activation domains of the acidic type form the majority of the activation domains so far identified in eukaryotic transcription factors from yeast to mammals, other types of activation domains have been identified in a number of different transcription factors in higher eukaryotes and these will be discussed in turn.

Glutamine-rich domains

Analysis of the constitutive transcription factor Sp1 (see Section 3.3.2) revealed that the two most potent activation domains contained approximately 25% glutamine residues and very few negatively charged residues (Courey and Tjian, 1988) (Figure 9.3). These glutamine-rich motifs are essential for the activation of transcription mediated by these domains since their deletion abolishes the ability to activate transcription. Most interestingly, however, transcriptional activation can be restored by substituting the glutamine-rich regions of Sp1 with a glutamine-rich region from the *Drosophila* homeobox transcription factor *Antennapedia*, which has no obvious sequence homology to the Sp1 sequence. Hence, as with the acidic activation domains, the activating ability of a glutamine-rich domain is not defined by its primary sequence but rather by its overall nature in being glutamine rich. In agreement with this, a continuous run of glutamine residues with no other amino acids has been shown to act as a transcriptional activation domain (Gerber *et al.*, 1994).

Similar glutamine-rich regions have been defined in transcription factors other than Sp1 and Antennapedia including the N-terminal activation domains of the octamer-binding proteins Oct-1 and Oct-2, the *Drosophila* homeobox proteins ultra-bithorax and zeste, and the yeast HAP1 and HAP2 transcription factors, indicating that this motif is quite widespread, being found in different transcription factors in different species (for a review, see Mitchell and Tjian, 1989).

Proline-rich domains

Studies on the constitutive factor CTF/NF1 which binds to the CCAAT box motif (see Section 3.3.3) defined a third type of activation domain distinct from those previously discussed. Thus the activation domain located at the C-terminus of CTF/NF1 is not rich in acidic or glutamine residues but instead contains numerous proline residues, forming approximately one-quarter of the amino acids in this region (Mermod *et al.*, 1989) (Figure 9.3). As with the other classes of activation domains, this region is capable of activating transcription when linked to the DNA-binding domains of other transcription factors. Moreover, as with the glutamine-rich domain, a continuous run of proline residues can mediate activation, indicating that the function of this type of domain depends primarily on its richness in proline (Gerber *et al.*, 1994). Similar proline-rich domains have now been identified in several other transcription factors such as the oncogene product Jun, AP2 and the C-terminal activation domain of Oct-2 (for a review, see Mitchell and Tjian, 1989). Thus, as with the glutamine-rich domains, proline-rich domains are not confined to a single factor whilst a single factor such as Oct-2 can contain two activation domains of different types.

In summary, therefore, it is clear that, as with DNA binding, several distinct protein motifs can activate transcription (Figure 9.3).

Functional relationship of the different activation domains

The existence of at least three distinct classes of activation domain raises the question of whether these three domains are functionally equivalent or whether they differ in their ability to activate transcription. This question was investigated by Seipel *et al.* (1992), who linked each of the activation domains to the DNA-binding domain of the GAL4 factor and tested the ability of these chimaeric proteins to activate transcription in mammalian cells when the GAL4 DNA binding site was placed at different positions relative to the start site of transcription (Figure 9.5). In these experiments, all three domains were able to activate transcription when the DNA-binding site was placed close to the start-site of transcription in the promoter region. In contrast, the glutamine-rich domain was unable to activate transcription when the binding site was located downstream of the transcription unit mimicking a position within an enhancer element (see Section 1.2.4). The acidic domain was strongly active from this enhancer position, whilst the proline-rich domain could also activate transcription from this position but only weakly.

These findings indicate therefore that clear differences exist in the abilities of the different activation domains to activate transcription

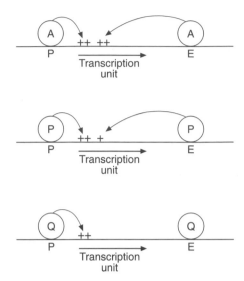

Figure 9.5. An acidic activation domain (A) can stimulate transcription when bound to DNA in the promoter (P) close to the transcriptional start site or when bound at a distant enhancer (E). In contrast, a proline-rich domain (P) stimulates only weakly from an enhancer position and a glutamine-rich domain (Q) does not stimulate at all from this position.

when bound to the DNA at different positions relative to the promoter. Such differences are likely to be important in determining the functional activity of different factors. In addition such differences in the activity of different activation domains are likely to reflect differences in the mechanisms by which these factors act. In agreement with this idea, acidic or proline-rich activation domains derived from mammalian factors can also activate transcription when introduced into yeast cells, whereas glutamine-rich domains cannot do so (Kinzler *et al.*, 1994).

In the next sections we will consider the mechanisms by which activation domains act, focusing particularly on the acidic domains where most information is available. Similarities and differences in the mode of action of the other activation domains will be discussed where this information is available (for reviews, see Ranish and Hahn, 1996; Stargell and Struhl, 1996).

9.2.3 Interaction of activation domains with the basal transcriptional complex

The widespread interchangeability of acidic activation domains from yeast, *Drosophila* and mammalian transcription factors, discussed above, strongly suggests that a single common mechanism may mediate transcriptional activation by acidic activation domains in a wide range of organisms. This idea is supported by the finding noted above that mammalian transcription factors carrying such domains, for example, the glucocorticoid receptor can activate a gene carrying their appropriate DNA-binding site in yeast cells, whilst the yeast GAL4 factor can do so in cells of *Drosophila*, tobacco plants and mammals (for reviews, see Guarente, 1988; Ptashne, 1988).

These considerations suggest that the target factor or factors with which these activators interact is likely to be highly conserved in evolution. A number of experiments have indicated that, in many cases, this target factor is a component of the basal transcription complex (see Section 3.2) required for the transcription of a number of different genes and not solely for that of the activated gene. Thus the over-expression of the yeast GAL4 protein, which contains a strong activation domain, results in the down regulation of genes which lack GAL4-binding sites such as the CYC1 gene as well as activating genes which do contain GAL4-binding sites. This phenomenon, which has been noted for a number of transcription factors with strong activation domains, is known as squelching (for a review, see Ptashne, 1988). Although the degree of squelching by any given factor is proportional to the strength of its activation domain, squelching differs from activation in that it does not require DNA-binding and can be achieved with truncated factors containing only the activation domain and lacking the DNA binding domain. This phenomenon can therefore be explained on the basis that a transcriptional activator when present in high concentration can interact with its target factor in solution as well as on the DNA. If this target factor is present at limiting concentrations, it will therefore be sequestered away from other genes which require it for transcription resulting in their inhibition (Figure 9.6).

The existence of squelching indicates therefore that, in many cases, the target factor for activation domains is likely to be a component of the basal transcriptional apparatus which is required for the transcription of a wide range of genes and which is conserved from yeast to mammals, allowing yeast activators to work in mammalian cells and vice versa. Obviously an activating factor could act by stimulating the binding of such a component so that the basal complex assembled more efficiently. Alternatively, it could act by interacting with a factor which had already bound so that the activity or stability of the assembled complex was stimulated. It appears that both these mechanisms are used and they will be discussed in turn.

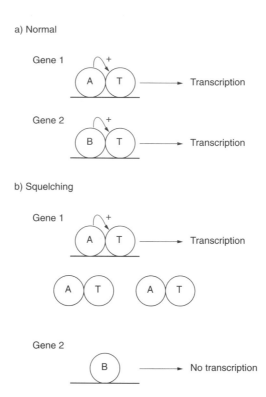

Figure 9.6 The process of squelching. In the normal case illustrated in (a), two distinct activator molecules A and B, involved in the activation of genes 1 and 2, respectively, both act by interacting with the general transcription factor T and both genes are transcribed. In squelching, illustrated in (b), factor A is present at high concentration and hence interacts with T both on gene 1 and in solution. Hence factor T is not available for transcription of gene 2 and therefore only gene 1 is transcribed, whilst transcription of gene 2 is squelched.

Factor binding

As described in Section 3.2.4, the basal transcriptional complex can assemble in a stepwise manner, with the binding of TFIID being followed by the binding of TFIIB and then the binding of RNA polymerase in association with TFIIF. Clearly an activator could increase the rate of complex assembly by enhancing any one of these assembly steps. Indeed, there is evidence that activators target several of these steps in the assembly process (Figure 9.7). Thus, for example, it appears that acidic activators interact directly with TFIID (see Section 3.2.4) to stimulate the binding of TFIID to the promoter (Figure 9.7a).

Although increased binding of TFIID to the promoter will directly enhance the assembly of the complex by allowing TFIIB to bind, there is

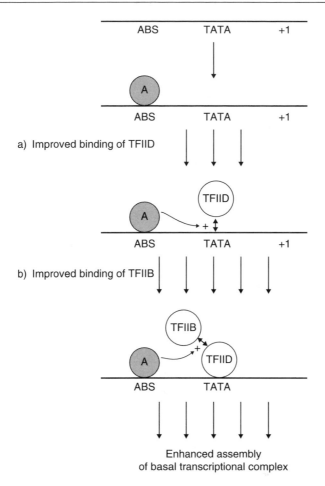

Figure 9.7. The binding of an activating molecule (A) to its binding site (ABS) can enhance both the binding of TFIID to the TATA box (a) and the recruitment of the TFIIB factor (b), so enhancing the rate of basal complex assembly and of transcription.

evidence that activators can also act directly to improve the recruitment of TFIIB independent of their effect on TFIID (Figure 9.7b). Thus it has been shown that both an acidic activator and glutamine- or proline-rich activators can greatly stimulate the binding of TFIIB to the promoter (Choy and Green, 1993). Hence activators can enhance the assembly of the basal transcriptional complex by independently enhancing the binding of both TFIID and TFIIB. This ability of activators to act at these two independent steps results in a strong synergistic activation of transcription in the presence of different activators targeting either TFIID or TFIIB (Gonzalez-Couto *et al.*, 1997).

As with TFIID, it has been shown that TFIIB interacts directly with activating molecules. Thus TFIIB can be purified on a column containing a bound acidic activator and interactions of TFIIB with non-acidic activators have also been reported. Moreover, mutations in the activator which abolish this interaction with TFIIB prevent it from activating transcription (for a review, see Hahn, 1993b). Thus the effect of activators on TFIIB is mediated via a direct protein–protein interaction, which is essential for their ability to stimulate transcription.

In addition to the stepwise pathway of complex assembly, it has also been proposed that the basal transcriptional complex can assemble in a much simpler manner with binding of TFIID being followed by binding of the RNA polymerase holoenzyme which contains the polymerase itself, TFIIB, TFIIF and TFIIH, as well as a number of other proteins (see Section 3.2.4). There is evidence that activators can also act in this pathway, not only by enhancing the recruitment of TFIID as described above but also by directly enhancing the binding of the RNA polymerase holoenzyme itself (Figure 9.8a). Thus, for example, if a DNA-binding domain is linked to the yeast protein Gal11, which is a component of the RNA polymerase holoenzyme, the holoenzyme is recruited to the DNA via this DNA-binding domain and transcription is activated (Figure 9.8b) (Barberis *et al.*, 1995). Hence the need for activators can be bypassed by

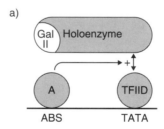

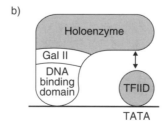

Figure 9.8 (a) Activators can act by enhancing the binding of the RNA polymerase holoenzyme complex following binding of TFIID. (b) In agreement with this idea, the need for an activator can be bypassed by artifically attaching a DNA-binding domain to the Gal11 component of the holoenzyme, so enhancing holoenzyme recruitment by allowing it to bind to DNA directly.

recruiting the RNA polymerase holoenzyme to DNA via an artifical DNA-binding domain.

Indeed, on the basis of experiments of this type, Ptashne and Gann (1997) have argued that the only role of activators is to enhance the assembly of the basal complex by interacting with one or other of its specific components, so facilitating their recruitment to the DNA. However, whilst such enhanced recruitment of specific components of the complex clearly plays a major part in the action of transcriptional activators and operates in both pathways of complex assembly, it is likely that other effects are also involved in the action of transcriptional activators. These effects are discussed in the next section.

Factor activity

In addition to their effects on complex assembly, it is clear that activators can also stimulate transcription at a subsequent step following assembly

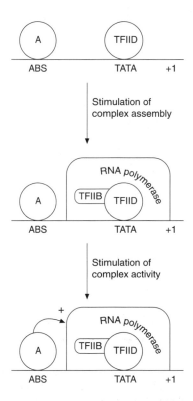

Figure 9.9. An activator can stimulate transcription both by promoting the assembly of the basal transcription complex and by stimulating its activity following assembly.

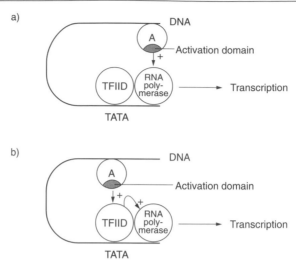

Figure 9.10 Two possible mechanisms by which an activating factor (A) could stimulate the activity of the basal trascriptional complex. This could occur via direct interaction with the RNA polymerase itself (a) or by interaction with another transcription factor such as TFIID, which in turn interacts with the polymerase (b).

of the complex, resulting in its enhanced stability or increased activity (Choy and Green, 1993) (Figure 9.9).

An obvious mechanism for activation would be for activating domains to interact directly with the RNA polymerase itself to increase its activity (Figure 9.10a). As discussed in Section 3.1, the largest subunit of RNA polymerase II contains at its C-terminus multiple copies of a sequence whose consensus is Tyr-Ser-Pro-Thr-Ser-Pro-Ser, which is highly conserved in evolution and is essential for its function. This motif is very rich in hydroxyl groups and lacks negatively charged acidic residues.

It has therefore been suggested that this motif could interact directly either with a negatively charged acidic domain or via hydrogen bonding with amide groups in a glutamine-rich activation domain. This would provide a mechanism for direct interaction between activating domains and RNA polymerase itself, whilst the evolutionary conservation of the target region within the polymerase would explain why yeast activators work in mammalian cells and vice versa. In agreement with this idea it has been shown that yeast mutants containing a reduced number of copies of the heptapeptide repeat in RNA polymerase II are defective in their response to activators such as GAL4.

Although these results are consistent with a direct interaction between transcriptional activators and the RNA polymerase, they are equally consistent with an indirect interaction in which the activator contacts another component of the transcriptional machinery which then interacts with the repeated motif in the polymerase. Indeed, it has recently been shown that the interaction of activators with RNA polymerase *in vitro* requires a mediator complex consisting of some 20 polypeptides rather than occurring directly (Bjorklund and Kim, 1996). Interestingly, the mediator has been shown to associate with the C-terminal domain of the RNA polymerase within the RNA polymerase holoenzyme complex (see Section 3.2.4), indicating that it could act to transduce the signal from activating molecules to this domain. Therefore despite the attractiveness of a model involving direct interaction between activating factors and the polymerase itself, it is unlikely to be correct and it appears that activators interact with the polymerase indirectly via other factors (Figure 9.10b).

As well as the mediator complex, TFIID is an obvious candidate for the component with which activating factors interact, since this factor is both required for the transcription of a wide variety of genes both with and without TATA boxes (see Section 3.2.5) and is highly conserved in evolution, with the yeast factor being able to promote transcription in mammalian cell extracts and vice versa. Evidence for an effect of activating factors on TFIID has been obtained in the case of the yeast acidic activating factor GAL4 (Horikoshi *et al.*, 1988). Thus, in the absence of GAL4, TFIID was shown to be bound only at the TATA box of a promoter containing both a TATA box and GAL4-binding sites. In contrast, in the presence of GAL4 bound to its upstream binding sites in the promoter, the conformation of TFIID was altered such that it now covered both the TATA box and the start site for transcription (Figure 9.11). Moreover, no change in TFIID conformation was observed in the presence of a truncated GAL4 molecule, which can bind to DNA but lacks the acidic activation domain. Hence, an acidic activator can produce a

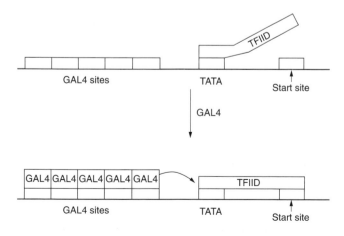

Figure 9.11 Effect of GAL4 binding on the binding of TFIID.

change in TFIID conformation, resulting in its binding to the start site for transcription, and this effect correlates with the ability of GAL4 to activate transcription rather than being a consequence of its binding to DNA.

It is clear therefore that activating molecules can alter the configuration of TFIID bound to the promoter by interacting directly with it. As well as interacting with TFIID to change its configuration, activators can also interact with TFIIB changing its conformation and enhancing its ability to recruit the complex of RNA polymerase II and TFIIF (Roberts and Green, 1994) (Figure 9.12). Hence activators appear to target both TFIID and TFIIB in two ways. Firstly, as described in the previous section, they enhance their binding to the promoter and, secondly, they alter their conformation so as to enhance their activity (Figure 9.13).

Together with TFIIB, TFIID constitutes a major target for transcriptional activators. Interestingly, however, other components of the basal complex, such as TFIIA (Ozer *et al.*, 1994), TFIIF (Joliot *et al.*, 1995) and TFIIH (Xiao *et al.*, 1994), have also been shown to interact with transcriptional activators. Hence a number of different factors within the basal transcriptional complex serve as targets for direct interactions with transcriptional activators. However, it is clear that, in some cases, activators interact with the basal complex only indirectly via other factors and such interactions are discussed in the next section.

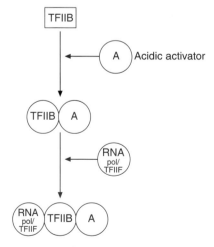

Figure 9.12 The binding of an acidic activator (A) to TFIIB produces a conformational change which enhances the ability of TFIIB to interact with the RNA polymerase–TFIIF complex, thereby enhancing its recruitment to the promoter.

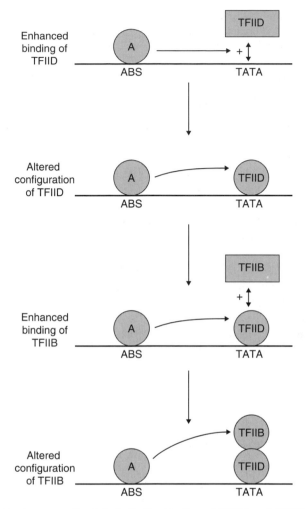

Figure 9.13 Activators can stimulate both the binding of TFIIB and TFIID and enhance their activity by altering their conformation (square to circle).

9.2.4 Interaction of activation domains with co-activator molecules

TAFs

As described in chapter 3 (section 3.2.5) TFIID consists of the TBP protein that binds to the TATA box and a number of other proteins known as TAFs (TBP-associated factors). In some cases, where activators interact with TFIID, such interactions can be reproduced with purified

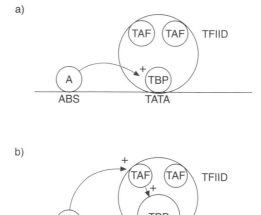

Figure 9.14 Interaction of an activator molecule with TBP can occur either directly (a) or indirectly (b) via an intermediate TBP-associated adaptor molecule (TAF).

TBP. Moreover, mutations in specific acidic activators which interfere with their ability to interact with TBP also abolish their ability to activate transcription, indicating an important functional role for these interactions.

Although there is thus evidence that the ability to interact with TBP appears to be essential for transcriptional activation in some cases (Figure 9.14a), there is also evidence that, in some circumstances, such activation requires interaction of the activator with the TAFs rather than with TBP. Thus, in many cases, stimulation of transcription *in vivo* by activator molecules does not occur with purified TBP but is dependent upon the presence of the TFIID complex and hence of the TAFs. This suggests a model in which the interaction of activators with TBP occurs indirectly via TAFs, with the TAFs being co-activator molecules linking the activators with the basal transcriptional complex (Figure 9.14b) (for reviews, see Verrijzer and Tjian, 1996; Tansey and Herr, 1997).

Interestingly, there is evidence that different classes of activation domain may interact with different TAFs (Chen *et al.*, 1994). Thus, whilst acidic activation domains have been shown to interact directly with $TAF_{II}31$ (also known as $TAF_{II}40$), the glutamine-rich domain of Sp1 interacts with $TAF_{II}110$, whilst multiple activators including proline-rich activators target $TAF_{II}55$. Hence different types of activation domains may have different targets within the TFIID complex (Figure 9.15).

In agreement with this idea, the acidic activation domain of VP16 is not capable of squelching gene activation by the non-acidic activation domain

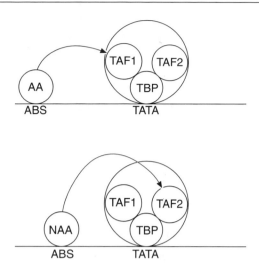

Figure 9.15. Acidic (AA) and non-acidic (NAA) activator molecules may interact with different TBP-associated factors (TAFs) within the TFIID complex.

of the oestrogen receptor, whereas the oestrogen receptor activation domain is capable of squelching gene activation mediated both by its own activation domain and by the acidic domain of VP16, indicating that they contact different molecules. Moreover, these findings suggest that a series

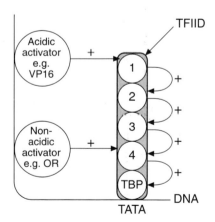

Figure 9.16 Interaction of different activator molecules with different adaptor molecules (1–4) which each activate each other and ultimately activate TBP. Note that the ability of the non-acidic activation domain of the oestrogen receptor to squelch activation by the acidic activation domain of VP16, but not vice versa, can be explained if the oestrogen receptor interacts with an adaptor molecule (4) closer to TBP in the series than that with which VP16 interacts (1).

of TAFs within TFIID may mediate activation, with the acidic activation domain of VP16 contacting a factor which is located earlier in the series than that contacted by the non-acidic activation domain of the oestrogen receptor (Figure 9.16). Hence the factor contacted by the activation domain of the oestrogen receptor would also be essential for activation by VP16 (factor 4 in Figure 9.16), whereas the factor contacted by the acidic activation domain of VP16 (factor 1 in Figure 9.16) would not be required for activation by the oestrogen receptor.

The functional differences which exist between different factors in their ability to activate transcription from different positions and in different species (see Section 9.2.2) are therefore paralleled by differences in their ability to interact with different TAFs. This ability of different activation domains to interact with different TAFs can produce a strong synergistic activation of transcription which is far stronger than the sum of that observed with either activation domain alone. Thus, the ability of different activators to bind to different TAFs in the TFIID complex would result in greatly enhanced recruitment of TFIID compared to the effect of either activator alone (Figure 9.17) (for a review, see Buratowski, 1995).

These findings thus suggest that the TAFs are of importance for transcriptional activation and mediate some of the interactions between activators and TFIID which were described in Section 9.2.3. However, it is clear that their importance varies between different species and on different promoters. Thus, whilst TAFs appear to be of central importance in transcriptional activation in higher eukaryotes such as man and *Drosophila*, they are not essential for transcriptional activation at most promoters in yeast (for reviews, see Chao and Young, 1996; Sauer and Tjian, 1997). Similarly, even in higher eukaryotes, specific TAFs appear to be of key importance at particular types of promoters. Thus mutation of $TAF_{II}250$ inhibits the expression of specific genes and results in cell-cycle arrest in mammalian cells without affecting the transcription of other genes (Wang and Tjian, 1994).

This idea that particular TAFs may play a critical role in mediating the response to activators at specific genes has recently been extended by

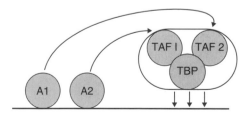

Figure 9.17 The ability of different activators (A1 and A2) to interact with different TAFs will result in a strong synergistic enhancement of TFIID recruitment and hence of transcriptional activation.

findings suggesting that TAFs also function in promoter selectivity. Thus it appears that TFIID complexes containing particular TAFs assemble preferentially at particular promoters. This effect may be mediated by particular TAFs binding preferentially to particular core promoters (see Section 1.2.1) containing different sequences between the TATA box and the start site of transcription (Figure 9.18). Thus, as noted above, most yeast genes do not require TAFs for the activation of transcription. However, a few genes involved in cell-cycle progression such as the cyclin genes have been shown to be dependent upon $TAF_{II}145$ for their transcription. This dependence upon $TAF_{II}145$ is not due to the nature of the activator sequences in the promoter but is dependent upon the nature of the core promoter (Shen and Green, 1997) (Figure 9.19). Although the yeast promoters used in this study contain a TATA box, the ability of TAFs to interact with specific core promoter sequences may be of particular importance on promoters lacking a TATA box and containing an initiator element where, as discussed in Section 3.2.5, TBP is

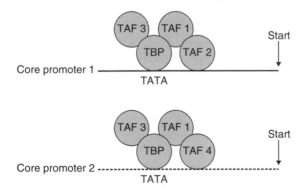

Figure 9.18 TFIID complexes containing different TAFs bind preferentially to different core promoters containing different sequences between the TATA box and the transcriptional start site.

Gene 1	UAS	Core	$TAF_{11}145$ dependent
Gene 2	UAS	Core	$TAF_{11}145$ independent
1/2	UAS	Core	$TAF_{11}145$ independent
2/1	UAS	Core	$TAF_{11}145$ dependent

Figure 9.19 The dependence of particular yeast promoters on $TAF_{II}145$ for transcription is determined by the nature of the core promoter not on the upstream activator binding sites (UAS).

brought to the promoter by factors binding to the initiator element rather than by TBP binding to the TATA box (for a review, see Roeder, 1996).

Thus, particular TFIID complexes containing specific combinations of TAFs may bind selectively to specific promoters rather than only responding to transcriptional activators following binding. This idea has been supported by the finding of a cell type-specific form of TAF$_{II}$130, known as TAF$_{II}$105, which is expressed only in B lymphocytes (Dikstein *et al.*, 1996). Hence different forms of TFIID containing different TAFs may exist in different tissues and may thus play a role in the cell type-specific regulation of gene expression. Indeed, a second form of TBP which is expressed specifically in the brain has recently been identified suggesting that cell type-specific forms of TBP as well as of the TAFs can exist (for a review, see Buratowski, 1997). Obviously, the different TFIID complexes formed in this manner may also differ in their responses to different transcriptional activators. Thus, for example, TAF$_{II}$30, which mediates transcriptional activation by the oestrogen receptor is found in only some TFIID complexes. In others, it is replaced by TAF$_{II}$18 which does not mediate activation by the receptor (for a review see Chang and Jaehning, 1997).

Hence the TAF factors play a key role in transcription, by acting as co-activators mediating the response to specific activators and by regulating the binding of TFIID to specific promoters containing particular sequences adjacent to the TATA box (Figure 9.20). This ability of the TAFs to act as an intemediate between the basal transcriptional complex

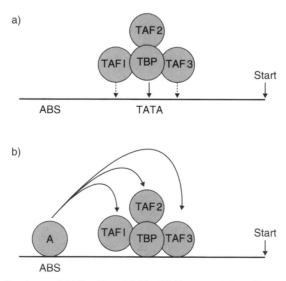

Figure 9.20 Mechanisms of TAF action. (a) The TAFs may act to enhance binding of TFIID to specific promoters by interacting with DNA sequences adjacent to the TATA box to which TBP binds. (b) The TAFs can mediate the response of TFIID to transcriptional activators.

and transcriptional activators is evidently paralleled by the role of the mediator complex, which acts as an intermediate between activators and the RNA polymerase itself within the RNA polymerase holoenzyme complex (see Section 9.2.3).

CBP and other co-activators

In addition to factors such as the TAFs and the mediator, which were originally defined via their association with the basal transcriptional complex, other co-activators exist which were originally defined on the basis of their essential role in transcriptional activation mediated by a specific transcriptional activator. Thus, as discussed in Section 4.3.2, the CBP factor was originally defined as a co-activator essential for cyclic AMP-stimulated transcription mediated via the CREB factor. Subsequently, however, it was shown that CBP and its close relative p300 are essential co-activators for a vast range of other factors such as the nuclear receptors (Section 4.4.2), MyoD (Section 5.3.1), AP1 (Section 7.2.1), p53 (Section 7.3.2) and a number of others (for a review, see Shikama *et al.*, 1997) (Figure 9.21).

This ability of CBP and p300 to interact with a vast array of transcription factors places them at the centre of a whole range of signalling pathways in the cell and they thus play a critical role in gene activation via these pathways. The relatively low abundance of CBP/p300 in the cell means that different signalling pathways compete for them and results in mutual antagonism between different competing pathways such

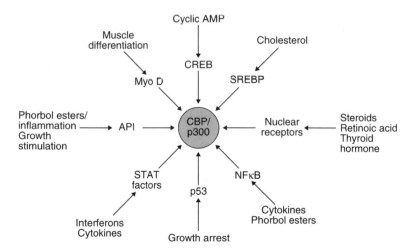

Figure 9.21 Some transcription factors which interact with the CBP/p300 co-activators and the signalling pathways which activate them.

as the inflammation mediated by the AP1 pathway and the anti-inflammatory effects of glucocorticoids (see Section 4.4.4) or the growth-promoting effects of the AP1 pathway compared to the growth-arresting effects of the p53 pathway (see Section 7.3.2). Interestingly, the activation domain of CREB undergoes a structural transition from a coiled structure to form two α-helices when it interacts with CBP (Radhakrishnan *et al.*, 1997). This evidently parallels the change in the activation domain of VP16 when it interacts with TAF$_{II}$31 (see Section 9.2.2), suggesting that the formation of a specific helical structure may be a general feature which occurs when many activation domains interact with their targets.

Although the p300/CBP proteins are the best defined co-activators, other co-activators have also been defined on the basis of their association with particular activators. Thus, for example, the nuclear receptors discussed in Section 4.4 interact not only with CBP but also with a range of other co-activators such as TIF1, TIF2, SRC-1 and Sug1 (for a review, see Glass *et al.*, 1997). Moreover, several of these co-activators associate with the receptors only after they have been activated by binding their ligand, indicating that they are likely to play a key role in the ability of the receptors to activate transcription only following ligand binding (see Section 4.4.2).

The key role of CBP/p300 and other co-activators obviously leads to the question of how they act. In the case of CBP/p300, it has been shown that they can interact with either TBP or TFIIB and can form part of the RNA polymerase holoenzyme complex (see Section 4.3.2). Hence, like the TAFs, they could act as classical co-activators linking an activating molecule and the basal transcriptional complex (see Figure 4.11). Alternatively, as described in Sections 4.3.2 and 4.4.3, several co-activators, including CBP and SRC-1, have histone acetyltransferase activity and may therefore activate transcription by acetylating histone molecules and thereby altering chromatin structure (see Figure 4.12) (see Section 1.4.3). Similar histone acetyltransferase activity is also observed for the TAF$_{II}$250 component of TFIID (see Section 3.2.5), indicating that some TAFs may also act via the alteration of chromatin structure. Such activation of transcription via changes in chromatin structure is discussed further in Section 9.2.5.

A multitude of targets for transcriptional activators

There thus exists an array of target factors which are contacted by transcriptional activators and these include components of the basal complex such as TFIIB and TBP, as well as the various TAFs and other co-activators with some factors being contacted by activators of all classes and others by activators of only one class. Even when the finding that some of these targets such as individual TAFs can interact with only one class of activation domain is taken into account, there still remains a bewildering number of targets within the basal complex. Thus, for example, in the most extreme case described so far, the acidic activation domain of VP16 has been reported to interact with TFIIB, TFIIH, TBP, TAF$_{II}$40, TAF$_{II}$31 and the RNA polymerase holoenzyme (for a review, see Chang and Jaehning, 1997). It should be noted, however, that the various possible targets are not mutually exclusive. Indeed, the ability of different molecules of the same factor or different activating factors to interact with different components within the basal transcriptional complex is likely to be essential for the strong enhancement of transcription which is the fundamental aim of activating molecules (Figure 9.22) (for a review, see Carey, 1998).

9.2.5 Other targets for transcriptional activators

Although the basal transcriptional complex which initiates transcription is the best characterized target for transcriptional activators and co-activators (as discussed in the preceding sections), at least two other stages of the transcriptional process can be targeted by such activators and these will be discussed in turn.

(a) Modulation of chromatin structure

As discussed in Section 1.4, the DNA molecule is associated with histones and other proteins to form particles known as nucleosomes which are the basic unit of chromatin structure. Prior to the onset of transcription, the chromatin structure becomes altered, thus allowing the subsequent binding of the factors which actually stimulate transcription. As discussed in previous chapters, this alteration in chromatin structure

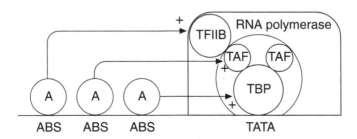

Figure 9.22 The ability of multiple activating molecules (A) to contact different factors allows strong activation of transcription.

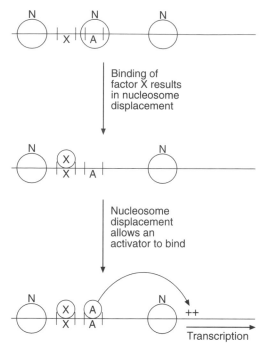

Figure 9.23 A transcription factor (X) can stimulate transcription by binding to DNA and displacing a nucleosome (N), so allowing a constitutively expressed activator (A) to bind and activate transcription.

can itself be produced by the binding of a specific transcription factor. This results in a change in the nucleosome pattern of DNA–histone association, thereby allowing other activating factors access to their specific DNA-binding sites (Figure 9.23) (for reviews, see Beato and Eisfeld, 1997; Hartzog and Winston, 1997; Li *et al.*, 1997a; Wu, 1997).

Thus, as described in Section 4.4.3, the binding of the activated glucocorticoid receptor to the mouse mammary tumour virus promoter results in an alteration in the nucleosomal organization of the DNA. This allows the ubiquitous NFI factor to bind to its binding site, which was previously masked by a nucleosome and hence allows it to activate transcription by interacting with the basal transcriptional complex. Similarly, as discussed in Section 4.2.1, the binding site for the heat shock factor (HSF) in the hsp70 gene promoter is unmasked by the prior binding of the GAGA factor to its binding site in the gene promoter. In non-heat-shock cells, however, HSF is present in an inactive non-DNA-binding form and therefore cannot bind to its site, even though it is accessible. Following heat shock, HSF is converted to a DNA-binding form and immediately binds to its site and activates transcription.

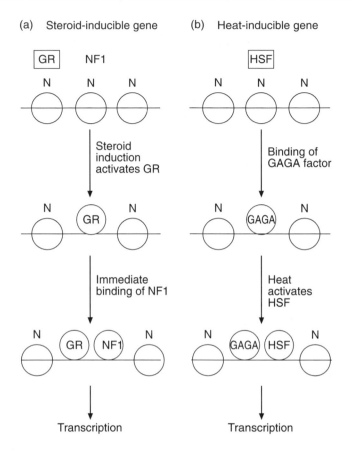

Figure 9.24. Two-stage induction of genes by steroid treatment or heat shock involving alteration of chromatin structure and the subsequent binding of an activator molecule. In the steroid case (a), the steroid activates the glucocorticoid receptor (GR), resulting in a change in chromatin structure and immediate binding of the constitutively expressed NF1 factor. In contrast in the heat-inducible case (b), the chromatin structure has already been altered prior to heat shock by binding of the GAGA factor but transcription only occurs when heat-shock activates the heat-shock factor (HSF) to a DNA-binding form.

These examples illustrate therefore how factors which alter chromatin structure can prepare the way for the binding of the factors which actually activate transcription with such binding occurring either immediately following the change in chromatin structure as in the glucocorticoid receptor/NFI case or following a subsequent stimulus as in the GAGA/HSF case (Figure 9.24). The importance of factors which can alter chromatin structure is illustrated by the finding (discussed in Section 4.2.1) that the mutational inactivation of the trithorax gene encoding the GAGA factor in *Drosophila* leads to reduced viability, eye abnormalities and female sterility. These effects result from the failure of the mutants to alter the chromatin structure of a number of homeobox genes from an inactive to an active state so preventing their transcription. Hence, the GAGA factor has a widespread role in remodelling the chromatin structure of a number of different genes (for reviews, see Schumacher and Magnusson, 1997; Wilkins and Lis, 1997).

Interestingly, as well as interacting with the basal transcriptional apparatus, activation domains have also been shown to be involved in the ability of specific factors to alter chromatin structure. Thus, the activation of the yeast PHO5 gene promoter following phosphate starvation is mediated by the binding of the PHO4 factor to the PHO5 promoter, resulting in nucleosome displacement. Suprisingly, a truncated PHO4 molecule lacking the acidic activation domain but retaining the DNA-binding domain is incapable of nucleosome displacement, whilst this ability can be restored by linking the truncated PHO4 molecule to the acidic activation domain of VP16 (Figure 9.25). Hence the ability of PHO4 to disrupt chromatin structure is dependent upon the acidic activation domain, which also interacts with the basal transcription complex to stimulate transcription (for reviews, see Lohr, 1997; Svaren and Horz, 1997). This dual function is also seen in the case of the glucocorticoid receptor which, as well as altering chromatin structure thereby facilitating NF1 binding and consequent transcriptional activation, can also itself directly stimulate transcription via interaction with other transcription factors (see Section 4.4.3).

These findings indicate therefore that the remodelling of chromatin structure can be achieved both by specific factors such as the GAGA factor and by the activation domains of other factors which can modulate

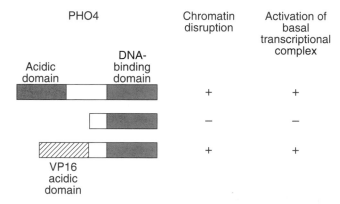

Figure 9.25 Both the disruption of chromatin and the activation of the basal transcription complex by the yeast PHO4 factor are dependent on its acidic activation domain. They are therefore lost when this domain is deleted but can be restored by addition of the acidic activation domain of VP16.

chromatin structure as well as activate transcription directly by interacting with the basal transcriptional apparatus. In both these types of cases, the ability to alter chromatin structure is likely to depend upon the ability of these factors to recruit other factors which then actually alter chromatin structure. Thus, for example, both the GAGA factor (see Section 4.2.1) and the glucocorticoid receptor (see Section 4.4.3) can bind the SWI/SNF complex which, as discussed in Section 1.4.2, can alter the nucleosomal structure by opening up the chromatin.

Hence chromatin opening by specific factors can occur via recruitment of the SWI/SNF complex (Figure 9.26a). In addition, however, many of the co-activator molecules which were described in Section 9.2.4 have histone acetyltransferase activity and can therefore produce a more open chromatin structure containing highly acetylated histones (see Section 1.4.3). Therefore, factors such as the glucocorticoid receptor, which recruit co-activators such as CBP with this activity, can remodel chromatin structure via these factors acetylating histones thereby producing a more open chromatin structure (Figure 9.26b). It is clear therefore that the alteration of chromatin structure by specific factors is of considerable importance in the control of transcription.

(b) Stimulation of transcriptional elongation

In most genes, once transcription has been initiated the RNA polymerase continues to transcribe the DNA until it has produced a complete RNA transcript. In some genes, however, such as the c-*myc* oncogene (see Section 7.2.3), some transcripts terminate prematurely and do not produce an RNA capable of encoding the appropriate protein (for a review, see Greenblatt *et al.*, 1993). Moreover, this process can be regulated. Thus, when the c-*myc* gene is transcribed in the pro-myeloid cell line HL-60, most transcripts terminate near the end of the first exon and do not produce a functional mRNA encoding the complete Myc protein. When the HL-60 cells are differentiated to form granulocytes, however, the majority of transcripts pass through this block and full-length mRNA is produced (Figure 9.27). Hence, in this case, an increased level of functional c-*myc* mRNA able to produce the Myc protein is obtained without an increase in transcriptional initiation (for a review, see Spencer and Groudine, 1990).

Although the regulatory factor which modulates c-*myc* elongation is unknown, a factor acting at the level of transcriptional elongation has been identified in the human immunodeficiency virus (HIV). Thus, in this case, only short prematurely truncated transcripts are produced from the HIV promoter in the absence of the viral Tat protein. When Tat is present, however, it stimulates both the rate of initiation of

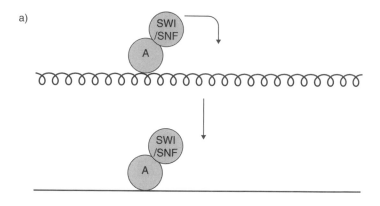

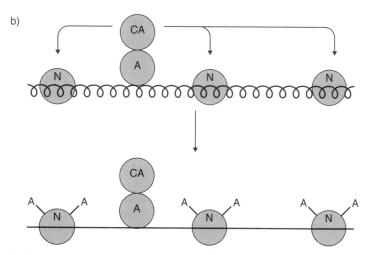

Figure 9.26 A transcriptional activator (A) can alter chromatin structure from a tightly packed (wavy line) to a more open (solid line) structure either (a) by recruiting the SWI/SNF chromatin remodelling complex or (b) by recruiting a co-activator (CA) with histone acetyltransferase activity. N = nucleosome.

transcription and also the proportion of full-length transcripts which are produced, so overcoming the block to transcriptional elongation (Figure 9.28) (for reviews, see Greenblatt *et al.*, 1993; Jones, 1997). A similar role for HSF in stimulating transcriptional elongation in the *hsp70* gene has also been proposed (for a review, see Lis and Wu, 1993). Hence activating factors can act to stimulate the proportion of full-length RNA transcripts that are produced.

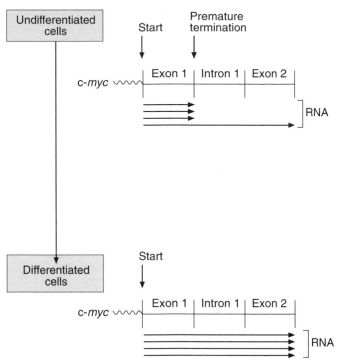

Figure 9.27 In undifferentiated HL-60 cells most transcripts from the c-*myc* gene terminate prematurely at the end of the first exon. In differentiated cells, however, this does not occur resulting in an increase in full-length functional transcripts without increased initiation.

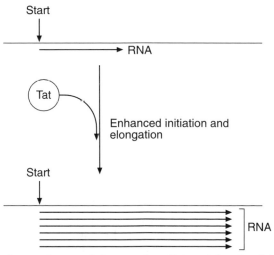

Figure 9.28. The human immunodeficency virus Tat protein stimulates transcriptional initiation so that more RNA is made and also enhances the proportion of full-length RNA species capable of encoding viral proteins.

Interestingly, as in the case of transcriptional initiation and the disruption of chromatin, there is evidence that the stimulation of transcriptional elongation involves activation domains. Thus when binding sites for the yeast transcriptional activator GAL4 were placed upstream of the c-*myc* promoter, both the rate of initiation and the proportion of full-length transcripts were greatly stimulated by the binding of hybrid transcription factors containing the DNA-binding domain of GAL4 linked to an acidic or non-acidic activation domain (Yankulon *et al.*, 1994). This effect was not observed in the presence of the GAL4 DNA-binding domain alone or when the GAL4 binding sites were deleted (Figure 9.29). Hence the ability of activating factors to act at the level of transcriptional elongation is dependent on the same activation domains which act to stimulate transcription at other stages.

These findings indicate therefore that, in the minority of genes where some transcripts terminate prematurely, it is possible for transcriptional activators to act by enhancing the proportion of full-length transcripts which are produced as well as by stimulating the number of transcripts

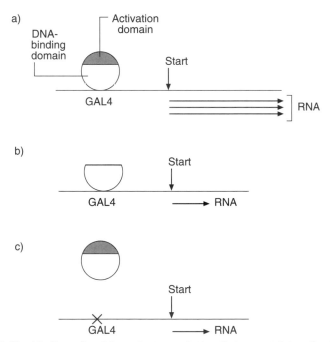

Figure 9.29 The binding of a chimaeric transcription factor containing the GAL4 DNA-binding domain linked to an activation domain (shaded) stimulates both transcriptional initiation and the proportion of full-length transcripts produced by a c-*myc* gene carrying a DNA-binding site for GAL4 (a). This effect is not observed when only the DNA-binding domain of GAL4 is present (b) or when the GAL4 binding sites are deleted so that the activator cannot bind (c).

which are initiated (for a review, see Bentley, 1995). Similarly, as discussed in Section 7.3.4, proteins such as the von Hippel–Lindau protein can inhibit transcription by interfering with transcriptional elongation.

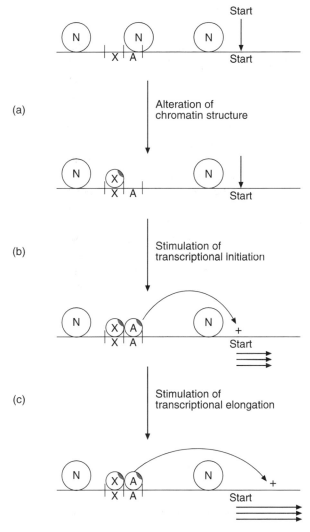

Figure 9.30 An activation domain (shaded) within an activating transcription factor can act: (a) by disrupting the nucleosome (N) arrangement of the DNA so that an activating factor can bind; (b) by stimulating the rate of transcriptional initiation and (c) by increasing the proportion of initiated transcripts which go on to produce a full-length RNA.

9.2.6 Multi-stage activation of transcription

It is clear therefore that activation of gene expression by transcription factors can occur at three distinct stages to stimulate transcription. Thus activating factors can disrupt the chromatin structure to allow other activating factors to bind, stimulate the rate of transcriptional initiation so that more RNA transcripts are initiated and, in some genes, can increase the proportion of full-length transcripts that are produced (Figure 9.30). Taken together, with the multiple targets for activating factors within the initiation complex, these effects allow transcriptional activators to fulfil their function and strongly stimulate transcription.

9.3 REPRESSION OF TRANSCRIPTION

Although the majority of transcription factors which have so far been described act in a positive manner, a number of cases have now been reported in which a transcription factor exerts an inhibitory effect on transcription. This effect can occur by indirect repression in which the repressor interferes with the action of an activating factor, so preventing it stimulating transcription (Figure 9.31a–d). Alternatively, it can occur via direct repression in which the factor reduces the activity of the basal transcriptional complex (Figure 9.31e). These two mechanisms will be discussed in turn (for reviews, see Cowell, 1994; Hanna-Rose and Hansen, 1996; Latchman, 1996).

9.3.1 Indirect repression

Several mechanisms exist by which an inhibitor can interfere with the action of an activator.

(a) Inhibition of activator binding by masking of its DNA-binding site

One means by which repression can occur is by the masking of the DNA-binding site for the factor, so preventing it binding to the DNA and activating transcription. By preventing the binding of the positively acting factor, the negatively acting factor effectively inhibits gene activation. This masking of the binding site can be achieved simply by the negatively acting factor binding to the same site as the positively

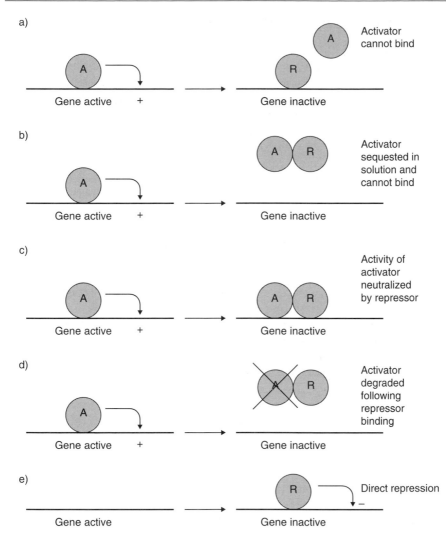

Figure 9.31 Potential mechanisms by which a transcription factor can repress gene expression. This can occur: by the repressor (R) binding to DNA and preventing an activator (A) from binding and activating gene expression (a); by the repressor interacting with the activator in solution and preventing its DNA binding (b), by the repressor binding to DNA with the activator and neutralizing its ability to activate gene expression (c), by the repressor promoting degradation of the activator (d) or by direct repression by an inhibitory transcription factor (e).

acting factor but failing to activate transcription (Figure 9.32a). This is seen, for example, in the case of the Sp3 factor, a factor related to the Sp1 factor discussed in Section 3.3.2. Thus the Sp3 factor binds to the same site as Sp1 but, unlike Sp1, it cannot activate transcription. It therefore

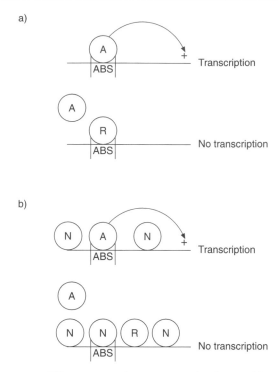

Figure 9.32. A repressor (R) can prevent gene expression by masking the binding sites (ABS) for an activator (A) either by binding directly to the site (a) or by reorganizing chromatin structure so that the binding site is masked by a nucleosome (N) (b).

blocks the Sp1-binding site, preventing Sp1 binding and activation of transcription (for a review, see Lania *et al.*, 1997).

Other examples of this mechanism have been discussed in previous chapters. Thus, the eng protein binds to the same site as the Ftz protein and hence Ftz cannot stimulate gene expression in the presence of eng (see Section 6.2.3). Similarly, the binding of the glucocorticoid receptor to an nGRE inhibits gene expression by preventing activator binding (see Section 4.4.4), whilst the CCAAT displacement protein prevents the CCAAT binding protein from binding to its site in the sperm histone H2B gene in embryonic tissues (see Section 3.3.3).

As well as directly masking the DNA-binding site of an activator by binding directly to the site, an inhibitor can also achieve such masking indirectly by reorganizing the nucleosome arrangement of the chromatin. This effect is evidently the reverse of the mechanism of gene activation which involves nucleosome displacement so unmasking an activator-binding site and allowing it to bind (see Section 9.2.5). Thus, if a factor reorganizes the pattern of nucleosomes so that the binding site of the activator is masked by a nucleosome, this will effectively repress gene activation (Figure 9.32b). An example of a factor which acts in this way is the polycomb repressor of *Drosophila*, which normally represses inappropriate expression of several homeotic genes by modulating their chromatin structure so that activating molecules cannot bind. When this factor is inactive, inappropriate expression of these genes in the wrong cell type is observed, leading to dramatic transformations in the nature of specific parts of the body. By directing the tight packing of specific genes and thereby preventing transcription, the polycomb factor evidently has the opposite effect to that of the GAGA/trithorax factor, which as discussed in Section 9.2.5, directs an open chromatin structure, allowing activator binding (for reviews, see Pirrotta, 1997; Schumacher and Magnusson, 1997).

This antagonism between factors which open the chromatin structure and those which close it up is also seen in the case of factors that alter chromatin structure by altering the acetylation of histones. Thus, as discussed in Section 9.2.5, several co-activator molecules including CBP have histone acetyltransferase activity and can therefore acetylate histones producing a more open chromatin structure. Conversely, many co-repressor molecules which bind to inhibitory transcription factors can deacetylate histones, producing the closed chromatin structure that is associated with deacetylated histones (see Sections 1.4.3 and 7.2.3) (for a review, of histone deacetylation, see Pazin and Kadonaga (1997)).

Hence the state of histone acetylation and the structure of chromatin can be controlled by the balance of deacetylases and acetylases which are bound to the DNA (Figure 9.33). In most cases the acetylating co-activators and deacetylating co-repressors will normally be brought to the DNA via interactions with distinct activating and inhibiting

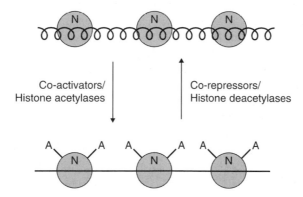

Figure 9.33 The balance between tightly packed chromatin (wavy line) and open chromatin (solid line) can be controlled by the balance between co-activating molecules, which acetylate histones, and co-repressors, which deacetylate histones in the nucleosome (N).

transcription factors, respectively. In the case of the thyroid hormone receptor, however, both types of factors bind to the same molecule. Thus, as discussed in Section 4.4.4, prior to exposure to hormone, the receptor binds co-repressors such as mRPD3 which have histone deacetylase activity whilst following hormone binding these factors dissociate and are replaced by co-activators such as CBP, which acetylate histones and allow the receptor to activate transcription.

Hence the inhibition of DNA binding by a specific activator via masking of its binding site is a major method of transcriptional inhibition. This operates both via specific factors, which bind directly to the binding site of the activator, and other more general factors, which organize a closed chromatin structure in which the binding site is masked along with the rest of that region of DNA.

(b) Inhibition of activator binding by formation of a non-DNA-binding complex

As well as preventing activator binding to DNA via masking its binding site, an inhibitor can also inhibit transcription via the formation of a non-DNA-binding complex with an activating factor (Figure 9.31b). Thus, as discussed in Section 5.3.2, gene activation by helix–loop–helix proteins can be inhibited by the Id protein, which can dimerize with them by means of its helix–loop–helix domain and then inhibit their DNA-binding, since it lacks the basic DNA-binding domain.

(c) Quenching of an activator

The cases of repression described so far all involve the inhibition of DNA binding either by blocking the binding site for a factor (Figure 9.31a) or by forming a non-DNA-binding protein–protein complex (Figure 9.31b). Since DNA binding is a necessary prerequisite for gene activation, this constitutes an effective form of repression. In addition, however, inhibition of transcription can also be achieved by interfering with transcriptional activation by a DNA-bound factor in a phenomenon known as quenching (Figure 9.31c). A simple example of this type is seen in the case of the negatively acting yeast factor GAL80, which inhibits gene activation by the positively acting GAL4 protein. This is achieved by the binding of GAL80 to DNA-bound GAL4, such binding occurring via the 30 amino acids located at the extreme C-terminus of the GAL4 molecule. As these amino acids are located close to the GAL4 activation domain, the binding of GAL80 to GAL4 masks the GAL4 activation domain and hence inhibits the activation of gene expression by GAL4. In response to treatment with

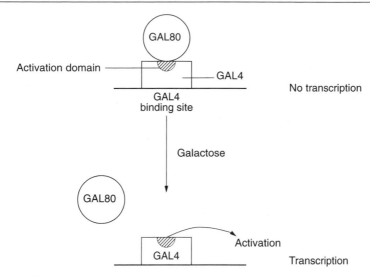

Figure 9.34 Galactose activates gene expression by removing the GAL80 protein from DNA-bound GAL4 protein, unmasking the activation domain on GAL4.

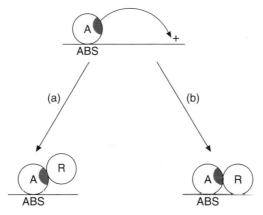

Figure 9.35 The ability of a bound activator (A) to stimulate transcription via its activation domain (shaded) can be inhibited by quenching of the activation domain by inhibitory factors (R), which either bind to the activator without binding to DNA (a) or which bind to a site adjacent to the activator (b).

galactose, GAL80 dissociates from GAL4, allowing GAL4 to fulfil its function of activating galactose-inducible genes. Hence this system provides an elegant means of modulating gene expression in response to galactose with the activating GAL4 factor being bound to DNA both prior to and after galactose addition, but being able to activate gene expression only following the galactose-induced dissociation of the quenching GAL80 factor (Johnston *et al.*, 1987) (Figure 9.34). A similar case in which one mechanism for repression by MDM2 involves the masking of the activation domain of p53 was discussed in Section 7.3.2.

A related example of quenching in which the inhibitory factor binds to a DNA sequence adjacent to the quenched factor, rather than only to the factor itself, is seen in the case of the c-*myc* promoter. Thus an inhibitory transcription factor, myc-PRF, binds to a site adjacent to that occupied by an activating factor, myc-CF1, and interferes with its ability to activate c-*myc* gene transcription (Kakkis *et al.*, 1989). Hence quenching can occur either by an inhibitory factor binding to the positively acting factor (Figure 9.35a) or by the inhibitory factor binding to DNA adjacent to the positive factor (Figure 9.35b). In both cases, however, this effect involves the inhibitor interfering with the ability of the activator's activation domain to stimulate transcription.

(d) Degradation of the activator

As well as interfering functionally with the action of the activator by preventing its DNA binding or quenching its activation domain, an indirect repressor can act by targeting an activator for degradation (Figure 9.31d). This is seen in the case of MDM2 which, as well as quenching the activation domain of p53 (see above), also targets it for degradation, so using multiple mechanisms to prevent p53 activating transcription (see Section 7.3.2).

In the case of p53, MDM2 is likely to do this by stimulating the recognition of p53 by protease enzymes which are present in the cell (Figure 9.36a). However, the AEBP1 transcription factor, which regulates adipocyte differentiation, actually itself has the ability to degrade other proteins and is therefore likely to act directly by degrading activators to which it binds (Figure 9.36b). Hence this factor combines

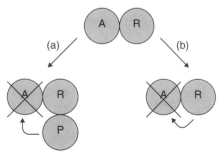

Figure 9.36 A repressor (R) can promote the degradation of an activator (A) either (a) indirectly by making it a target for a protease (P) or (b) directly by itself degrading the activator.

the ability to bind to DNA with the ability to degrade other factors with which it comes into contact (He *et al.*, 1995).

Thus the degradation of activators mediated directly or indirectly by inhibiting factors appears to be an important mechanism of transcriptional repression. Interestingly, however, it is also possible for the reverse to occur with a factor activating transcription because it directs the degradation of a repressor. Thus, in *Drosophila*, the PHYL and SINA factors produce the degradation of the TTK88 transcription repressor hence producing activation of transcription (Li *et al.*, 1997b; Tang *et al.*, 1997).

9.3.2 Direct repression

In the cases described so far, a negative factor exerts its effect by neutralizing the action of a positively acting factor by preventing its DNA binding (Figure 9.31a and b), inhibiting its activation of transcription following such binding (Figure 9.31c) or promoting its degradation (Figure 9.31d). In other cases, however, the inhibitory effect of a particular factor can be observed in the absence of any activating factors. This indicates that these inhibitory factors inhibit transcription directly by interacting with the basal transcriptional complex to reduce its activity (Figure 9.31e). Several factors of this type have now been shown to bind to specific DNA-binding sites within their target genes and reduce the activity of the basal transcriptional complex (Figure 9.37a). This effect is evidently similar in nature but opposite in effect to the

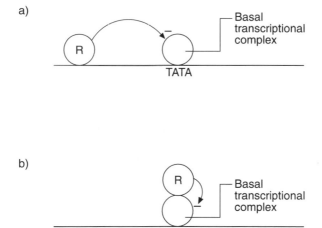

Figure 9.37 An inhibitory factor (R) can reduce the activity of the basal transcriptional complex either by binding to DNA and then interacting with the complex (a), or by binding directly to the complex by protein–protein interaction (b).

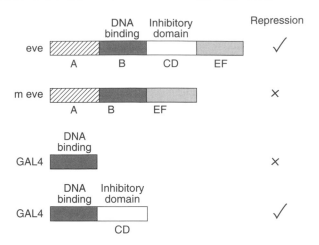

Figure 9.38 A specific region (CD) of the eve factor, which is distinct from the DNA-binding domain (B), acts as a transferrable inhibitory domain. Thus its deletion from the eve protein results in a loss of the ability to repress transcription, whilst its linkage to the DNA-binding domain of GAL4 generates a functional repressor.

stimulation of the basal complex by the binding of activating factors to specific DNA sequences in the promoter.

A number of factors capable of inhibiting the basal initiation complex in this way have been described in previous chapters including the yeast α2 factor (see Section 5.4.3), the *Drosophila* eve protein (see Section 6.2.2) and the mammalian thyroid hormone receptor, which can directly inhibit the basal complex in the absence of thyroid hormone but stimulates its activity in the presence of the hormone (see Section 4.4.4). Indeed, the number of such directly inhibitory factors is growing steadily and now includes some which were previously thought to function only in an indirect manner. Thus, the MDM2 factor, which was thought to function solely by masking the activation domain of p53, has now been shown to function as a direct repressor of transcription as well as targeting p53 for degradation (see Section 7.3.2), whilst the Rb-1 protein, which was originally thought to function solely by inhibiting E2F, is now known to also act as a direct repressor (see Section 7.3.3).

Interestingly, inhibitory factors of the directly acting type generally contain a small domain which can confer the ability to repress gene expression upon the DNA-binding domain of another factor when the two are artifically linked (see, for example, Han and Manley, 1993; Lillycrop *et al.*, 1994) (Figure 9.38). Hence these directly repressing factors contain specific inhibitory domains paralleling the existence of specific activation domains in activating transcription factors. By analogy with activation domains, inhibitory domains are likely to inhibit either the assembly of

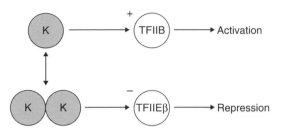

Figure 9.39 The Kruppel factor (K), when present as a monomer can interact with TFIIB to stimulate transcription. At high concentrations when it forms a dimer, it interacts with TFIIEβ to repress transcription.

the basal transcriptional complex or reduce its activity and/or stability after it has assembled. In agreement with this idea, the inhibitory domain of the thyroid hormone receptor has been shown to interact directly with TFIIB (Baniahmad *et al.*, 1993). Similarly, the inhibitory domain of the Kruppel repressor in *Drosophila* has been shown to interact with another component of the basal transcriptional complex, TFIIEβ (Sauer *et al.*, 1995). Interestingly, Kruppel can also act as an activator by interacting with TFIIB to stimulate its activity. This interaction with TFIIB is seen in the monomeric Kruppel factor, which hence acts as an activator, whereas the Kruppel dimer, which forms at high concentrations, inhibits transcription by interacting with TFIIEβ. Hence Kruppel can act as activator or repressor depending on its concentration in the cell, which results in its being present as an activating monomer or an inhibitory dimer (Figure 9.39).

Although inhibitory domains can thus clearly act by interacting with the basal transcriptional complex, it is likely that, as with activation domains (see Section 9.2.5), they can also act by altering chromatin structure via alteration of histone acetylation. Thus, as discussed in Sections 4.4.4 and 9.3.1, the thyroid hormone receptor inhibitory domain can recruit co-repressor molecules with the ability to deacetylate histones. Hence the activity of inhibitory domains appears to parallel that of activation domains in terms of their dual abilities to target the basal transcriptional complex and recruit other factors capable of modulating chromatin structure.

Interestingly, the inhibitory domain in the human Wilms' tumour anti-oncogene product and those from several *Drosophila* inhibitory factors including eve appear to share the common features of proline richness and an absence of charged residues (Han and Manley, 1993), suggesting that these factors have a common inhibitory domain. However, other inhibitory domains, such as those in the mammalian factors Oct-2

(Lillycrop *et al.*, 1994) and E4BP4 (Cowell and Hurst, 1994), are distinct both from this common domain and from each other indicating that, as with activation domains, several types of inhibitory domain may exist.

Whatever the precise number of inhibitory domains, it is clear that a number of factors possess such domains and can therefore inhibit transcription by binding to upstream DNA sequences and inhibiting the activity of the basal transcriptional complex. The binding of such factors is likely to be of vital importance in producing the inhibitory effect of many of the silencer elements, which were described in Section 1.3, the silencer element in the chicken lysozyme gene, for example, having been shown to act by binding the inhibitory thyroid hormone receptor.

As well as interfering with the basal complex by binding to distinct DNA-binding sites (Figure 9.37a), it is also possible for inhibitory factors to bind to the complex itself by protein–protein interaction and thereby interfere with its activity or assembly (Figure 9.37b) (for a review, see Drapkin *et al.*, 1993). An example of this is provided by the Dr1 protein, which inhibits the assembly of the basal transcriptional complex by binding to TBP and preventing TFIIB from binding (Figure 9.40) (Inostroza *et al.*, 1992). As the recruitment of TFIIB to the promoter by interaction with the TBP component of TFIID is an essential step in the

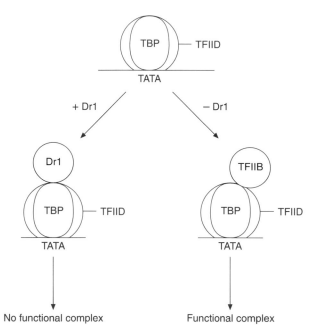

Figure 9.40 The Dr1 inhibitory factor can interact with the TBP component of TFIID, thereby preventing it binding TFIIB and thus inhibiting the assembly of the basal transcriptional complex.

assembly of the basal transcriptional complex, this effectively inhibits transcription (see Section 3.2.4).

As noted in Section 3.2.5, TBP is a component of the initiation complexes of all three polymerases and, in each case, acts by recruiting other factors to the promoter. It has been shown (White *et al.*, 1994) that Dr1 can inhibit this ability of TBP to recruit other factors within the RNA polymerase II and III initiation complexes but not within the RNA polymerase I complex. It therefore inhibits transcription by RNA

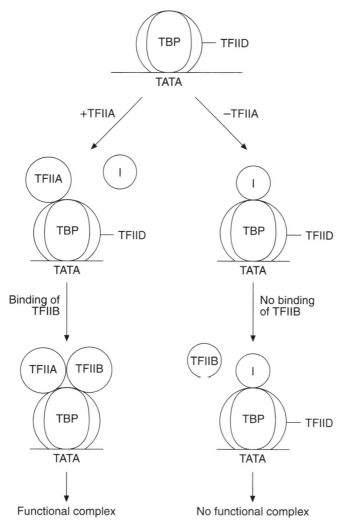

Figure 9.41 Binding of TFIIA to TBP prevents inhibitory molecules (I) from binding, but still allows the binding of TFIIB and thereby promotes the assembly of the basal transcriptional complex.

polymerase II and III but not by RNA polymerase I. Thus Dr1 may play a critical role in regulating the balance of transcriptional activity between the ribosomal genes which are the only genes transcribed by RNA polymerase I and all the other genes in the cell.

In addition to Dr1, other factors which bind to TBP and inhibit the assembly of the RNA polymerase II basal complex, such as NC1, NC2 and Dr2, have been described and are likely to be important in controlling the rate of transcription (for a review, see Drapkin *et al.*, 1993). Indeed, as well as interacting with activating factors (see Section 9.2.3), the TFIIA factor also appears to be able to bind to TBP, preventing these inhibitors from binding, thus allowing the recruitment of TFIIB (Figure 9.41) (see Section 3.2.4). This indicates that the activity of inhibitory molecules, which act by interacting with the basal transcriptional complex, can be regulated by activating factors, and once again illustrates that the balance between transcriptional activators and repressors is of central importance in the control of transcription.

9.4 CONCLUSIONS

The fundamental property of transcription factors is their ability to influence the transcription of specific genes either positively or negatively. In this chapter, we have discussed how the activation of transcription is dependent on discrete activation domains whose structure and mechanism of action are gradually being understood, whilst the repression of transcription can be produced either by direct repression or by the neutralization of a positively acting factor. These properties offer ample scope for gene regulation in different cell types or in different tissues. Thus, in addition to the simple activation of gene expression by a positively acting factor present in only one cell type, the effect of a positively acting factor present in several different cell types can be affected by the presence or absence of a negatively acting factor which is active in only one cell type and which inhibits its activity. Similarly, a single factor may act either positively or negatively depending on the gene involved (as in the case of the glucocorticoid receptor) or depending on whether a specific hormone is present (as in the case of the thyroid hormone receptor).

Ultimately, however, all such potential mechanisms of gene regulation in response to specific stimuli or in specific cell types are dependent upon mechanisms which control the synthesis or activity of specific transcription factors in different cell types or in response to specific stimuli. These mechanisms are discussed in the next chapter.

REFERENCES

Baniahmad, A., Ha, I., Reinberg, D., Tsai, S., Tsai, M.J. and O'Malley, B.W. (1993). Interaction of human thyroid hormone receptor β with transcription factor TFIIB may mediate target gene derepression and activation by thyroid hormone. *Proceedings of the National Academy of Sciences USA* **90**, 8832–8836.

Barberis, A., Pearlberg, J., Simkovich, N., Farrell, S., Reinagel, P., Bamdad, C., Sigal, G. and Ptashne, M. (1995). Contact with a component of the polymerase II holoenzyme suffices for gene activation. *Cell* **81**, 359–368.

Beato, M. and Eisfeld, K. (1997). Transcription factor access to chromatin. *Nucleic Acids Research* **25**, 3559–3563.

Bentley, D. (1995). Regulation of transcriptional elongation by RNA polymerase II. *Current Opinion in Genetics and Development* **5**, 210–216.

Bjorklund, S. and Kim, Y.J. (1996). Mediator of transcriptional regulation. *Trends in Biochemical Sciences.* **21**, 335–337.

Buratowski, S. (1995). Mechanisms of gene activation. *Science* **270**, 1773–1774.

Buratowski, S. (1997). Multiple TATA-binding factors come back into style. *Cell.* **91**, 13–15.

Carey, M. (1998). The enhancesome and trascriptional synergy. *Cell* **92**, 5–8.

Chang, M. and Jaehning, J.A. (1997). A multiplicity of mediators: alternative forms of transcription complexes communicate with transcriptional regulators. *Nucleic Acids Research* **25**, 4861–4865.

Chao, D.M. and Young, R.A. (1996). Activation without a vital ingredient. *Nature* **383**, 119–120.

Chen, J-L., Attardi, L.D., Verrijzer, C.P., Yokomori, K. and Tjian, R. (1994). Assembly of recombinant TFIID reveals different co-activator requirements for different transcriptional activators. *Cell* **79**, 93–105.

Choy, B. and Green, M.R. (1993). Eukaryotic activators function during multiple steps of pre-initiation complex assembly. *Nature* **366**, 531–536.

Courey, A.J. and Tjian, R. (1988). Analysis of Sp1 *in vivo* reveals multiple transcriptional domains including a novel glutamine-rich activation motif. *Cell* **55**, 887–898.

Cowell, I.G. (1994). Repression versus activation in the control of gene transcription. *Trends in Biochemical Sciences* **19**, 38–42.

Cowell, I.G. and Hurst, H.C. (1994). Transcriptional repression by the human bZIP factor E4BP4: definition of a minimal repressor domain. *Nucleic Acids Research* **22**, 59–65.

Dickstein, R., Zhou, S. and Tjian, R. (1996). Human TAF$_{II}$105, is a cell type-specific TFIID subunit related to TAF$_{II}$130. *Cell* **87**, 137–146.

Drapkin, R., Merino, A. and Reinberg, D. (1993). Regulation of RNA polymerase II transcription. *Current Opinion in Cell Biology* **5**, 469–476.

Gerber, H.P., Seipel, K., Georgiev, O., Hofferer, M., Hug, M., Rusioni, S. and Schaffner, W. (1994). Transcriptional activation modulated by homopolymeric glutamine and proline residues. *Science* **263**, 808–811.

Glass, C.K., Rose, D.W. and Rosenfeld, M.G. (1997). Nuclear receptor coactivators. *Current Opinion in Cell Biology* **9**, 222–232.

Goding, C.R. and O'Hare, P. (1989). Herpes simplex virus Vmw65–octamer binding protein interaction: a paradigm for combinatorial control of transcription. *Virology* **173**, 363–367.

Gonzalez-Couto, E.K., Klages, N. and Strubin, M. (1997). Synergistic and promoter-selective activation of transcription by recruitment of transcription factors TFIID and TFIIB. *Proceedings of the National Academy of Sciences USA* **94**, 8036–8041.

Greenblatt, J., Nodwell, J.R. and Mason, S.W. (1993). Transcriptional anti-termination. *Nature* **364**, 401–406.

Guarente, L. (1988). USAs and enhancers: common mechanism of transcriptional activation in yeast and mammals. *Cell* **52**, 303–305.

Hahn, S. (1993a). Structure (?) and function of acidic transcription activators. *Cell* **72**, 481–483.

Hahn, S. (1993b). Efficiency in activation. *Nature* **363**, 672–673.

Hai, T., Horikoshi, M., Roeder, R.G. and Green, M.R. (1988). Analysis of the role of the transcription factor ATF in the assembly of a functional preinitiation complex. *Cell* **54**, 1043–1051.

Han, K. and Manley, J.L. (1993). Transcriptional repression by the *Drosophila* even skipped protein: definition of a minimal repressor domain. *Genes and Development* **7**, 491–503.

Hanna-Rose, W. and Hansen, U. (1996). Active repression mechanisms of eukaryotic transcriptional repressors. *Trends in Genetics* **12**, 229–234.

Hartzog, G.A. and Winston, F. (1997). Nucleosomes and transcription: recent lessons from genetics. *Current Opinion in Genetics and Development* **7**, 192–198.

He, G.P., Muise, A., Li, A.W. and Ro, H.S. (1995). A eukaryotic transcriptional repressor with carboxypeptidase activity. *Nature* **378**, 92–96.

Hollenberg, S.M. and Evans, R.M. (1988). Multiple and cooperative trans-activation domains of the human glucocorticoid receptor. *Cell* **55**, 899–906.

Hope, I.A. and Struhl, K. (1986). Functional dissection of a eukaryotic transcriptional activator, GCN4 of yeast. *Cell* **46**, 885–894.

Horikoshi, M., Carey, M.F., Kakidani, H. and Roeder, R.G. (1988).

Mechanism of action of a yeast activator: direct effect of GAL4 derivatives on mammalian TFIID promoter interactions. *Cell* **54**, 665–669.

Inostroza, J.A., Mermelstein, F.H., Ha, I., Lane, W.S. and Reinberg, D., (1992). Dr1, a TATA-binding protein-associated phosphoprotein and inhibitor of class II gene transcription. *Cell* **70**, 477–489.

Johnston, S.A., Salmeron, J.M. and Pincher, S.E. (1987). Interaction of positive and negative regulatory proteins in the galactose regulon of yeast. *Cell* **50**, 143–146.

Joliot, V., Demma, M. and Prywes, R. (1995). Interaction with RAP74 subunit of TFIIF is required for transcriptional activation by serum response factor. *Nature* **373**, 632–635.

Jones, K. (1997). Taking a new TAK on Tat transactivation. *Genes and Development* **11**, 2593–2599.

Kakkis, E., Riggs, K.J., Gillespie, W. and Calame, K. (1989). A transcriptional repressor of C-myc. *Nature* **339** 718–721.

Kinzler, M., Braus, G.H., Georgiev, O., Seipel, K. and Schaffner, W. (1994). Functional differences between mammalian transcriptional activation domains at the yeast GAL1 promoter. *EMBO Journal* **13**, 641–645.

Lania, L., Majello, B. and DeLuca, P. (1997). Transcriptional regulation by the Sp family proteins. *International Journal of Biochemistry and Cell Biology* **29**, 1313–1323.

Latchman, D.S. (1996). Inhibitory factors. *International Journal of Biochemistry and Cell Biology* **28**, 965–974.

Li, Q., Wrange, O. and Eriksson, P. (1997a). The role of chromatin in transcriptional regulation. *International Journal of Biochemistry and Cell Biology* **29**, 731–742.

Li, S., Li, Y., Carthew, R.W. and Lai, Z.C. (1997b). Photoreceptor cell differentiation requires regulated proteolysis of the transcriptional repressor tramtrack. *Cell* **90**, 469–478.

Lillycrop, K.A., Dawson, S.J., Estridge, J.K., Gester, T., Matthias, P. and Latchman, D.S. (1994). Repression of a herpes simplex virus immediate-early promoter by the Oct-2 transcription factor is dependent upon an inhibitory region at the N-terminus of the protein. *Molecular and Cellular Biology* **14**, 7633–7642.

Lis, J. and Wu, C. (1993). Protein traffic on the heat shock promoter: parking, stalling and trucking along. *Cell* **74**, 1–4.

Lohr, D. (1997). Nucleosome transactions on the promoters of the yeast GAL and PHO genes. *Journal of Biological Chemistry* **272**, 26795–26798.

Mermod, N., O'Neil, E.A., Kelley, T.J. and Tjian, R. (1989). The proline-rich transcriptional activator of CTF/NF-1 is distinct from the replication and DNA binding domain. *Cell* **58**, 741–753.

Mitchell, P.J. and Tjian, R. (1989). Transcriptional regulation in mammalian cells by sequence specific DNA binding proteins. *Science* **245**, 371–378.

Ozer, J., Moore, P.A., Bolden, A.H., Lee, A., Rosen, C.A. and Lieberman, P.M. (1994). Molecular cloning of the small (α) sub-unit of human TFIIA reveals functions critical for activated transcription. *Genes and Development* **8**, 2324.

Pazin, M.J. and Kadonaga, J.T. (1997). What's up and down with histone deacetylation and transcription. *Cell* **89**, 325–328.

Pirrotta, V. (1997). PcG complexes and chromatin silencing. *Current Opinion in Genetics and Development* **7**, 249–258.

Ptashne, M. (1988). How eukaryotic transcriptional activators work. *Nature* **335**, 683–689.

Ptashne, M. and Gann, A. (1997). Transcriptional activation by recruitment. *Nature* **386**, 569–577.

Radhakrishnan, I., Perez-Alvardo, G.C., Parker, D., Dyson, H.J., Montminy, M.R. and Wright, P. (1997). Solution structure of the KIX domain of CBP bound to the transactivation domain of CREB a model for activator: co-activator interactions. *Cell* **91**, 741–752.

Ranish, J.A. and Hahn, S. (1996). Transcription: basal factors and activation. *Current Opinion in Genetics and Development* **6**, 151–158.

Roberts, S.G.E. and Green, M.R. (1994). Activator-induced conformational change in general transcription factor TFIIB. *Nature* **371**, 717–720.

Roeder, R.G. (1996). The role of general initiation factors in transcription by RNA polymerase II. *Trends in Biochemical Sciences* **21**, 327–334.

Sauer, F., Fondell, J.D., Ohkuma, Y., Roeder, R. and Jackle, H. (1995). Control of transcription by Kruppel through interactions with TFIIB and TFIIEβ. Nature, 375, 162–164.

Sauer, F. and Tjian, R. (1997). Mechamisms of transcriptional activation: differences between yeast and man. *Current Opinion in Genetics and Development* **7**, 176–181.

Schumacher, A. and Magnusson, T. (1997). Murine polycomb and trithorax group genes regulate homeotic pathways and beyond. *Trends in Genetics* **13**, 167–170.

Seipel, K., Georgiev, O. and Schaffner, W. (1992). Different activation domains stimulate transcription from the remote ('enhancer') and proximal ('promoter') positions. *EMBO Journal* **11**, 4961–4968.

Shen, W.C. and Green, M.R. (1997). Yeast TAF$_{II}$145 functions as a core promoter selctivity factor, not a general co-activator. *Cell* **90**, 615–624.

Shikama, N., Lyon, J. and La Thangue, N.B. (1997). The p300/CBP family: integrating signals with transcription factors and chromatin. *Trends in Cell Biology* **7**, 230–236.

Spencer, C.A. and Groudine, M. (1990). Transcriptional elongation and eukaryotic gene regulation. *Oncogene* **5**, 777–785.

Stargell, L.A. and Struhl, K. (1996). Mechamisms of transcriptional activation *in vivo*: two steps forward. *Trends in Genetics* **12**, 311–315.

Svaren, J. and Horz, W. (1997). Transcription factors vs nucleosomes: regulation of the PH05 promoter in yeast. *Trends in Biochemical Sciences* **22**, 93–97.

Tang, A.H., Neufeld, T.P., Kwan, E. and Rubin, G.M. (1997). PHYL acts to down regulate TTK88, a transcriptional repressor of neuronal cell fates by a SINA-dependent mechanism. *Cell* **90**, 459–467.

Tansey, W.P. and Herr, W. (1997). TAFs: gilt by association. *Cell* **88**, 729–732.

Triezenberg, S.J. (1995). Structure and function of transcriptional activation domains. *Current Opinion in Genetics and Development* **5**, 190–196.

Uesugi, M., Nyanguile, O., La, H., Levine, A.J. and Verdine, G.L. (1997). Induced α-helix in the VP16 activation domain upon binding to a human TAF. *Science* **277**, 1310–1313.

Verrijzer, C.P. and Tjian, R. (1996). TAFs mediate transcriptional activation and promoter selectively. *Trends in Biochemical Sciences* **21**, 338–345.

Wang, E.H. and Tjian, R. (1994). Promoter selective defect in cell cycle mutant ts13 rescued by hTAF$_{II}$250 *in vitro*. *Science* **263**, 811–814.

White, R.J., Khoo, B.C.-E., Inostroza, J.A., Reinberg, D. and Jackson, S.P. (1994). Differential regulation of RNA polymerases I, II and III by the TBP-binding repressor Dr1. *Science* **266**, 448–450.

Wilkins, R.C. and Lis, J.T. (1997). Dynamics of potentiation and activation: GAGA factor and its role in heat shock gene regulation. *Nucleic Acids Research* **25**, 3963–2968.

Wu, C. (1997). Chromatin remodelling and the control of gene expression. *Journal of Biological Chemistry* **272**, 28171–28174.

Xiao, H., Pearson, A., Coulombe, B., Truant, R., Zhang, S., Regier, J.L., Triezenberg, S.J., Reinberg, D., Flores, O., Ingles, C.J. and Greenblatt, J. (1994). Binding of basal transcription factor TFIIH to the acidic activation domains of VP16 and p53. *Molecular and Cellular Biology* **14**, 7013–7024.

Yankulov, K., Blau, J., Purton, T., Roberts, S. and Bentley, D.L. (1994). Transcriptional elongation by RNA polymerase II is stimulated by transctivators. *Cell* **77**, 749–759.

CHAPTER TEN

What regulates the regulators?

10.1 INTRODUCTION

As we have discussed in previous chapters, transcription factors, in addition to their role in constitutive gene expression (Chapter 3) also play a central role in producing the induction of specific genes in response to particular stimuli (Chapter 4) and in the cell type-specific (Chapter 5) or developmentally regulated (Chapter 6) expression of other genes. The ability to bind to DNA (chapter 8) and influence the rate of transcription either positively or negatively (Chapter 9) are clearly features of many transcription factors that regulate gene expression in response to specific stimuli or in specific cell types. Most importantly, however, such factors must also have their activity regulated such that they only become active in the appropriate cell type or in response to the appropriate stimulus, thereby producing the desired pattern of gene expression. The means by which this occurs have already been discussed in the appropriate chapters dealing with specific transcription factors. The aim of this chapter is to categorize the various means by which this regulation of transcription factor activity can be achieved and to illustrate these processes both with examples that have already been discussed and with additional cases involving factors which have not so far been considered.

Two basic mechanisms by which the action of transcription factors can be regulated have been described. These involve either controlling the synthesis of the transcription factor so that it is made only when necessary (Figure 10.1a) or alternatively, regulating the activity of the factor so that pre-existing protein becomes activated when required (Figure 10.1b). These two mechanisms will be discussed in turn.

10.2 REGULATION OF SYNTHESIS

10.2.1. Evidence for the regulated synthesis of transcription factors

Regulating the synthesis of transcription factors such that they are only made when the genes which they regulate are to be activated is an obvious mechanism of ensuring that specific genes become activated only at the appropriate time and place. This mechanism is widely used, therefore, particularly for transcription factors which regulate the expression of cell type-specific or developmentally regulated genes. Thus, as discussed in Section 5.2.2, the octamer-binding protein Oct-2, is detectable in B cells, but is absent from many other cell types such as HeLa cells. Such cell type-specific expression of the Oct-2 protein is paralleled by the presence of the Oct-2 mRNA in B cells and not in HeLa cells. Hence the cell type-specific synthesis of the Oct-2 protein is controlled by regulating the synthesis of the Oct-2 mRNA. Similarly, the MyoD transcription factor and its corresponding mRNA are detectable

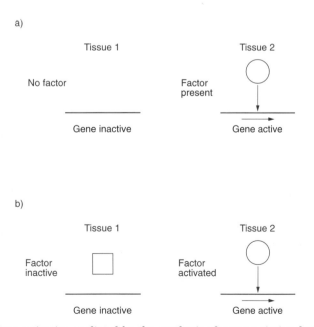

Figure 10.1 Gene activation mediated by the synthesis of a transcription factor only in a specific tissue (a) or its activation in a specific tissue (b).

only in cells of the skeletal muscle lineage in which MyoD plays a critical role in regulating muscle-specific gene expression (Davis *et al.*, 1987; see Section 5.3.2).

In addition to its role in controlling cell type-specific gene expression, regulation of transcription factor synthesis is also widely used in the control of developmentally regulated gene expression. Thus numerous studies of the *Drosophila* homeobox transcription factors discussed in Section 6.2, using both immunofluorescence with specific antibodies and *in situ* hybridization, have revealed highly specific expression patterns for individual factors and the mRNAs which encode them indicating that their role in regulating gene expression in development is dependent, at least in part, on the regulation of their synthesis (for a review, see Ingham, 1988) (Figure 10.2). In agreement with this idea, alteration of the levels of the homeobox protein Bicoid within individual cells of the *Drosophila* embryo alters the phenotype of the cell to that characteristic of cells which normally contain the new level of the Bicoid protein (Driever and Nusslein-Volhard, 1988). Hence, in a number of cases, where a factor must be active in a particular cell type or at a specific point in development, this is achieved by the factor being present only in the particular cells where it is required. Clearly, such regulated synthesis of a specific transcription factor could be achieved by any of the methods which are normally used to regulate the production of individual proteins such as the regulation of gene transcription, RNA splicing or translation of the mRNA (Figure 10.3) (for a review of the levels at which gene regulation can occur, see Latchman (1998)). Several of these mechanisms

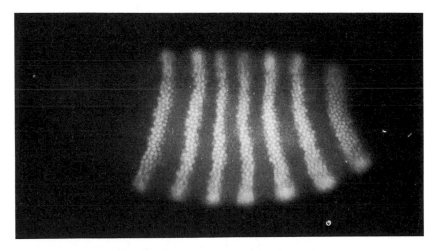

Figure 10.2 Localization of the Ftz protein in the *Drosophila* blastoderm embryo using a fluorescent antibody which reacts specifically with the protein. The anterior end of the embryo is to the left and the dorsal surface to the top of the photograph. Note the precise pattern of seven stripes of Ftz-expressing cells around the embryo.

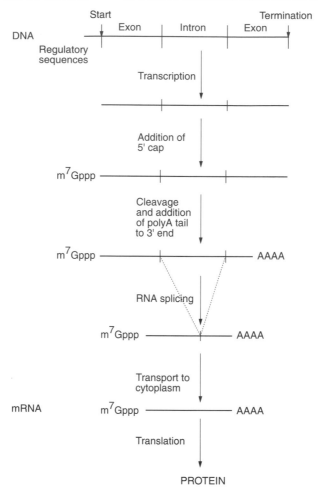

Figure 10.3 Potential regulatory stages in the expression of a gene encoding a transcription factor.

of gene regulation are utilized in the case of individual transcription factors and these will be discussed in turn.

10.2.2 Regulation of transcription

As discussed above, a number of cases where the cell type-specific expression of a transcription factor is paralleled by the presence of its corresponding mRNA in the same cell type have now been described. In turn, this cell type-specific expression of the transcription factor mRNA is likely to result from the regulated transcription of the gene encoding the transcription factor. Unfortunately, the low abundance of many transcription factors has precluded the direct demonstration of the regulated transcription of the genes which encode them. This has been achieved, however, in the case of the CCAAT box-binding factor C/EBP (Section 3.3.3), which regulates the transcription of several different liver-specific genes such as transthyretin and α-1-antitrypsin. Thus by using nuclear run-on assays to measure transcription of the gene encoding C/EBP directly, Xanthopoulos *et al.* (1989) were able to show that this gene is transcribed at high levels only in the liver, paralleling the presence of C/EBP itself and the mRNA encoding it at high levels only in this tissue (Figure 10.4). Hence the regulated transcription of the C/EBP gene in turn controls the production of the corresponding protein which, in turn, directly controls the liver-specific transcription of other genes such as α-1-antitrypsin and transthyretin.

Interestingly, as well as being used to regulate the relative amounts of a particular factor produced by different tissues, transcriptional control can also be used to regulate factor levels within a specific cell type. Thus the levels of the liver-specific transcription factor DBP are highest in rat hepatocytes in the afternoon and evening, with the protein being undetectable in the morning. This fluctuation is produced by regulated transcription of the gene encoding DBP, which is highest in the early evening and undetectable in the morning, whereas the C/EBP gene is transcribed at equal levels at all times. In turn, the alterations in DBP level produced in this way produce similar diurnal fluctuations in the transcription of the albumin gene, which is dependent on DBP for its transcription (Wuarin and Schibler, 1990).

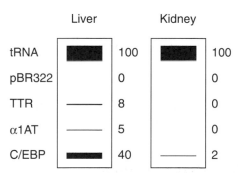

Figure 10.4 Nuclear run-on assay of transcription in the nuclei of kidney and liver. Values indicate the degree of transcription of each gene in the two tissues. Note the enhanced transcription in the liver of the gene encoding the transcription factor C/EBP as well as of the genes encoding the liver-specific proteins transthyretin (TTR) and α-1 antitrypsin (α1AT). The positive control transfer RNA gene is, as expected, transcribed at equal levels in both tissues, whilst the negative control, pBR322 bacterial plasmid, does not detect any transcription.

Although regulated transcription of the genes encoding the transcription factors themselves is likely therefore to constitute an important means of regulating their synthesis, it is clear that this process simply sets the problem of gene regulation one stage further back. Thus it will be necessary to have some means of regulating the specific transcription of the gene encoding the transcription factor itself, which in turn may require other transcription factors that are synthesized or are active only in that specific cell type. It is not surprising therefore that the synthesis of transcription factors is often modulated by post-transcriptional control mechanisms not requiring additional transcription factors. These mechanisms will now be discussed.

10.2.3 Regulation of RNA splicing

Numerous examples have now been described in eukaryotes where a single RNA species transcribed from a particular gene can be spliced in two or more different ways to yield different mRNAs encoding proteins with different properties (for reviews, see Latchman, 1998; McKeown, 1992; Rio, 1993). This process is also used in several cases of genes encoding specific transcription factors, for example, in the case of the *era-1* gene, which encodes a transcription factor that mediates the induction of gene expression in early embryonic cells in response to retinoic acid. In this case, two alternatively spliced mRNAs are produced, one of which encodes the active form of the molecule, whilst the other produces a protein lacking the homeobox region. As the homeobox mediates DNA binding by the intact protein (see Section 6.2.2), this truncated form of the protein is incapable of binding to DNA and activating gene expression (Larosa and Gudas, 1988). A similar use of alternative splicing to create mRNAs encoding proteins with and without the homeobox has also been reported for the Hoxb6 (2.2) gene (Shen *et al.*, 1991).

Hence in these cases where one of the two proteins encoded by the alternatively spliced mRNAs is inactive, alternative splicing can be used in the same way as the regulation of transcription in order to control the amount of functional protein that is produced.

Interestingly, however, unlike transcriptional regulation, alternative splicing can also be used to regulate the relative production of two distinct functional forms of a transcription factor that have different properties. This is seen in the case of the Pax8 factor which is a member of the Pax family (see Sections 6.3.2 and 8.2). In this case, alternative splicing results in the insertion of a single serine residue in the recognition helix of the paired domain, which is critical for DNA binding (Figure 10.5). This alters the DNA-binding properties of the factor so that it recognizes different DNA sequences to the form of Pax8, which lacks this residue (Kozmik *et al.*, 1997). Hence alternative splicing can introduce a subtle, single amino-acid change in a transcription factor, which results in the existence of two forms of the factor with different DNA-binding specificities.

As well as affecting DNA-binding specificity, alternative splicing can also produce forms of a transcription factor with distinct effects on transcription. This is seen in the case of the CREM factor which, as discussed in Section 4.3.3, exists in different forms generated by alternative splicing. This results in the existence of specific forms of the protein which can bind to DNA but lack the transcriptional activation domain and hence inhibit transcriptional activation by the full-length protein. Thus, the repressor molecules generated by alternative splicing of the CREM transcript, act as indirect repressors which passively prevent DNA binding by the positively acting forms containing a functional activation domain (see Section 9.3.1).

Alternative splicing can also occur however, in factors which contain a specific inhibitory domain and which can therefore function as direct repressors interfering with the activity of the basal transcriptional complex (see Section 9.3.2). Thus, although it is transcribed in B cells and not in most other cell types (see Section 10.2.1), the gene encoding the Oct-2 transcription factor is also transcribed in neuronal cells. In neuronal cells, the Oct-2 RNA is spliced so that the protein it encodes does not contain the C-terminal activation domain which allows it to activate transcription. It does, however, retain the N-terminal inhibitory domain discussed in Section 9.3.2 as well as the DNA-binding domain, and can therefore act as a direct inhibitor of gene expression (Lillycrop *et al.*, 1994). In contrast in B cells, alternative splicing produces an mRNA

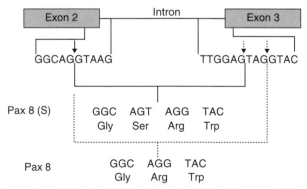

Figure 10.5 Alternative splicing in the Pax8 gene involving the use of different splice sites in exon 3 (dotted arrows) together with the same splice site in exon 2 (solid arrow) generates different forms of the protein with and without an additional serine residue and thus having different DNA-binding specificities.

which encodes a protein containing both the inhibitory domain and the stronger activation domain and which therefore activates transcription (Figure 10.6). Hence, in this case, alternative splicing produces different forms of a factor in different cell types which have opposite effects on the activity of their target promoters.

Such alternative splicing is also seen in the case of another transcription factor containing an inhibitory domain, namely the thyroid hormone receptor (see Section 9.3.2). Thus, as discussed in Section 4.4.4, alternative splicing produces two forms of the receptor, one of which lacks the ligand-binding domain and therefore cannot bind thyroid hormone (Figure 4.30). Although it cannot therefore respond to thyroid hormone, this α2 form of the protein still contains the DNA-binding domain and can therefore bind to the specific binding site for the receptor in hormone-responsive genes. By doing so, it acts as a dominant repressor of gene activation mediated by the normal receptor in response to hormone binding. Hence these two alternatively spliced forms of the transcription factor, which are made in different amounts in different tissues, mediate opposing effects on thyroid hormone-dependent gene expression.

As well as affecting the actual properties of a transcription factor, alternative splicing can also affect its degradation by proteases and therefore determine how much of the protein accumulates. This is seen in the case of the Hac1p protein, which is a member of the basic-leucine zipper transcription factor family discussed in Section 8.4. This factor accumulates at an increased level in the presence of unfolded proteins in

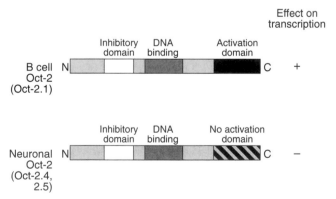

Figure 10.6 In B lymphocytes the predominant form of Oct-2 (Oct-2.1) contains the C-terminal activation domain as well as the DNA-binding domain and an inhibitory domain. As the activation domain overcomes the effect of the inhibitory domain, this form is able to activate transcription. In contrast, the predominant neuronal forms of Oct-2 (Oct-2.4 and −2.5) contain different C-terminal regions and lack the activation domain. As they retain both the inhibitory domain and the DNA-binding domain, however, they can bind to specific DNA-binding sites and inhibit gene expression.

the cell and then activates the expression of genes, which assist other proteins to fold properly. This increased accumulation of Hac1p is controlled by an alternative splicing event, which replaces the C-terminal tail of the Hac1p protein with a different peptide that renders it more resistant to degradation, allowing it to accumulate to increased levels (Cox and Walter, 1996) (Figure 10.7). Hence, in this case, alternative splicing alters the stability of the transcription factor rather than its activity.

The examples of alternative splicing discussed above thus illustrate the potential of this process in generating different forms of a particular transcription factor which, because of differences in the regions which mediate DNA binding, transcriptional activation or protein degradation have different properties or stability that result in differences in their effects on gene expression.

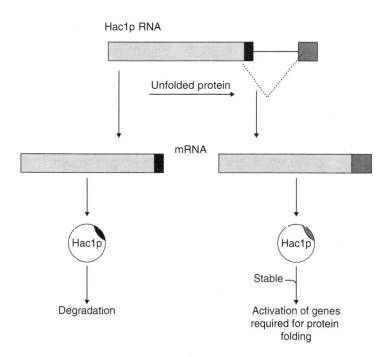

Figure 10.7 The Hac1p protein exists in different forms with different C-termini (solid and shaded boxes) which are generated by alternative splicing of its mRNA and have different stabilities. The presence of unfolded proteins in the cell stimulates the alternative splicing event, producing the more stable form of the protein so stimulating the expression of genes whose proteins products are needed to fold the unfolded proteins properly.

10.2.4 Regulation of translation

The final stage in the expression of a gene is the translation of its corresponding mRNA into protein. In theory, therefore, the regulation of synthesis of a particular transcription factor could be achieved by producing its mRNA in all cell types but translating it into active protein only in the particular cell type where it was required. However, the observed parallels between the cell type-specific expression of a particular transcription factor and the cell type-specific expression of its corresponding mRNA discussed above (Section 10.2.1) indicate that this cannot be the case for the majority of transcription factors. None the less, this mechanism is used to control the synthesis of at least one transcription factor in yeast.

Thus the yeast GCN4 transcription factor controls the activation of several genes in response to amino-acid starvation and the factor itself is synthesized in increased amounts following such starvation allowing it to mediate this effect. This increased synthesis of GCN4 following amino-acid starvation is mediated via increased translation of pre-existing GCN4 mRNA (for a review, see Hinnebusch, 1997). This translational regulation is dependent upon short sequences within the 5' untranslated region of the GCN4 mRNA, upstream of the start point for translation of the GCN4 protein.

Most interestingly, such sequences are capable of being translated to produce short peptides of two or three amino acids (Figure 10.8). Under conditions when amino acids are plentiful, these short peptides are synthesized and the ribosome fails to reinitiate at the start point for GCN4 production resulting in this protein not being synthesized. Following amino-acid starvation, however, the production of the small peptides is supressed and the production of GCN4 is correspondingly enhanced. Hence this mechanism ensures that GCN4 is synthesized only in response to amino-acid starvation and then activates the genes encoding the enzymes required for the biosynthetic pathways necessary to make good this deficiency.

Interestingly, the use of distinct translational start sites is also seen in the case of the C/EBPβ transcription factor expressed in the mammalian liver (see Section 3.3.3). In this case, however, the two start sites of translation result in two different forms of the C/EBPβ protein. One of these, known as liver activator protein (LAP) contains an activation domain as well as a basic DNA-binding domain and leucine zipper. The other is produced by translational initiation from a downstream start site and therefore lacks the activation domain, although it retains both the basic domain and the leucine zipper (Figure 10.9). This protein is therefore known as liver inhibitory protein (LIP), since it can bind to the same sites as LAP and inhibit gene activation (Descombes and Schibler, 1991).

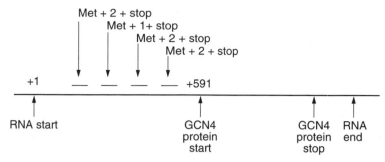

Figure 10.8 Presence of short open reading frames capable of producing small peptides in the 5′ untranslated region of the yeast GCN4 RNA. Translation of the RNA to produce these small proteins supresses translation of the GCN4 protein. The position of the methionine residue beginning each of the small peptides is indicated together with the number of additional amino acids incorporated before a stop codon is reached.

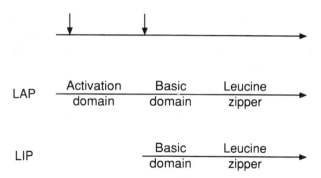

Figure 10.9 The use of different translational initiation codons (vertical arrows) produces the LAP protein, which possesses an activation domain and the LIP protein, which lacks this domain and therefore inhibits gene activation by LAP.

Hence, as with the regulation of splicing, the regulation of translation can be used to control the amount of an active factor which is produced as well as to regulate the balance between two functionally antagonistic factors encoded by the same gene.

10.2.5 Role of regulated synthesis

Regulating the synthesis of a transcription factor constitutes a metabolically inexpensive way of controlling its activity. Thus, in situations where the activity of a particular factor is not required, no energy is expended on making it in an inactive form. Such regulation probably takes place predominantly at the level of transcription so that no energy is expended on the production of an RNA, its splicing, transport, etc. However, even in cases where regulation occurs at later stages such as splicing or translation, the system is relatively efficient in terms of energy usage, since the step in gene expression which requires the most energy is the final one of translation.

In view of its metabolic efficiency, it is not surprising therefore that the regulation of their synthesis is widely used to control the activity of the factors which mediate cell type-specific gene regulation where differences in the activity of a given factor in different cell types are maintained for long periods of time. Similarly, alternative splicing or use of different translational initiation codons is used to produce different forms of the same factor which often have antagonistic effects on gene expression.

The regulation of factor activity by regulating its synthesis does suffer, however, from the defect that a change in the level of activity of a factor which is controlled purely by a change in its actual amount can take some time to occur. Thus in response to a signal that induces new transcription of the gene encoding a particular factor, it is necessary to go through all the stages illustrated in Figure 10.3, before the production of active factor which is capable of activating the expression of other genes in response to the inducing signal. It is not surprising therefore that, although some factors such as GCN4 which mediate inducible gene expression are regulated by the regulation of their synthesis, the majority of such factors are regulated by post-translational mechanisms which activate pre-existing transcription factor protein in response to the inducing signal. Thus, although mechanisms of this type are metabolically expensive in that they require the synthesis of the factor in situations where it is not required, they have the necessary rapid response time required for the regulation of inducible gene expression. Moreover, unlike transcriptional regulation, they constitute an independent method of gene regulation rather than requiring the activation of other transcription factors in order to activate the transcription of the gene encoding the factor itself.

10.3 REGULATION OF ACTIVITY

10.3.1 Evidence for the regulated activity of transcription factors

In a number of cases, it has been shown that a particular transcription factor pre-exists in an inactive form prior to its activation and the consequent switching on of the genes which depend on it for their activity. Thus, whilst the activation of heat-inducible genes by elevated temperature is dependent on the activity of the heat-shock transcription factor (HSF), this induction can be achieved in the presence of the protein synthesis inhibitor cycloheximide (Zimarino and Wu, 1987). Hence this process cannot be dependent on the synthesis of HSF in response to heat but rather must depend on the heat-induced activation of pre-existing inactive HSF (see Section 4.2.2). Similarly, as discussed in Section 9.3.1, the yeast GAL4 transcription factor pre-exists in cells prior to galactose treatment, which activates it by causing the dissociation of the inhibitory GAL80 protein.

Although for the reasons discussed above (Section 10.2.4) the activation of pre-existing transcription factors is predominantly used to modulate transcription factors involved in controlling inducible rather than cell type-specific gene expression, it has also been reported for factors involved in regulating cell type-specific gene expression. Thus, as discussed in Section 5.2.2, the transcription factor NFκB which is a heterodimer of two subunits p50 and p65, plays an important role in the B cell-specific expression of the immunoglobulin κ gene (for reviews, see Thanos and Maniatis, 1995; Baeuerle and Baltimore, 1996; Barnes, 1998). However, both subunits of NFκB are expressed in a wide variety of cell types and the factor is present in an inactive form both in pre-B cells and in a wide variety of other cell types such as T cells and HeLa cells, which do not express the immunoglobulin genes. This pre-existing form of NFκB can be activated by treatment with substances such as lipopolysaccharides or phorbol esters. These treatments therefore result in the activation of the immunoglobulin κ gene in pre-B cells, which do not normally express it, and the expression of other NFκB-dependent genes, such as the interleukin-2 α-receptor in T cells. This mechanism, in which pre-existing NFκB becomes activated both during B-cell differentiation and by agents such as phorbol esters which activate T cells, therefore, allows NFκB to play a dual role both in B-cell specific gene expression and in the expression of particular genes in response to T-cell activation by various agents (Figure 10.10). This effect would otherwise require a complex pattern of regulation in which NFκB was synthesized both in response to B-cell maturation and to agents which activate T cells.

Hence modulating the activity of a transcription factor represents a rapid and flexible means of activating a particular factor. Moreover, unlike transcriptional control, such mechanisms allow a direct linkage between the inducing stimulus and the activation of the factor, rather than requiring the regulated activity of other transcription factors which in turn activate transcription of the gene encoding the regulated factor. Hence they represent a highly efficient means of allowing specific cellular signalling pathways to produce changes in cellular transcription factor

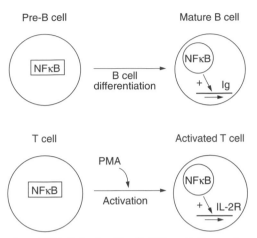

Figure 10.10 Activation of NFκB during B cell differentiation or by agents such as PMA, which activate T cells, allows it to activate expression of the immunoglobulin κ-chain gene in B cells and the interleukin-2 receptor gene in activated T cells.

activity and hence affect gene expression (for reviews, see Karin, 1994; Hill and Treisman, 1995; Treisman, 1996).

In the most extreme example of the linkage between signalling pathways and transcription factors, the signalling molecule and the transcription factor are identical. Thus, in response to microbial infection, mammalian neutrophils secrete the protein lactoferrin into the medium. It has been shown that the lactoferrin protein can be taken up by other cells of the immune system. It then enters the nucleus of the cells and binds to specific DNA sequences activating genes whose protein products are required for the cells to neutralize the microbial infection (He and Furmanski, 1995). Hence, in this case, the signalling factor and the transcription factor are the same protein (for a discussion, see Baeuerle, 1995). In most cases, however, the signalling molecule acts indirectly to produce a change in the activity of a distinct transcription factor which pre-existed within the cell in an inactive form prior to exposure to the signal. Four basic means by which such mechanisms can regulate factor activity have been described (Figure 10.11) and these will be discussed in turn.

10.3.2 Regulation by protein-ligand binding

As discussed above, one of the principal advantages of regulating the activity of a factor in response to an inducing stimulus is that it allows a direct interaction between the inducing stimulus and the activation of

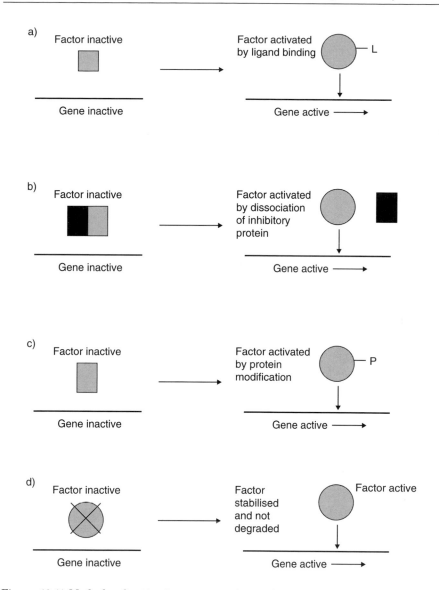

Figure 10.11 Methods of activating a transcription factor in response to an inducing stimulus. This can occur by a ligand-mediated conformational change (a), by removal of an inhibitory protein (b), by a modification to the protein such as phosphorylation (c) or by stablizing the factor so that it is not degraded (d).

the factor, ensuring a rapid response. The simplest method for this is for an inducing ligand to bind to the transcription factor and alter its structure so that it becomes activated (Figure 10.11a).

An example of this effect is seen in the case of the ACE1 factor, which mediates the induction of the yeast metallothionein gene in response to copper. In this case, the transcription factor undergoes a major conformational change in the presence of copper, which converts it to an active form which is able to bind to its appropriate binding sites in the metallothionein gene promoter and activate transcription (Figure 10.12) (for a review, see Thiele *et al.*, 1992).

A similar example of such ligand-induced regulation that occurs in mammalian cells is seen in the case of the thyroid hormone receptor. Thus, as discussed in Section 4.4.4, this receptor binds to DNA in the absence of thyroid hormone and inhibits gene expression via a specific inhibitory domain. Upon binding of thyroid hormone, the receptor undergoes a conformational change, which exposes its activation domain and allows it to bind co-activating molecules and activate transcription (see Figure 4.28). The importance of this conversion from repressor to activator is seen in the case of mutant forms of the thyroid hormone receptor which cannot undergo this conformational change because they do not bind thyroid hormone. This is observed not only in the v-*erbA* oncogene, as discussed in Section 7.2.2, but also in patients with generalized thyroid hormone resistance. Thus these patients have been shown to produce forms of the receptor which can repress gene expression but which cannot activate genes in response to thyroid hormone. Most interestingly, the presence of these dominant negative forms of the receptor results in impairment of physical and mental

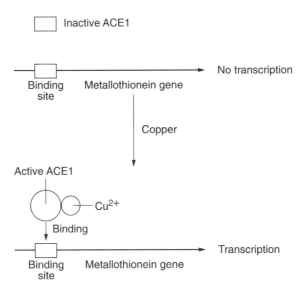

Figure 10.12 Activation of the ACE1 factor in response to copper results in transcription of the metallothionein gene.

development, which is much more severe that that observed if the receptor is absent completely (Baniahmad *et al.*, 1992).

Although the receptors for steroid hormones such as the glucocorticoids belong to the same family of receptors as the thyroid hormone receptor, they are regulated by their ligand in a somewhat different manner. Thus, prior to hormone addition, the glucocorticoid receptor is prevented from binding to DNA by its association with the 90 kDa heat-shock protein (hsp90) (see Section 4.4.2). Steroid treatment induces a conformational change in the receptor resulting in its dissociation from hsp90. This step is essential for DNA binding by the receptors which, although inherently able to bind to DNA, cannot do so until they are released from the complex with hsp90 (Figure 4.19) (for reviews, see Pratt, 1997; Pratt and Toft, 1997). Most interestingly, the association of hsp90 with the glucocorticoid receptor occurs via the C-terminal region of the receptor, which also contains the steroid-binding domain. It has been suggested therefore that, by associating with the C terminal region of the receptor, hsp90 masks adjacent domains whose activity is necessary for gene activation by the receptor, for example, those involved in receptor dimerization or subsequent DNA binding thereby preventing DNA binding from occurring. Following steroid treatment, however, the steroid binds to the C-terminus of the receptor displacing hsp90 and thereby unmasking these domains and allowing DNA binding to occur (Figure 10.13). Hence, activation of the steroid receptors involves a ligand-induced conformational change which, unlike the case of the thyroid hormone receptor, results in the dissociation of an inhibitory protein.

The association and subsequent steroid-induced dissociation of the receptors from hsp90 therefore appears to play an important role in their activation. As discussed in Section 4.4.2, however, steroid treatment also involves a conformational change in the receptor which occurs upon steroid binding and which unmasks a steroid-dependent activation domain in the receptor. Activation of the receptors by steroid therefore involves both ligand-induced conformational changes analogous to that produced in the ACE1 factor by copper and in the thyroid hormone receptor by thyroid hormone as well as the dissociation of an inhibitory protein, and thus combines the mechanisms illustrated in Figure 10.11a and b.

10.3.3 Regulation by protein–protein interactions

As described above, the glucocorticoid receptor is regulated by its interaction with hsp90 which prevents it binding to DNA and activating transcription in the absence of steroid hormone. A similar mechanism is used in the case of the NFκB factor which, as discussed above, only

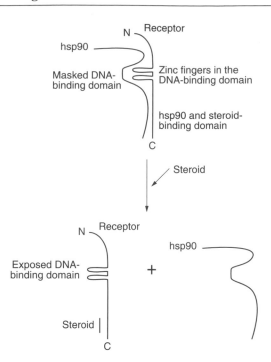

Figure 10.13 Interaction of hsp90 and the glucocorticoid receptor. Hsp90 binds to the receptor via the C-terminal region of the receptor, which also binds steroid and may mask regions of the receptor necessary for dimerization or DNA binding. When steroid is added, it binds to the receptor at the C-terminus, displacing hsp90 and exposing the masked regions.

activates transcription in mature B cells or in other cell types following treatment with agents such as lipopolysaccharides or phorbol esters. In agreement with this, no active form of NFκB capable of binding to DNA can be detected in DNA-mobility shift assays (see Section 2.2.1), using either cytoplasmic or nuclear extracts prepared from pre-B cells or non-B-cell types. Interestingly, however, such activity can be detected in the cytoplasm but not the nucleus of such cells following denaturation and subsequent renaturation of the proteins in the extract. Hence NFκB exists in the cytoplasm of pre-B cells and other cell types in an inactive form, which is complexed with another protein known as IκB that inhibits its activity (for reviews, see Thanos and Maniatis, 1995; Baeuerle and Baltimore, 1996; Perkins, 1997). The release of NFκB from IκB by the denaturation–renaturation treatment therefore results in the appearance of active NFκB capable of binding to DNA (Figure 10.14a).

These findings suggested therefore that treatments with substances such as lipopolysaccharides or phorbol esters do not activate NFκB by interacting directly with it in a manner analogous to the activation of the ACE1 factor by copper. Rather they are likely to produce the dissociation of NFκB from IκB resulting in its activation. In agreement with this idea, phorbol ester treatment of cells prior to their fractionation eliminated the latent NFκB activity in the cytoplasm and resulted in the appearance of active NFκB in the nucleus (Figure 10.14b). These substances act therefore by releasing NFκB from IκB allowing it to move to the nucleus where it can bind to DNA and activate gene expression. Hence this constitutes an example of the activation of a factor by the dissociation of an inhibitory protein (Figure 10.11b).

Such a mechanism is used to regulate the activity of many different transcription factors. Thus, apart from the NFκB/IκB and glucocorticoid receptor–hsp90 interactions, other examples of inhibitory interactions discussed in previous chapters including those between DNA-binding helix–loop–helix proteins and Id (Section 5.3.2) and p53 and the MDM2 protein (Section 7.3.2). Hence inhibitory interactions of this type are widely used to regulate the activity of specific transcription factors.

In contrast, other transcription factors may be inactive alone and may need to complex with a second factor in order to be active. This is seen in the case of the Fos protein which cannot bind to DNA without first forming a heterodimer with the Jun protein (see Section 7.2.1). A similar mechanism also operates in the case of the Myc factor which cannot bind to DNA except as a complex with the Max protein (see Section 7.2.3). Hence protein–protein interactions between transcription factors can result in either inhibition or stimulation of their activity. The need for Fos and Myc to interact with another factor prior to DNA binding arises from their inability to form a homodimer, coupled with the need for factors of this type to bind to DNA as dimers. Hence they need to form heterodimers with another factor prior to DNA binding (for further discussion, see Section 8.4).

Even in the case of factors such as Jun which can form DNA-binding homodimers, the formation of heterodimers with another factor offers the potential to produce a dimer with properties distinct from those of either homodimer. Thus, the Jun homodimer can bind strongly to AP1 sites but only weakly to the cyclic AMP response element (CRE). In contrast, a heterodimer of Jun and the CREB factor binds strongly to a CRE and more weakly to an AP1 site. Heterodimerization can therefore represent a means of producing multi-protein factors with unique properties different from that of either protein partner alone (for reviews, see Jones, 1990; Lamb and McKnight, 1991).

Hence, as well as stimulating or inhibiting the activity of a particular factor, the interaction with another factor can also alter its properties, directing it to specific DNA-binding sites to which it would not normally bind. This effect is also seen in the yeast mating-type system where, as

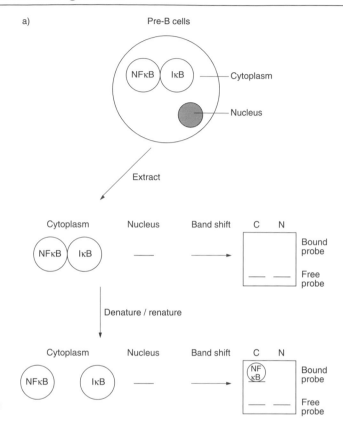

Figure 10.14

discussed in Section 5.4.3, the α2 repressor is directed to different DNA-binding sites in different genes following its interaction with either the a1 or MCM1 proteins. Similarly, as discussed in Section 6.2.3, the *Drosophila* extradenticle protein changes the DNA-binding specificity of the Ubx protein so that it binds to certain DNA-binding sites with high affinity in the presence of extradenticle and with low affinity in its absence.

Although several examples of one transcription factor altering the DNA-binding specificity of another have thus been defined, such protein–protein interactions can also change the specificity of a transcription factor in at least one other way. This is seen in the case of the *Drosophila* dorsal protein which is related to the mammalian NFκB factors. Thus this factor is capable of both activating and repressing specific genes. Such an ability is not due, for example, to the production of different forms by alternative splicing since both activation and repression take place in the same cell type. Rather, it appears to depend on the existence of a DNA sequence (the ventral repression element or VRE) adjacent to

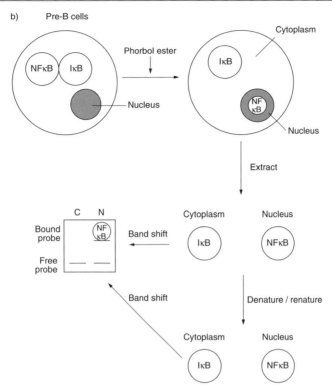

Figure 10.14 Regulation of NFkB. (a) In pre-B cells NFκB is located in the cytoplasm in an inactive form which is complexed to IκB. DNA mobility band shift assays do not therefore detect active NFκB. If a cytoplasmic extract is first denatured and renatured, however, active NFκB will be released from IκB and will be detected in a subsequent band-shift assay. (b) In mature B cells, NFκB has been released from IκB and is present in the nucleus in an active DNA-binding form. It can therefore be detected in a DNA mobility shift assay without a denaturation–renaturation step which has no effect on the binding activity.

the dorsal binding site in genes such as zen, which are repressed by dorsal, whereas the VRE sequence is absent in genes such as twist, which are activated by dorsal.

It has been shown that DSP1 (dorsal switch protein), a member of the HMG family of transcription factors (see Section 8.5), binds to the VRE and interacts with the dorsal protein, changing it from an activator to a repressor. Hence in genes such as twist, where DSP1 cannot bind, dorsal activates expression, whereas in genes such as zen, which DSP1 can bind, dorsal represses expression (Figure 10.15) (for a review, see Ip, 1995). It has been shown that DSP1 can interact with the basal transcriptional complex and disrupt the association of TFIIA with TBP (Kirov *et al.*, 1996). It therefore acts as an active transcriptional repressor interfering with the assembly of the basal transcriptional complex (for further discussion of this repression mechanism, see Section 9.3.2).

Interestingly, it has recently been shown that, like DSP1, the *Drosophila* groucho protein can switch dorsal from activator to repressor, indicating that multiple proteins can mediate this effect (Dubnicoff *et al.*, 1997). Moreover, a similar negative element to the VRE is associated with the NFκB binding site in the mammalian β-interferon promoter and the *Drosophila* DSP1 protein can similarly switch NFκB from activator to repressor when DSP1 is artificially expressed in mammalian cells. This mechanism may thus not be confined to *Drosophila* and a mammalian homologue of DSP1 may regulate NFκB activity in a similar manner (for a discussion, see Thanos and Maniatis, 1995).

Protein–protein interactions between different factors can thus either stimulate or inhibit their activity or alter that activity, either in terms of DNA-binding specificity or even from activator to repressor. It is likely that the wide variety of protein–protein interactions and their diverse effects allow the relatively small number of transcription factors that exist to produce the complex patterns of gene expression that are required in normal development and differentiation.

10.3.4 Regulation by protein modification

Many transcription factors are modified extensively following translation by the addition of O-linked monosaccharide residues (Jackson and Tjian, 1988), acetylation (Gu and Roeder, 1997) or by phosphorylation (for reviews, see Hunter and Karin, 1992; Hill and Treisman, 1995; Treisman, 1996). Such modifications represent obvious targets for agents that induce gene activation. Thus, such agents could act by altering the

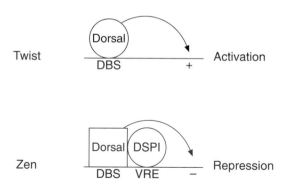

Figure 10.15 The interaction of DSPI bound at the ventral repression element (VRE) with the dorsal protein bound at its adjacent binding site (DBS) in the zen promoter results in dorsal acting as a repressor of transcription, whereas in the absence of binding sites for DSPI as in the twist promoter, it acts as an activator.

activity of a modifying enzyme, such as a kinase. In turn, this enzyme would modify the transcription factor, resulting in its activation and providing a simple and direct means of activating a particular factor in response to a specific signal (Figure 10.11c).

The most direct example of such an effect is seen in the case of gene activation by the interferons α and γ. Thus these molecules bind to cell surface receptors which are associated with factors having tyrosine kinase activity. The binding of interferon to the receptor stimulates the kinase activity and results in the phosphorylation of transcription factors known as STATs. In turn, this results in the dimerization of the signal transducers and activators of transcription (STAT) proteins, allowing them to move to the nucleus where they bind to DNA and activate interferon-responsive genes (Figure 10.16) (for reviews, see Ihle and Kerr, 1995; Darnell, 1997; Horvath and Darnell, 1997).

Another example of this type is provided by the CREB factor, which mediates the induction of specific genes in response to cyclic AMP treatment following phosphorylation by the protein kinase A enzyme, which is activated by cyclic AMP (see Section 4.3). Hence, in this case, the activation of a specific enzyme by the inducing agent allows the transcription factor to activate transcription and hence results in the activation of cyclic AMP-inducible genes. Similarly, the phosphorylation of the heat-shock factor (HSF) following exposure of cells to elevated temperature increases the activity of its activation domain leading to increased transcription of heat-inducible genes (see Section 4.2.2), whilst

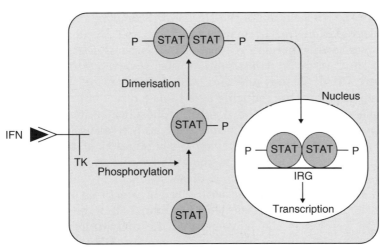

Figure 10.16 Binding of interferon (IFN) to its receptor results in activation of an associated tyrosine kinase (TK) activity leading to phosphorylation of a STAT transcription factor, allowing it to dimerize and move to the nucleus and stimulate interferon-responsive genes (IRG).

the ability of the retinoic acid receptor to stimulate transcription is enhanced by phosphorylation of its activation domain by the basal transcription factor TFIIH (see Section 3.2.4).

In contrast to these effects on transcriptional activation ability, phosphorylation of the serum response factor (SRF), which mediates the induction of several mammalian genes in response to growth factors or serum addition, increases its ability to bind to DNA rather than directly increasing the activity of its activation domain. Interestingly, SRF normally binds to DNA in association with an accessory protein p62^{TCF}. The ability of p62^{TCF} to associate with SRF is itself stimulated by phosphorylation. Similarly, as discussed in Section 4.3.2, phosphorylation of CREB by protein kinase A allows it to stimulate transcription because it allows it to assosciate with the CBP co-activator. In contrast, as discussed in Section 7.3.3, the association of the Rb-1 protein with other transcription factors such as E2F and MyoD is dependent upon the dephosphorylation of Rb-1. Hence the phosphorylation state of a transcription factor can control its ability to associate with other factors and regulate their activity as well as its ability to enter the nucleus, bind to DNA or stimulate transcription.

The effect of phosphorylation on protein–protein interactions is also involved in the dissociation of NFκB and its associated inhibitory protein IκB, which was discussed above (Section 10.3.3). In this case, however, the target for phosphorylation is the inhibitory protein IκB rather than the potentially active transcription factor itself. Thus, following treatment with phorbol esters or other stimuli such as tumour necrosis factor or interleukin 1, IκB becomes phosphorylated. Such phosphorylation results in the dissociation of the NFκB–IκB complex and targets IκB for rapid degradation (for a review, see Pahl and Baeuerle, 1996). This breakdown of the complex results in NFκB being free to move to the nucleus and activate transcription (for reviews, see Baeuerle and Baltimore, 1996; Thanos and Maniatis, 1995; Perkins, 1997). Hence, in this case, as before, the inducing agent has a direct effect on the activity of a kinase enzyme but the resulting phosphorylation inactivates an inhibitory factor rather than stimulating an activating factor.

This example therefore involves a combination of two of the post-translational activation mechanisms we have discussed, namely, protein modification (Figure 10.11c) and dissociation of an inhibitory protein (Figure 10.11b). Moreover, as with the glucocorticoid receptor and its dissociation from hsp90, discussed in Section 10.3.2, the net effect of the activation process is the movement of the activating factor from the cytoplasm to the nucleus where it can bind to DNA. Thus regulatory processes can activate a transcription factor by changing its localization in the cell as well as altering its inherent ability to bind to DNA or to activate transcription (for a review, see Vandromme *et al.*, 1996).

Clearly a key role in the regulation of the NFκB pathway will therefore be played by the enzymes which actually phosphorylate IκB in response to specific stimuli. Such an IκB kinase has recently been identified and shown to be activated following treatment with substances which stimulate NFκB activity (for reviews, see Israel, 1997; Stancovski and Baltimore, 1997; Verma and Stevenson, 1997). Hence such stimuli act by activating the IκB kinase, resulting in phosphorylation of IκB leading to its degradation and thus activation of NFκB (figure 10.17a).

In contrast, other stimuli such as glucocorticoid hormone treatment can inhibit NFκB activity. Although this may involve a direct inhibitory interaction between the activated glucocorticoid receptor and the NFκB protein itself (de Bosscher *et al.*, 1997), it is also likely to involve the ability of glucocorticoid to induce enhanced IκB synthesis, resulting in inhibition of NFκB (for a review, see Marx, 1995) (Figure 10.17b). Hence the ability of IκB to interfere with NFκB is modulated both by processes which alter the activity of IκB by phosphorylating it (Figure 10.17a) and by altering its rate of synthesis (Figure 10.17b).

In addition to its activation of NFκB, treatment with phorbol esters also results in the increased expression of several cellular genes which contain specific binding sites for the transcription factor AP1. As discussed in Section 7.2.1, this transcription factor in fact consists of a complex mixture of proteins including the proto-oncogene products Fos and Jun. Following treatment of cells with phorbol esters, the ability of Jun to bind to AP1 sites in DNA is stimulated and this effect, together with the increased levels of Fos and Jun produced by phorbol ester treatment results in the increased transcription of phorbol ester-inducible genes. As with the activation of NFκB, phorbol esters appear to increase DNA binding of Jun by activating protein kinase C.

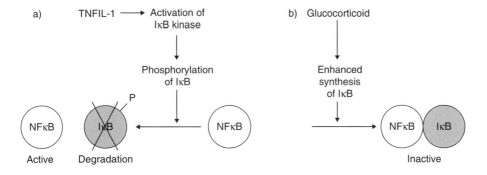

Figure 10.17 Regulation of NFκB activity by IκB can be modulated by stimuli, which result in its phosphorylation and degradation, leading to activation of NFκB (a), or by stimuli, which enhance its synthesis thereby inactivating NFκB (b).

Paradoxically, however, it has been shown (Boyle *et al.*, 1991) that the increased DNA-binding ability of Jun following phorbol ester treatment is mediated by its dephosphorylation at three specific sites located immediately adjacent to the basic DNA-binding domain, indicating that protein kinase C acts by stimulating a phosphatase enzyme, which in turn dephosphorylates Jun (Figure 10.18).

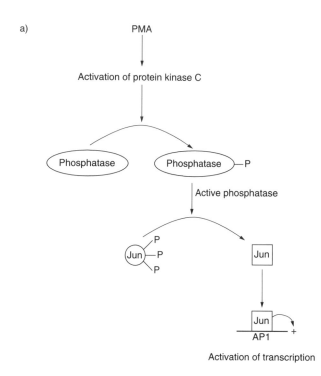

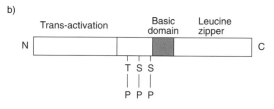

Figure 10.18 (a) Activation of Jun binding to DNA by dephosphorylation. The dephosphorylation of Jun protein following PMA treatment increases its ability to bind to AP1 sites and activate PMA-responsive genes. This is likely to be mediated via the PMA-dependent activation of protein kinase C, which in turn phosphorylates a phosphatase enzyme allowing it to dephosphorylate Jun. (b) Position in the Jun protein of the two serine (S) and one threonine (T) residues, which are dephosphorylated in response to PMA. Note the close proximity to the basic domain (shaded) which mediates DNA binding. The positions of the transactivation domain and leucine zipper are also indicated.

Such an inhibitory effect of phosphorylation on the activity of a transcription factor is not unique to the Jun protein, a similar effect of phosphorylation in reducing DNA-binding activity having also been observed in the Myb proto-oncogene protein, which was discussed in Section 7.2.4 (Luscher *et al.*, 1990). Moreover, DNA-binding ability is not the only target for such inhibitory effects of phosphorylation. Thus phosphorylation of the bicoid protein reduces its ability to activate transcription without affecting its DNA binding activity, presumably by inhibiting the activity of its activation domain (Ronchi *et al.*, 1993).

Hence protein modification by phosphorylation can have a wide variety of effects on transcription factors, either stimulating or inhibiting their activity and acting via a direct effect on the ability of the factor to enter the nucleus, bind to DNA, associate with another protein or activate transcription or by an indirect effect affecting the activity of an inhibitory protein. The directness and rapidity of this means of transcription factor activation suggests that the other forms of modification which have been observed for specific transcription factors are also likely to be the targets for regulatory processes.

Indeed, the first example in which the activity of a transcription factor is regulated by its acetylation has recently been described (Gu and Roeder, 1997). Thus, the addition of acetyl residues to the C-terminal domain of the p53 protein (Section 7.3.2) increases the DNA-binding ability of the protein, possibly by preventing an inhibitory interaction of this C-terminal domain with the DNA-binding domain (Figure 10.19). Interestingly, this acetylation of p53 is carried out by the p300 co-activator molecule which, as described in Section 9.2.5, associates with p53 as well as with a wide variety of other transcription factors. This finding indicates that as well as acetylating histones and thereby modifying chromatin structure (see Section 1.4.3), p300 and the related CBP co-activators may also use their acetyltransferase activity to acetylate specific transcription factors and thereby modify their activity (Figure 10.20).

10.3.5 Regulation of protein degradation and processing

One of the simplest mechanisms of regulating the activity of a particular transcription factor is to ensure that it is stable and can fulfil its function in one situation, but is rapidly degraded in another situation and so cannot fulfil its function (for a review, see Pahl and Baeuerle, 1996) (Figure 10.11d). Indeed, many of the regulatory processes which we have previously described act in this manner. Thus, the alternative splicing of the Hac1p factor mRNA resulted in different forms of the protein with different stabilities (Section 10.2.3), whilst the phosphorylation of the

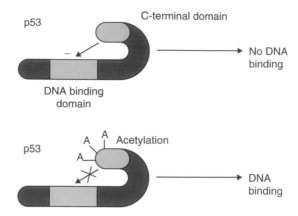

Figure 10.19 Acetylation (A) of the p53 C-terminus activates its DNA binding ability possibly by preventing an inhibitory interaction between the C-terminus and the *DNA-binding domain.*

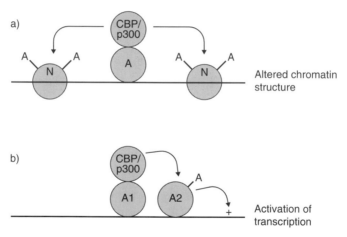

Figure 10.20 Possible mechanisms of action of CBP/p300. Following recruitment to DNA by an activating molecule (A1) the acetyl transferase activity of CBP/p300 may either (a) acetylate histones producing a more open chromatin structure or (b) acetylate another activating transcription factor (A2), allowing it to stimulate transcription.

IκB factor resulted in its degradation, thereby releasing active NFκB (section 10.3.4). Similarly, the p53 transcription factor is targeted for degradation by association with the MDM2 factor (see Section 7.3.2), whilst the TTK 88 transcriptional repressor is degraded following association with the PHYL and SINA proteins (see Section 9.3.1).

Hence regulating the stability of a transcription factor so that it is different in different situations is an important means of regulating transcription factor activity (Figure 10.21a). In addition, however, proteolysis can also be used to activate a transcription factor. This can be achieved by cleaving an inactive precursor to produce an active form of the transcription factor (Figure 10.21b). This form of regulation is also seen in the NFκB family. Thus, an NFκB-related protein p105 is synthesized as a single molecule in which the NFκB portion is linked to an IκB-like region, which inhibits its activity resulting in an inactive precursor protein. Following exposure to an activating stimulus, the IκB portion is phosphorylated and is cleaved off to release active NFκB (for a review, see Thanos and Maniatis, 1995). This mechanism evidently resembles the regulation of NFκB by IκB described in Section 10.3.4, except that, in this case, the NFκB and IκB-like activities are contained in the same molecule rather than in different molecules (Figure 10.22).

This regulatory mechanism is also seen in the case of the SREBP transcription factors which activate gene expression in response to removal of cholesterol (for a review, see Brown and Goldstein, 1997). In the presence of cholesterol, these factors are anchored in the endoplasmic reticulum by a specific region of the protein. When cells are deprived of cholesterol, this region of the protein is cleaved off, allowing the protein to move to the nucleus and switch on genes whose protein products are required for cholesterol biosynthesis (Figure 10.23). Interestingly, both of the cases of regulation by proteolytic cleavage we have described result in a change in localization of the transcription factor with the NFκB portion of the p105 moving from the cytoplasm to the nucleus and the

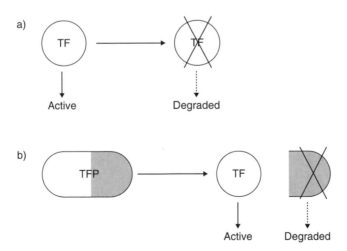

Figure 10.21 Proteolytic cleavage of a transcription factor can be used either (a) to degrade the factor, so preventing it from acting, or (b) to cleave an inactive precursor molecule to produce an active factor.

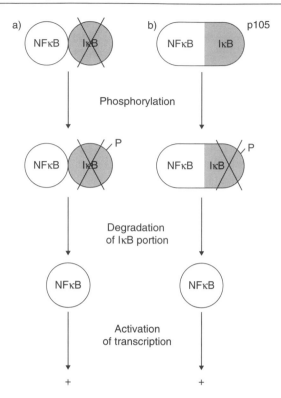

Figure 10.22 In the NFκB family, activation of NFκB can be achieved either (a) by phosphorylation and degradation of an associated IκB protein, or (b) by phosphorylation of the IκB portion of a large precursor protein (p105), resulting in its proteolytic processing to release active NFκB.

activated SREBP factor moving from the endoplasmic reticulum membrane to the nucleus. This further underlies the importance of changes in transcription factor localization brought about by regulatory processes (see also Section 10.3.4).

10.3.6 Role of regulated activity

In addition to its ability to produce a very rapid activation of gene expression, modification of the activity of a pre-existing protein also allows specific targets for modification to be used in different cases. Thus the various regulating processes discussed above affect the activity of transcription factors at a wide variety of different stages. For example, in the case of phosphorylation (Section 10.3.4), we have seen how in different cases a single process can alter the DNA-binding ability of a

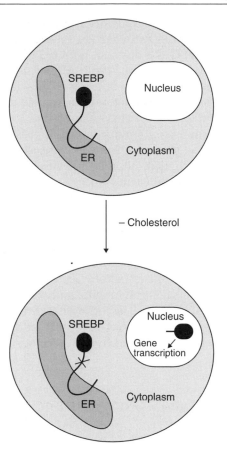

Figure 10.23 In the presence of cholesterol, the SREBP factor is anchored in the membrane of the endoplasmic reticulum and hence cannot enter the nucleus. On removal of cholesterol, the SREBP precursor is cleaved, releasing the active form of the protein which can move to the nucleus and activate the expression of genes involved in cholesterol biosynthesis.

factor, its localization within the cell, its trans-activation ability, its ability to associate with another protein or its degradation.

Clearly, therefore, post-translational mechanisms for activating pre-existing protein could be used to stimulate independently either the DNA binding or the transcriptional activation activities of a single factor in different situations within a complex regulatory pathway. Indeed, such a combination of mechanisms is actually used to regulate the activity of the yeast GAL4 transcription factor. Thus, as discussed in Section 9.3.1, the activation of transcription of galactose-inducible genes by GAL4 is mediated by the galactose-induced dissociation of the inhibitory GAL80 protein which exposes the activation domain of DNA bound GAL4.

Interestingly, however, this effect only occurs when the cells are grown in the presence of glycerol as the main carbon source. By contrast, however, in the presence of glucose, GAL4 does not bind to DNA and the addition of galactose has no effect (Giniger *et al.*, 1985). Hence, by having a system in which glucose modulates the DNA binding of the factor and galactose modulates the activation of bound factor, it is possible for glucose to inhibit the stimulatory effect of galactose. This ensures that the enzymes required for galactose metabolism are only induced in the presence of glycerol and not in the presence of the preferred nutrient glucose (Figure 10.24).

Such a system in which two different activities of a single factor are independently modulated could clearly not be achieved by stimulating the *de novo* synthesis of the factor which would simply result in more of it being present. Hence, in addition to its rapidity, the activation of pre-existing factor has the advantage of flexibility in potentially being able to generate different forms of the factor with different activities. It should be noted, however, that this effect can also be achieved, for example, by alternative splicing of the RNA encoding the factor (Section 10.2.3), which can, for example, generate forms of the protein with and without the DNA-binding domain, as in the case of the Era-1 factor, with and without the activation domain, as in the case of CREM or Oct-2, or with and without the ligand-binding domain, as in the case of the thyroid hormone receptor.

10.4 CONCLUSIONS

In this chapter we have discussed how the regulation of gene expression by transcription factors is achieved both by the regulated synthesis or by the regulated activity of these factors. Although there are exceptions, the

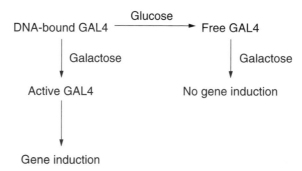

Figure 10.24 Effects of glucose and galactose on GAL4 activity. Note that whilst galactose stimulates the ability of DNA-bound GAL4 to activate transcription, this effect does not occur in the presence of glucose, which results in the release of GAL4 from DNA.

regulation of synthesis of a particular factor is used primarily in cases of factors which mediate tissue-specific or developmentally regulated gene expression where a factor is only required in a small proportion of cell types and is never required in most cell types. In contrast, however, the rapid induction of transcription in response to inducers of gene expression is primarily achieved by the activation of pre-existing inactive forms of transcription factors that are present in most cell types, since this process, although more metabolically expensive, provides the required rapidity in response.

Although these two processes have been discussed separately, it should not be thought that a given factor can only be regulated either at the level of synthesis or at the level of activity. In fact, in many cases of inducible gene expression which involve activation of pre-existing factors, such activation is supplemented by the slower process of synthesizing new factors in response to the inducing agent. Thus, in the case of the stimulation of genes containing AP1 sites by phorbol esters discussed above (Section 10.3.4), the phorbol ester-induced increase in the DNA binding of pre-existing Jun protein is supplemented by increased synthesis of both Fos and Jun following phorbol ester treatment, and such newly synthesized Fos and Jun will clearly eventually become a major part of the increased AP1 activity observed following phorbol ester treatment (see Section 7.2.1). Similarly, the activation of NFκB by dissociation from IκB following treatment with substances such as phorbol esters which activate T cells (see Sections 10.3.3 and 10.3.4) has been shown to be supplemented by increased synthesis of NFκB and its corresponding mRNA following T-cell activation, whilst increased synthesis of IκB itself occurs in response to glucocorticoid (Section 10.3.4). Hence, in many cases, the rapid effects of post-translational processes in activating gene expression are supplemented by *de novo* synthesis of the factor which, although slower, will enhance and maintain the effect.

This combination of regulated synthesis and regulated activity is also seen in the case of factors which mediate tissue-specific gene expression and which are synthesized in only a few cell types. Thus, in the case of the MyoD factor which regulates muscle-specific genes, the factor and its corresponding mRNA are synthesized only in cells of the muscle lineage (see Section 5.3.2). The activation of MyoD-dependent genes which occurs when myoblast cells within this lineage differentiate into myotubes is not, however, mediated by new synthesis of MyoD, which is present at equal levels in both cell types. Rather, it occurs due to the decline in the level of the inhibitory protein Id, resulting in the post-transcriptional activation of pre-existing MyoD and the transcription of MyoD-dependent genes. Hence, in this case, regulation of synthesis is used to avoid the wasteful production of MyoD in cells of non-muscle lineage, whilst the activation of pre-existing MyoD ensures a rapid response to agents which induce differentiation within cells of the muscle lineage. Thus, in a number of cases, a combination of both regulated synthesis and regulated activity allows the precise requirements of a particular response to be fulfilled rapidly but with minimum unnecessary wastage of energy.

Similarly, in many situations an alteration in the activity of one transcription factor produced by a post-translational change results in the activation of the gene encoding another transcription factor resulting in its enhanced synthesis. This is seen, for example, in the case of the B-cell specific factors Oct-2 and NFκB, which were discussed earlier in this chapter as examples, respectively, of regulated synthesis (Section 10.2.1) and regulated activity (Section 10.3.1). Thus recent evidence indicates that the presence of activated NFκB is essential for the transcription of the Oct-2 gene in B lymphocytes directly linking the cell type-specific regulation of NFκB activity with the cell type-specific regulation of Oct-2 synthesis (Bendall *et al.*, 1997).

In summary, therefore, the different properties of regulated synthesis and regulated activity allow these two processes, both independently and in combination, to regulate the complex processes of inducible, tissue-specific and developmentally regulated gene expression efficiently.

REFERENCES

Baeuerle, P.A. (1995). Enter a polypeptide messenger. *Nature* **373**, 661–662.

Baeuerle, P.A. and Baltimore, D. (1996). NF-κB: ten years after. *Cell* **87**, 13–20.

Baniahmad, A., Tsai, S.Y., O'Malley, B.W. and Tsai, M.J. (1992). Kindred S thyroid hormone receptor is an active and constitutive silencer and a repressor for thyroid hormone and retinoic acid responses. *Proceedings of the National Academy of Sciences USA* **89**, 10633–10637.

Barnes, P.J. (1998). NFκB. *International Journal of Biochemistry and Cell Biology* (in press).

Bendall, H.H., Scherer, D.C., Edson, C.R., Ballard, D.W. and Oltz, E.M. (1997). Transcription factor NF-κB regulates inducible Oct-2 expression in precursor B lymphocytes. *Journal of Biological Chemistry* **272**, 28826–28828.

Boyle, W.J., Smeal, T., Defize, L.H.K., Angel, P., Woodgett, J.R., Karin, M. and Hunter, T. (1991). Activation of protein kinase C decreases phosphorylation of c Jun at sites that negatively regulate its DNA binding activity. *Cell* **64**, 573–584.

Brown, M.S. and Goldstein, J.L. (1997). The SREBP pathway: regulation of cholesterol metabolism by proteolysis of a membrane-bound transcription factor. *Cell* **89**, 331–340.

Cox, J.S. and Walter, P. (1996). A novel mechanism for regulating the activity of a transcription factor that controls the unfolded protein response. *Cell* **87**, 391–404.

Darnell, J.E. Jr (1997). STATs and gene regulation. *Science* **277**, 1630–1635.

Davis, H.L., Weintraub, H. and Lassar, A.B. (1987). Expression of a single transfected cDNA converts fibroblasts to myoblasts. *Cell* **51**, 987–1000.

de Bosscher, K., Schmitz, M.L., Vanden Berghe, W., Plaisance, S., Fiers, W. and Haegeman, G. (1997). Glucocorticoid-mediated repression of nuclear factor-κB-dependent transcription involves direct interference with transactivation. *Proceedings of the National Academy of Sciences USA* **94**, 13504–13509.

Descombes, P. and Schibler, U. (1991). A liver transcriptional activator protein LAP and a transcriptional inhibitory protein LIP are translated from the same mRNA. *Cell* **67**, 569–580.

Driever, W. and Nusllein-Volhard, C. (1988). The bicoid protein determines position in the *Drosophila* embryo in a concentration-dependent manner. *Cell* **54**, 95–104.

Dubnicoff, T., Valentine, S.A., Chen, G., Shi, T., Lengyel, J.A., Parough, Z. and Courey, A.J. (1997). Conversion of Dorsal from an activator to a repressor by the global corepressor Groucho. *Genes and Development* **11**, 2952–2957.

Giniger, E., Varnam, S.M. and Ptashne, M. (1985). Specific DNA binding of GAL4, a positive regulatory protein of yeast. *Cell* **40**, 767–774.

Gu, W. and Roeder, R.G. (1997). Activation of p53 sequence-specific DNA binding by acetylation of the p53 C-terminal domain. *Cell* **90**, 595–606.

He, J. and Furmanski, P. (1995). Sequence specificity and transcriptional activation in the binding of lactoferrin to DNA. *Nature* **373**, 721–724.

Hill, C.S. and Treisman, R. (1995). Transcriptional regulation by extracellular signals: mechanisms and specificity. *Cell* **80**, 199–211.

Hinnebusch, A.G. (1997). Translational regulation of GCN4. *Journal of Biological Chemistry* **272**, 21661–21664.

Horvath, C.M. and Darnell, J.E. Jr (1997). The state of the STATs: recent developments in the study of signal transduction to the nucleus. *Current Opinion in Cell Biology* **9**, 233–239.

Hunter, T. and Karin, M. (1992). The regulation of transcription by phosphorylation. *Cell* **70**, 375–387.

Ihle, J.N. and Kerr, I.M. (1995). Jaks and Stats in signalling by the cytokine receptor super-family. *Trends in Genetics* **11**, 69–74.

Ingham, P.W. (1988). The molecular genetics of embryonic pattern formation in *Drosophila*. *Nature* **335**, 25–34.

Ip, Y.T. (1995). Converting an activator into a repressor. *Current Biology* **5**, 1–3.

Israel, A. (1997). IκB kinase all zipped up. *Nature* **388**, 519–521.

Jackson, S.P. and Tjian, R. (1988). O-Glycosylation of eukaryotic transcription factors: implications for mechanisms of transcriptional regulation. *Cell* **55**, 125–133.

Jones, N. (1990). Transcriptional regulation by dimerization: two sides to an incestuous relationsip. *Cell* **61**, 9–11.

Karin, M.(1994). Signal transduction from the cell surface to the nucleus through the phosphorylation of transcription factors. *Current Opinion in Cell Biology* **6**, 415–424.

Kirov, N.C., Lieberman, P.M. and Rushlow, C. (1996). The transcriptional co-repressor DSP1 inhibits activated transcription by disrupting TFIIA-TBP complex formation. *EMBO Journal* **15**, 7079–7087.

Kozmik, Z., Czerny, T. and Busslinger, M. (1997). Alternatively spliced insertions in the paired domain restrict the DNA sequence specificity of Pax6 and Pax8. *EMBO Journal* **16**, 6793–6803.

Lamb, P. and McKnight, S.L. (1991). Diversity and specificity in transcriptional regulation: the benefits of heterotypic dimerization. *Trends in Biochemical Sciences* **16**, 417–422.

Larosa, G.J. and Gudas, L.J. (1988). Early retinoic acid-induced F9 teratocarcinoma stem cell gene ERA-1: alternative splicing creates transcripts for a homeobox-containing protein and one lacking the homeobox. *Molecular and Cellular Biology* **8**, 3906–3917.

Latchman, D.S. (1998). *Gene Regulation: A Eukaryotic Perspective*, 3rd edn. London, New York: Chapman and Hall.

Lillycrop, K.A., Dawson, S.J., Estridge, J.K., Gerster, T., Matthias, P. and Latchman, D.S. (1994). Repression of a herpes simplex virus immediate-early promoter by the Oct-2 transcription factor is dependent upon an inhibitory region at the N-terminus of the protein. *Molecular and Cellular Biology* **14**, 7633–7642.

Luscher, B., Christenson, E., Litchfield, D.W., Knebs, E.G. and Eiserman, R.N. (1990). *Myb* DNA binding inhibited by phosphorylation at a site deleted during oncogenic activation. *Nature* **344**, 517–522.

McKeown, M. (1992). Alternative mRNA splicing. *Annual Review of Cell Biology* **8**, 133–155.

Marx, J. (1995). How the glucocorticoids suppress immunity. *Science* **270**, 232–233.

Pahl, H.L. and Baeuerle, P.A. (1996). Control of gene expression by proteolysis. *Current Opinion in Cell Biology* **8**, 340–347.

Perkins, N. (1997). Achieving transcriptional specificity with NFκB. *International Journal of Biochemistry and Cell Biology* **29**, 1433–1448.

Pratt, W.B. (1997). The role of the hsp90-based chaperone system in signal transduction by nuclear receptors and receptor signalling via

MAP kinase. *Annual Review of Pharmacology and Toxicology* **37**, 297–326.

Pratt, W.B. and Toft, D.O. (1997). Steroid receptor interactions with heat shock protein and immunophilin chaperones. *Endocrinology Reviews* **18**, 306–360.

Rio, D.C. (1993). Splicing of pre mRNA. Mechanisms, regulation and role in development. *Current Opinion in Genetics and Development* **3**, 574–584.

Ronchi, E., Treisman, J., Dostatri, N., Struhl, G. and Desplan, C. (1993). Down-regulation of the *Drosophila* morphogen bicoid by the torso receptor mediated signal transduction cascade. *Cell* **74**, 347–355.

Shen, W.F., Detmer, K., Simonitch-Easton, T., Lawrence, H.J. and Largman, C. (1991). Alternative splicing of the Hox 2.2 homeobox gene in human hematopoetic cells and murine embryonic and adult tissues. *Nucleic Acids Research* **19**, 539–545.

Stancovski, I. and Baltimore, D. (1997). NF-κB activation: the IκB kinase revealed. *Cell* **91**, 299–302.

Thanos, D. and Maniatis, T. (1995). NFκB: a lesson in family values. *Cell* **80**, 529–532.

Thiele, D.J. (1992). Metal regulated transcription in eukaryotes. *Nucleic Acids Research* **20**, 1183–1191.

Treisman, R. (1996). Regulation of transcription by MAP kinase cascades. *Current Opinion in Cell Biology*, **8**, 205–215.

Vandromme, M., Gauthier-Rouviere, C., Lamb, N. and Fernadez, A. (1996). Regulation of transcription factor localisation: fine tuning of gene expression. *Trends in Biochemical Sciences* **21**, 59–64.

Verma, I.M. and Stevenson, J. (1997). IκB kinase: beginning not the end. *Proceedings of the National Academy of Sciences USA* **94**, 11758–11760.

Wuarin, J. and Schibler, U. (1990). Expression of the liver-enriched transcriptional activator protein DBP follows a stringent circadian rhythm. *Cell* **63**, 1257–1269.

Xanthopoulos, K.G., Mirkovitch, J., Decker, T., Kuo, C.G. and Darnell, J.E. Jr (1989). Cell-specific transcriptional control of the mouse DNA binding protein mC/EBP. *Proceedings of the National Academy of Sciences USA* **86**, 4117–4121.

Zimarino, V. and Wu, C. (1987). Induction of sequence-specific binding of *Drosophila* heat-shock activator protein without protein synthesis. *Nature* **327**, 727–730.

CHAPTER ELEVEN

Conclusions and future prospects

At the time the first edition of this book was published (1991), enormous progress had been made in understanding the nature and role of transcription factors. Thus the roles of specific factors in processes such as constitutive (Chapter 3), inducible (Chapter 4), tissue-specific (Chapter 5) and developmentally regulated (Chapter 6) gene expression had been defined as had their involvement in diseases such as cancer (Chapter 7). Moreover, by studying these factors in detail, it proved possible to analyse how they fulfil their function in these processes by binding to specific sites in the DNA of regulated genes (Chapter 8) and activating or repressing transcription (Chapter 9), as well as the regulatory processes which result in their doing so only at the appropriate time and place (Chapter 10). Moreover, the regions of individual factors which mediate these effects and the critical amino acids within them which are of importance had been identified in a number of cases.

In the intervening years up to the current publication of the third edition, much further progress has been made in these areas. In addition, the ability to prepare 'knock-out' mice in which the gene encoding an individual factor has been inactivated has allowed the *in vivo* functional role of many factors to be assessed directly, whilst numerous studies have elucidated the structure of specific factors either in isolation or bound to DNA, as illustrated in the colour plate section. It has become increasingly clear, however, that the activity of a particular factor cannot be considered in isolation. Thus, very often the activity of a factor can be stimulated either positively or negatively by its interaction with another factor. For example, the Fos protein needs to interact with the Jun protein to form a DNA-binding complex (see Sections 7.2.1 and 8.4).

Conversely, the DNA-binding ability of the glucocorticoid receptor is inhibited by its association with hsp90 (see Sections 4.4.2 and 10.3.2), whilst that of the MyoD factor is inhibited by its association with Id (Section 5.3.2).

Additionally, however, it has become clear that, as well as stimulating or inhibiting factor activity, such protein–protein interactions can also alter the specificity of a factor. This may involve altering its DNA-binding specificity and hence the target genes it affects as in the case of the interactions between the yeast a1 or MCM1 proteins and the α2 protein (see Section 5.4.3). Alternatively, it may completely change the factor from activator to repressor as in the case of the dorsal/DSP1 interaction (see Section 10.3.3). Indeed, differences in the ability to interact with other proteins result in factors with identical DNA-binding specificities having entirely different functional effects as in the case of Ubx and Antp (see Section 6.2.3). Hence, by altering the specificity of particular factors, interactions of this type are likely to play a crucial role in the complex regulatory networks which allow a relatively small number of transcription factors to control highly complex processes such as development (see Chapter 6).

As well as such regulatory interactions between different factors, it has become increasingly clear in recent years that many activating transcription factors need to interact with other factors, known as co-activators, in order to stimulate transcription. The most important of such co-activators, CBP, was originally characterized as being required for transcriptional activation in response to cyclic AMP treatment, mediated via the CREB transcription factor (see Section 4.3.2). It is also involved, however, in transcriptional activation mediated via a number of other transcription factors activated by different signalling pathways (see Section 9.2.4). In turn, because of the limiting amounts of CBP in the cell, the different transcription factors and signalling pathways compete for CBP resulting in mutual antagonism between, for example, the signalling pathways mediated by AP1 and the glucocorticoid receptor (see Section 4.4.4).

Thus, the critical dependence of many activating factors on a specific co-activator can result in a functional link between two different factors which do not themselves interact but which compete for the same co-activator. Moreover, the activity of a transcription factor can be regulated by controlling its ability to interact with its co-activator. For example, only the phosphorylated form of CREB can interact with CBP and therefore activate transcription, whereas the non-phosphorylated form does not interact with CBP and is thus inactive (see Section 4.3.2). Similarly, in the absence of thyroid hormone, the thyroid hormone receptor has an inhibitory effect on transcription because it binds co-repressor molecules, which act to inhibit transcription. Following exposure to thyroid hormone, however, the receptor undergoes a conformational change which allows it to bind co-activator molecules and hence activate transcription (see Section 4.4.4).

Hence the interaction between activators and co-activators plays a critical role in the activation of transcription and its regulation. Although co-activators are likely to act in some cases by interacting with the basal transcriptional complex (see Sections 9.2.3 and 9.2.4), the finding that many co-activators have histone acetyltransferase activity (see Sections 9.2.4 and 9.2.5) indicates that they may stimulate transcription via altering chromatin structure. Hence such factors could act by acetylating histones, thereby altering the chromatin structure to a more open structure able to support active transcription (see Section 1.4.3).

Conversely, co-repressors which have histone deacetylase activity may act by producing a more closed chromatin structure incompatible with transcription (see Sections 4.4.4 and 9.3.2). Interestingly, as described in Section 10.3.4, it has been shown that the p53 transcription factor exhibits enhanced DNA-binding activity following acetylation by the p300 co-activator, which is closely related to CBP. It is therefore possible that the acetyltransferase activity of co-activators may mediate transcriptional activation by acetylating other transcription factors as well as the histones, providing a further aspect to the activities of these molecules.

Ultimately, therefore, the understanding of transcription factor function will require a knowledge of the nature and effect of interactions between different transcription factors, their co-activators and co-repressors, which is as good as that now available for individual factors. Moreover, it will be necessary to establish how such changes modulate the activity of the basal transcriptional complex and alter chromatin structure. Clearly much work remains to be done before this is achieved. The rapid progress since the first edition of this work was published suggests, however, that an eventual understanding in molecular terms of the manner in which transcription factors control highly complex processes such as *Drosophila* and even mammalian development can ultimately be achieved.

Index